INSIGHTS TO NEUROIMMUNE BIOLOGY

INSIGHTS TO NEUROIMMUNE BIOLOGY

Second Edition

Edited by

ISTVAN BERCZI

Department of Immunology, Faculty of Medicine, The University of Manitoba, Winnipeg, Manitoba, Canada; Departamento de Fisiología y Farmacología, Centro de Ciencias Básicas, Universidad Autónoma de Aguascalientes, Aguascalientes, Estado Libre y Soberano de Aguascalientes, Mexico City, Mexico

ELSEVIER

Amsterdam • Boston • Heidelberg • London • New York • Oxford
Paris • San Diego • San Francisco • Singapore • Sydney • Tokyo

Elsevier
Radarweg 29, PO Box 211, 1000 AE Amsterdam, Netherlands
The Boulevard, Langford Lane, Kidlington, Oxford OX5 1GB, UK
50 Hampshire Street, 5th Floor, Cambridge, MA 02139, USA

British Library Cataloguing-in-Publication Data
A catalogue record for this book is available from the British Library

Library of Congress Cataloging-in-Publication Data
A catalog record for this book is available from the Library of Congress

ISBN: 978-0-12-801770-8

Typeset by Thomson Digital

TABLE OF CONTENTS

LIST OF CONTRIBUTORS

Istvan Berczi
Department of Immunology, Faculty of Medicine, The University of Manitoba, Winnipeg, Manitoba, Canada; Departamento de Fisiología y Farmacología, Centro de Ciencias Básicas, Universidad Autónoma de Aguascalientes, Aguascalientes, Estado Libre y Soberano de Aguascalientes, Mexico City, Mexico

Rafael Campos-Rodríguez
Departamento Bioquímica, ESM, IPN, México

Roberto Chavira-Ramírez
Departamento de Hormonas esteroideas, ICMyN "SZ", Mexico City, México

Kalman Kovacs
Department of Pathology and Laboratory Medicine, St. Michael's Hospital, University of Toronto, Toronto, Ontario, Canada

Dora Lippai
Second Department of Internal Medicine, Semmelweis University, Budapest, Hungary

Alejandro Organista-Esparza
Departamento de Fisiología y Farmacología, Centro de Ciencias Básicas, Universidad Autónoma de Aguascalientes, Aguascalientes, Estado Libre y Soberano de Aguascalientes, México City, México

Andrés Quintanar-Stephano
Departamento de Fisiología y Farmacología, Centro de Ciencias Básicas, Universidad Autónoma de Aguascalientes, Aguascalientes, Estado Libre y Soberano de Aguascalientes, México City, México

Nicolas Rohleder
Department of Psychology, Brandeis University, Waltham, MA, USA

Fabio Rotondo
Department of Pathology and Laboratory Medicine, St. Michael's Hospital, University of Toronto, Toronto, Ontario, Canada

Gyongyi Szabo
Department of Medicine, University of Massachusetts Medical School, Worcester, MA, USA

FOREWORD

In this book, the neuroimmune regulation by neuropeptides is discussed. It was a pleasure to work on this book. It is clear that *Medicine* and animal medication will receive an actual face-lift from the new capabilities, which have been offered by these mediators. Now, there is a candidate peptide to treat almost any disease condition, such as inflammation, immunological problems, organ–tissue damage, infectious diseases, etc. You name it and there is a peptide to treat, and this is just the beginning. I am sure that more mediators will be discovered to resolve greater problems. We were present during the entire development of *Neuroimmune Biology* and we knew for long time that this book was coming. Finally it has come. Initially, the body has its own immunity against infections, hence the individuals are healthy. When things get astray, we might develop a disease. But now, salvation is on the way! Science will pay back all the due investment. Besides, we should also remember the results that have been obtained till date.

Istvan Berczi, Fabio Rotondo, Kalman Kovacs

PREFACE

The participation of the nervous system in inflammatory reactions was suggested over a century ago. Yet, as potentially important contributors to inflammatory mechanisms, nervous elements have been largely ignored for a long time. However, discovery of the highly specific pharmacological effects of capsaicin in the 1950s and the selective neurotoxic action of vanilloid compounds in the 1970s, on nociceptive sensory neurons, have revived interest in the study of neurogenic inflammatory processes. Early observations furnished firm evidence for the existence of an inflammatory response induced by a purely neurogenic route, resulting from the activation of nociceptive afferents. Morphological studies utilizing the neurotoxic/neurodegenerative actions of capsaicin demonstrated a widespread system of peptidergic and nonpeptidergic capsaicin-sensitive afferent nerves, which innervate the skin, mucous membranes, as well as most of the visceral organs and tissues. From studies conducted upon different organs, tissues, and cells, a complex system of primary sensory neurons has emerged, which parallels the autonomic nervous system not only in its extent, but also probably in its significance. Afferent nerves, once believed to serve merely as sensory receptors conveying impulses generated by noxious stimuli, have evolved into key players in a complicated local regulatory system that participates in the contractile, glandular, vascular, inflammatory, immune, protective, restorative, and trophic functions, of somatic and visceral tissues. Neuropeptides released from nociceptive afferents in response to tissue injury or to a wide variety of chemical stimuli, involving inflammatory mediators, tissue, and mast cell-derived agents, play a pivotal role in these processes and are potent modulators of inflammatory reactions. New facets of capsaicin-sensitive neuronal and cellular responses were revealed, following the cloning of the capsaicin/vanilloid receptor, now known as the *transient receptor potential vanilloid type 1 receptor* (*TRPV1*), which is primarily expressed in nociceptive primary sensory neurons, but also in some other neurons and cells. A number of contributions in this volume focus on the characterization and functional traits of nociceptor neurons and on the mechanisms, which activate them. Further chapters deal with the parts played by primary sensory neurons in inflammatory reactions and in the regulation/modulation of the functions of various organs and tissues under physiological and pathological conditions. Certain chapters touch upon the therapeutic implications offered by the use of vanilloids, novel nonpeptide antagonists of peptide and vanilloid receptors. It is the editor's hope that this volume will contribute to and initiate new interest in the understanding of diverse roles fulfilled by primary afferent neurons, in the functioning of the body during health and disease.

G. Jancsó
Szeged, Hungary
May 26, 2008

PART 1

Introduction: Neuroimmune Regulation in Health and Disease

CHAPTER 1

Neuroimmune Regulation in Health and Disease

Istvan Berczi
Department of Immunology, Faculty of Medicine, The University of Manitoba, Winnipeg, Manitoba, Canada; Departamento de Fisiología y Farmacología, Centro de Ciencias Básicas, Universidad Autónoma de Aguascalientes, Aguascalientes, Estado Libre y Soberano de Aguascalientes, Mexico City, Mexico

1 INTRODUCTION

1.1 Health

The French scientist Claude Bernard was the first to recognize the fact that living organisms maintain their physiological and biochemical parameters within defined (normal) levels, which show only minor variations.[1] The scientific name of this condition is *homeostasis.* Every parameter in a healthy body that may be measured will stay within normal (physiological) levels. When we are healthy, we feel well with positive emotions and in general with a healthy mind, we are able to overcome and triumph over life's challenges. In *Medicine* there are books of physiology, internal medicine, pathology, pharmacology, and gynecology, which are the major subject areas that deal with *health and disease.* Information available on these subjects is incredibly large, and it definitely needs computer management to help with accessibility.

It is the basic information in *Physiology* that doing your work requires energy and activity, so the basic values of heart rate, metabolism, and the function of our entire system will increase. But it should not go higher than the upper normal level. Age, sex, and other parameters influence normal levels. Physical work requires more effort than mental work.

Realistically the Yin-Yang hypothesis is valid in most cases. Many people live with minor, not life-threatening problems; they feel and act like a healthy person because they are only a "little bit sick." This means that little problems can be overlooked. However, the opposite condition is when someone is a "little bit healthy," but most of the time is miserable. A sick person is unable to function normally. The bodily parameters and functions are abnormal. This is a *disease*! Everybody has met people who used to be

Insights to Neuroimmune Biology
http://dx.doi.org/10.1016/B978-0-12-801770-8.00001-X

sick, acted like one, or looked like one. However, their disease could not be diagnosed. We come across such people less frequently with the advance of *Medicine*, but such patients still exist. Mental conditions more often cause problems with diagnosis of the patients.

Today, we know that higher animals and man have developed the neuroimmune super system (NISS),[2] whereby the central nervous system (CNS), endocrine system (ES), and immune system (IS) are joined together to create a central regulatory circuitry. NISS has to deal with everything happening in the body, right from conception to death. In healthy people NISS must function normally, and tissue and organs must be healthy. "Healthy body, healthy mind" expresses folk wisdom about the importance of mind–body interaction for being healthy.

2 DISEASE

2.1 A Brief History

In the ancient cultures of Persia, Greece, and in the Roman Empire, fever was believed to posses healing powers.[3] This sentiment was maintained until modern times, when Boivin[4] isolated the first pyrogen, lipopolysaccharide (LPS), or endotoxin, from Gram-negative bacteria. LPS is an outstanding pyrogen. Countless studies on LPS have indicated that it is capable of inducing a disease known as "endotoxin shock," which may have lethal consequences.[5] Later it was discovered that a subtoxic doses of endotoxin actually provided protection to animals against infectious diseases and other harmful infections. This condition was known as "endotoxin tolerance." LPS, especially in the detoxified form, could be used as an immunological adjuvant and was effective for the treatment of radiation diseases; drug-induced immunosuppression and bone marrow failure had a beneficial effect on a number of other pathological conditions.[6]

Now we have recognized that LPS is an evolutionally preserved cross-reacting antigen (e.g., all Gram-negative bacteria have LPS), which may be called homologous epitopes (or "*Homotopes*" for short).[7] LPS is an antigen for innate immunity (INIM). Throughout millennia, the INIM system evolved to recognize homotopes within infectious agents. Since these are cross-reacting antigens, recognition may be economized. For example, the INIM antigen receptor, toll-like receptor-4 (TLR4) is one receptor, which recognizes LPS, and with this single receptor the INIM system is able to control infections from all the species of Gram-negative bacteria. This is characteristic of the innate immune recognition.[7]

Table 1.1 Neuroendocrine responses to endotoxins

HPT and GLH	Responses	HPA axis	Responses	Sex hormones	Responses
TRH	↓	CRF	↑	LH	↑↓
TSH	0↓	AVP	↑	FSH	↓
T4	↓	ACTH	↑	E2	↑↓
T3	↓	GC	↑	TS	↑↓
PRL	↑↓	α-MSH	↑	DHEA	↓?
GH	↑↓	β-END	↑	PS	↑↓
IGF-I	↓	CAT	↑		
INS	↑				
GLU	↑				
LEP	↑				

HPT, hypothalamus; GLH, growth and lactogenic hormones; TRH, thyroxin-releasing hormone; T4, thyroxin; T3, tri-iodo-thyronin; PRL, prolactin; GH, growth hormone; IGF-I, insulin-like growth factor-1; GLU, glucagon; LEP, leptine; HPA axis, hypothalamus–pituitary–adrenal axis; CRF or rather CRH, corticotrophin; AVP, arginine–vasopressin; ACTH, adrenocorticotropic hormone; GC, glucocorticoids; α-MSH, alpha-melanocyte-stimulating hormone; β-END, beta-endorphine; CAT, catecholamine; sex hormones; LH, luteinizing hormone; FSH, follicle-stimulating hormone; E2, estradiol; TSH, thyroid-stimulating hormone; DHEA, dehydroepiandrosterone; PS, progesterone.

From the table, it will be obvious that the HPA axis is activated in endotoxin shock, which is analogous to acute febrile illness or the APR. Every hormone is stimulated in the HPA axis. Interestingly insulin, glucagon, and leptin are also stimulated. These hormones must play an important role in the APR. The rest of the hormones are flatly inhibited or after temporary stimulation are inhibited (e.g., GH and PRL). Modified from Ref. [7].

When LPS is given to animals, it is capable of eliciting the syndrome of the infectious diseases caused by Gram-negative bacteria. There are neuroendocrine, metabolic, and immune alterations, just like in the disease (Table 1.1).[7]

If we carefully examine the data given in Table 1.1, we notice that during a disease, hormones of the hypothalamus–pituitary–adrenal (HPA) axis are all activated. This activation is very characteristic of the stress response, as described by Hans Selye[8] and today it is very clear that activation of the HPA axis follows infection, trauma, all sorts of injuries, and even mental illness. What does this mean? What does the HPA axis do? As stated, LPS can induce acute febrile illness, such as septicemia for instance. This is an emergency situation where bacteria are growing in the blood. This emergency situation cannot be handled by the *adaptive immune system* (ADIM) as here, lymphocytes have to grow and differentiate first and only after 5–7 days can respond in an immunological manner. On the other hand, LPS activates the INIM system, which produces interleukin (IL)-1-beta, IL-6, tumor necrosis (TNF)-alpha, and granulocyte–macrophage colony stimulating factor

(GMCSF). These cytokines activate the HPA axis, which are the major mediators of this system. The hypothalamus secretes corticotrophin (CRH) and vasopressin (VP); the adrenal secretes adrenocorticotropic hormone (ACTH), which rapidly releases glucocorticoids (GCs) and catecholamines (CAT). In turn GC and CAT stimulate INIM.[9] This system starts up immediately after the noxious stimulus and becomes fully active within 1 day, exerting maximum protection. So the IS is converted from a dual reactivity to the single INIM response. GC and CAT stimulate suppressor-regulatory T lymphocytes (Tsr), which play an important role in suppressing ADIM. The HPA axis is also suppressive. It can be noticed from Table 1.1 that after a transient initial stimulation, prolactin (PRL) and growth hormone (GH), will be suppressed during acute illness (*acute phase response* (APR)). These hormones maintain ADIM, and now they are not needed. But the INIM system is certainly up to the task of handling host defense. Most of us suffer numerous times febrile illness in a lifetime and normally we recover. This is how well INIM does the job![9]

Insulin, glucagon, and leptin are also activated during APR. These hormones play an important role in sick organisms. Thyroid hormones and sex steroid hormones are suppressed.[9]

The hypothalamic corticotrophin-stimulating hormone (CRH) controls APR, and VP supports CRH at this stage. However, when the acute phase subsides and the disease gets chronic, it is VP and not CRH that regulates the chronic inflammation. VP is capable of bringing about healing and recovery as it restores PRL and stimulates the HPA axis. So VP can create the homeostatic balance that is necessary for *healing*.[10]

The APR is an emergency host defense reaction. This is generally accepted today. APR is mediated by the central nervous system (CNS), ES, and IS, which form the NISS, which is the supreme regulator and protector of the host organism. NISS is with us for a lifetime.[11] Now we compare APR with the *stress reaction*. Stress was claimed to be host response to noxious (nocuous) agents, and brain, and the HPA axis, were involved. Selye himself recognized that stress causes a *"general adaptation syndrome,"*[8] which definitely indicates that the host handles difficulties of adapting to some noxious agents, infection, trauma, and toxins. The IS perfected adaptation an indeed INIM belongs to the stress system, as we know it today. So if stress and APR are identical and APR is a host defense reaction, this means that stress is also a host defense response. The HPA axis, whether mobilized by stress or by APR, is the first to respond and it is activated every time when there is disease or any other noxious agents are around. The host is responding to control the situation.[11] Now we will examine the diseases where the HPA axis is activated.

2.2 HPA Axis and Disease

2.2.1 New Observations Showing that Stress may be Beneficial

In contrast to the general view that stress is harmful, we were unable to find a harmful effect of chronic stress on the internal diseases (gastric ulceration and angina pectoris).[12] Increasing physical activity altered the measures of obesity in farm men and women but *did not affect mental health measures or cortisol secretion.*[13]

In experimental models of *neurodegenerative diseases*, chronic stress or GC treatment was found to exacerbate both clinical symptoms and neurodegenerative processes. However, recent evidence also shows that glucocorticoid receptor (GR) can *exert neuroprotective effects*. Thus, for any potential therapeutic strategy in these neurodegenerative diseases, we need to understand the precise modifications both in HPA axis and in GR activity, and find ways to harness their protective actions.[14]

Results support the notion of altered HPA axis regulation in chronically work-stressed teachers, with differential patterns of *hyper- and hyporeactivity* depending on individual stress conditions and the tested functional level of the HPA axis.[15]

Chronic psychosocial stressor exposure *impairs in vitro* ACTH responsiveness of both the left and right adrenal glands, *whereas in vivo adrenal responsiveness increases to an acute heterotypic stressor.* This suggests that an additional factor present during acute stressor exposure *in vivo* rescues left and right adrenal. ACTH sensitivity, or ACTH itself acts as a CORT secretagogue in chronically stressed chronic subordinate colony housing mice.[16]

2.2.2 Early Life Programming by GCs

The fetus is susceptible to internal and external stimuli that can lead to adverse long-term health consequences. *GCs* are important developmental *switches*, driving changes in gene regulation that are necessary for *normal growth and maturation*. The fetal-HPA axis is particularly susceptible to *long-term programming by GCs*; these effects can persist throughout the life of an organism. *Dysfunction of the HPA axis*, as a result of fetal programming, has been associated with impaired brain growth, altered behavior and increased susceptibility to chronic diseases (such as, metabolic and cardiovascular diseases). Moreover, the effects of GC-mediated programming are evident in subsequent generations, and transmission of these changes can occur through both maternal and paternal lineages.[17]

Low birth weight, a marker of an *adverse in utero environment*, is associated with cardiometabolic disease and brain disorders in adulthood. The adaptive changes made by the fetus in response to the intrauterine environment result in permanent changes in physiology, structure, and metabolism, a phenomenon termed *early life programming*. We have carried out detailed studies in men and women showing that high levels of endogenous GCs, or treatment with exogenous GCs, is associated with an adverse metabolic profile, increased cardiovascular disease, altered mood, and cognitive decline. Studies in humans have now demonstrated that *high maternal cortisol* in pregnancy and/or *inhibition of HSD2* are associated with programmed outcomes in childhood including higher blood pressure, behavioral disorders as well as altered brain structure. Alterations in DNA methylation of genes, important in regulating cortisol levels, tissue GC action, blood pressure, and fetal growth, are present in adulthood in association with both early life parameters and cardiometabolic risk factors.[18]

This is the first evidence for long-lasting effects of *antenatal synthetic GC exposure* on HPA-axis reactivity in term-born children. These findings may bear important implications regarding the vulnerability for stress-related physical and psychiatric disorders, for which *dysregulation of the HPA axis* has been discussed as a potential causal factor.[19]

Early-life stress can impact health in later stages but less is known about how early-life stress impairs HPA axis activity, contributing to *maladaptation* of the stress–response system. Early-life *stress exposure* (either prenatal or in the early postnatal period) can impact developmental pathways resulting in lasting structural and regulatory changes that *predispose to adulthood disease*. Epidemiological, clinical, and experimental studies have demonstrated that early-life stress produces *long-term hyperresponsiveness to stress* with exaggerated circulating *GCs,* and enhanced *anxiety and depression-like behaviors*. Recently, evidence has emerged on early-life stress-induced *metabolic derangements*, for example, *hyperinsulinemia* and altered *insulin sensitivity* on exposure to a high-energy diet later in life.[20]

Pregnancy and *long-term adverse "programming" effects* on the offspring. The consequences of social stress exposure depends on whether during pregnancy the stress occurs, and many of the effects on the offspring are sex specific. Stress during early pregnancy tends to result in pregnancy loss. Stress exposure later in pregnancy results in programmed offspring of low birth weight: a risk factor for various adulthood diseases. Neuroendocrine and behavioral responses to stress in the offspring are particularly sensitive to *fetal programming by prenatal stress,* indicated by enhanced *HPA axis responses* and increased anxiety behavior, which results from permanent changes in the offspring's brain. The hypothalamic–pituitary–gonadal axis may also be affected. Prenatal social stress also *programs future maternal behavior.*[21]

Male stress exposure occurs either throughout puberty or in adulthood. Remarkably, offspring of sires from both paternal stress groups displayed significantly *reduced HPA stress axis responsiveness.* Gene set enrichment analysis in offspring stress-regulating brain regions, the *paraventricular nucleus (PVN) and bed nucleus of the stria terminalis* revealed global pattern changes in transcription suggestive of *epigenetic reprogramming* and consistent of altered offspring stress response, including *increased expression of GC-responsive genes in the PVN.* By examining potential epigenetic mechanisms of germ cell transmission, we found robust changes in *sperm microRNA* (miR) content.[22]

Infants exposed to *chorioamnionitis with funisitis* had a significantly different patterns of cortisol across the samples compared with infants with *chorioamnionitis alone* or no prenatal inflammation ($F(4,139) = 7.3996$, $P < 0.0001$). Postnatal infections, necrotizing enterocolitis, and chronic lung disease were not significantly associated with the cortisol pattern at 18 months CA.[23]

2.2.3 Schizophrenia

Selective abnormalities of GC receptor mRNA (GR mRNA) expression in the *lateral orbitofrontal cortex (OFC) in psychiatric illness*, which are more specific and may be less influenced by NR3C1 genotype than those of the dorsolateral prefrontal cortex reported previously. Our results suggest that the *GRα-D1 protein isoform may be upregulated widely across the frontal cortex in psychiatric illness.*[24]

Abnormal HPA axis function, as indexed by elevated diurnal cortisol levels and/or a blunted *cortisol awakening response (CAR)*, has been observed among patients with first episode psychosis and associated with neurocognitive deficits in this population. *Family history of illness* (FHx) in children, but not *multiple antecedents of schizophrenia* (ASz) children, was characterized by a blunted CAR relative to *typically developing* (TD) *peers* (effect size = −0.73, $P = 0.01$) that was not explained by psychosocial stress exposure or by distress relating to these experiences. Neither FHx nor ASz children were characterized by elevated diurnal cortisol. *Between FHx and ASz children, more pronounced HPA axis function abnormalities (i.e., higher diurnal cortisol levels and greater blunting of the CAR) were associated with poor performance on test of verbal memory and executive function. Hormonal abnormality precedes a disease!*[25]

HPA activity may be chronically elevated; in melancholic depression, panic disorder, obsessive–compulsive disorder, and schizophrenia. The HPA axis may be more reactive to stress in social anxiety disorder and autism spectrum disorders. In contrast, HPA activity is more likely to be low in PTSD and atypical depression. Antidepressants are widely considered to inhibit HPA activity, although inhibition is not unanimously reported in the literature. There is evidence, also uneven, that the mood stabilizers, *lithium*, and *carbamazepine* have the potential to augment HPA measures, while *benzodiazepines,* atypical antipsychotics, and to some extent, typical antipsychotics have the potential to inhibit HPA activity.[26]

2.2.4 Alzheimer's Disease (AD)

Given the capacity of GCs and corticotropin-releasing hormone to induce AD-associated pathologies, a role has been suggested for circadian cortisol hypersecretion in the initiation of sporadic AD; and a temporal mechanism for AD development featuring neuroinflammation-mediated suppression of central GR signaling, has been proposed. The latter may represent a critical phase in AD development, where the density of functional GR is proposed to underlie the "cognitive reserve."

Supporting evidence for this mechanism is drawn from the brain regional locations of *AD neuropathologies*, and from risk factors for AD development (aging, ApoE-4 genotype, and hypertension). Thus, it is argued that *basal hypercortisolemia* merits further scrutiny regarding AD causation and development.[27]

Intracerebroventricular (i.c.v.) injection of amyloid-β(25–35) peptide (Aβ(25–35)) in rat, is a validated acute model of AD. Aβ(25–35) induces memory impairment, alteration of anxiety responses, HPA axis hyperactivity, GR, and mineralocorticoid receptor (MR) increases in brain regions related to HPA axis functions. GRs are progressively translocated into neurons' nucleus, while membrane version of MR is evidenced in all structures considered. The MR/GR ratio was modified in all structures considered.[28]

2.2.5 Huntington's Disease (HD)

The early *Huntington's disease (HD)* group had significantly lower morning cortisol levels relative to pre-HD and controls. In contrast, the early-HD group with at least mild or greater levels of depression symptoms had a comparable cortisol concentration to pre-HD and controls.[29]

Of these two classification approaches, HD motor sign severity was more strongly associated with *high evening cortisol* levels and both reduced information encoding and memory retrieval. Separately, there was also a trend of *higher cortisol levels in pre-HD.* The findings suggest hypercortisolism and the underlying pathological changes may begin many years before a clinical diagnosis is made! But the memory decline associated with HPA axis disturbance may only become detectable once motor signs become pronounced.[30]

We have examined HPA axis response of R6/1 mice following *acute stress* and have found evidence of a *female-specific dysregulation of the HPA axis in R6/1 mice*, which we further isolated to a *hyperresponse of adrenal cortical cells to stimulation* by adrenocorticotrophin hormone. Interestingly, the adrenal pathophysiology was not detected in mice that had been housed *in environmentally enriching conditions*, an effect of enrichment that was also reproduced *in vitro.* This constitutes the first evidence that environmental enrichment can in fact exert a lasting influence on peripheral organ function. *Cognitive stimulation* may therefore not only have benefits for mental function, but also for overall physiological wellbeing.[31]

We found that *CORT consumption did not alter rotarod performance of R6/1 HD or wild-type (WT) littermates. However, the onset of hippocampal-dependent Y-maze deficits was accelerated in male R6/1 mice by 5 days of CORT treatment,* whereas short-term memory of WT and female R6/1 mice was unaffected. We then further investigated the male HD susceptibility to CORT by measuring TrkB activation, BDNF and GR expression, as well as the level of cell proliferation in the hippocampus. *CORT treatment increased the levels of phosphorylated TrkB in male R6/1 mice only.*[32]

2.2.6 Multiple Sclerosis (MS)

Cortisol release as well as *glucocorticoid sensitivity* (GCS) were strongly correlated with time since diagnosis but not with neurological disability. Patients with shorter disease duration (2–12 months) expressed a significantly higher cortisol stress response while MS patients with longer disease duration (14–36 months) showed a significantly diminished HPA response as well as lower poststress GCS.[33]

HPA axis activation in untreated MS drifts from hypothalamo–pituitary to more adrenal activation, consistent with adrenal sensitization or hypertrophy due to chronic HPA axis activation. HPA system regulation remains more stable in MS patients on DMT.[34] High cortisol levels were associated with slower disease progression, especially in females with secondary progressive MS. Interestingly, normal-appearing white matter (NAWM) of patients with high cortisol levels displayed elevated expression of *GC-responsive genes, such as CD163*, and decreased expression of proinflammatory genes, such as *tumor necrosis factor-α (TNF-α)*. Thus, HPA axis hyperactivity in MS coincides with *low inflammation and/or high neurodegeneration,* and may impact on lesion pathology and molecular mechanisms in NAWM and thereby be of great importance for suppression of disease activity.[35]

Circadian cortisol release, in particular *CAR,* shows a course-specific pattern with most pronounced release in *relapsing–remitting* MS (RRMS). There is also some evidence for greater CAR in RRMS patients with *Kurtzke's Expanded Disability Status Scale* (EDSS) progression. As a consequence, *CAR might be of predictive value in terms of neurological disability in RRMS patients.* The possible role of neuroendocrine–immune interactions in MS pathogenesis is further discussed.[36]

2.2.7 Anxiety and Depression

Increased *dyslipidemia* and (abdominal) obesity risk in patients with more severe anxiety disorders and *depression* may be partly explained by *chronic low-grade inflammation* and *smoking.* TCAs may increase metabolic risk through *enhanced sympathetic* and *decreased parasympathetic ANS activity.* That the *HPA axis had no impact* in our sample may reflect the possibility that the HPA axis only plays a role in acute stress situations rather than under basal conditions.[37]

Among patients with depressive or anxiety disorders, a *lower CAR* – which may be indicative of underlying exhaustion of the HPA axis – predicted an *unfavorable course trajectory.*[38]

2.2.8 Major Depressive Disorder

Depressed women exhibited *flatter diurnal cortisol rhythms* and *more impaired suppression of cortisol following dexamethasone administration* than nondepressed women over three sampling days. In addition, *flatter diurnal cortisol* slopes were associated with reduced cortisol response to dexamethasone treatment, both for all women and for depressed women, when considered separately. Finally, greater self-reported depression severity was associated with flatter diurnal cortisol slopes and with less dexamethasone-related cortisol suppression for depressed women.[39]

2.2.9 Various Other Forms of Depression

Our study provides further support for the predictive role *over 5.5 years of HPA axis dysregulation, that is, lower morning cortisol levels, of recurrence in recurrently depressed patients. Childhood trauma* is associated to having lower cortisol levels. It might have long-term consequences for dealing with stress and the HPA axis.[40]

Results indicate that *microglial activation is significantly increased in the infralimbic, cingulate, and medial orbital cortices; nucleus accumbens; caudate putamen; amygdala; and hippocampus of the mouse brain* as a function of *unpredictable chronic mild stress (UCMS)* while levels of proinflammatory cytokines did not differ among the groups. This finding suggests that *neuroinflammation occurs in depression* and may be implicated in the subject's behavioral response. They also suggest that UCMS could be a potentially reliable model to study depression-induced *neuroinflammation*.[41]

Elevated long-term cortisol levels might play a role in a subgroup of patients with bipolar disorder (BD). There may be differences in pathogenesis of younger and older onset BD suggesting two different disease entities.[42]

These studies support the presence of *HPA-axis hyperactivity and a blunted HPA-axis response to stress at the onset of psychosis*.[43]

2.2.10 Inflammation

The *anti-inflammatory role of androgens*, loss of androgens (*androgen drain*), bimodal role of *estrogens (support B cells and inhibit macrophages and T cells)*, increased conversion of androgens to estrogens in inflammation (androgen drain), *disturbances of the gonadal axis, inadequate amount of HPA* axis hormones relative to inflammation (*disproportion principle*), biologics partly improve neuroendocrine axes, anticorticotropin-releasing hormone therapies improve inflammation (*antalarmin*), bimodal role of the *sympathetic nervous system (SNS) (proinflammatory early, anti-inflammatory late;* most probably due to catecholamine-producing local cells), *anti-inflammatory role of alpha-melanocyte-stimulating hormone* (α-MSH), *vasoactive intestinal peptide, and the vagus nerve via* α*7 nicotinergic receptors. Circadian rhythms* of hypothalamic origin are responsible for circadian rhythms of symptoms (neuroimmune link revealed). Important new pain-sensitizing immunological pathways were found in the last decade.[44]

Coronary artery disease *pre- and postpercutaneous coronary intervention* (PCI), states were associated with increased percentage of activated/degranulated neutrophils, as indicated by elevated lactoferrin parameters. The 1d-PCI declines those associated with plasma ACTH in both groups. The correlation of plasma cortisol with plasma lactoferrin in the extremely stressed *acute coronary syndrome* (ACS), before stenting. However, it suggests an association of cortisol with neutrophil activation.[45]

Previous studies have shown altered expression of central and peripheral HPA axis hormones in chronic inflammatory skin diseases and skin tumors, and that hyperactive lesional HPA axis hormones may negatively feedback to the central HPA axis and interact with some cytokines and neuropeptides, leading to symptom deterioration. This provides an evidence-based understanding of the activation of the central and peripheral HPA axis in common skin diseases and its association with disease activity.[46]

2.2.11 Various Other Conditions

HPA axis contributes to the regulation of T-cell *activation in AIDS*. This may represent an important pathway, through which psychological states and the HPA axis influence progression of HIV.[47]

In type 2 diabetic patients, the HPA axis is clearly hyperactive as evident in increased urinary free cortisol, diminished cortisol suppression after dexamethasone and increased ACTH-induced cortisol levels. Abdominal obesity and the presence of chronic complications increased the HPA axis hyperactivity in type 2 diabetes. Augmentation of positive feedback is associated with insulin resistance and negative feedback with abdominal obesity.[48]

In the hippocampus of aged rats with spatial *memory impairments*, GR protein level was decreased in the nucleus but not in the cytosol, and levels of GC response elements binding activity was decreased.[49]

African women have been previously presented with an increased vascular reactivity. Conversely, Caucasian women have displayed an increased central cardiac reactivity. We included African (n = 102) and Caucasian (n = 115) women in the study, matched for age and body mass index. Elevated CRP levels were observed in African women compared to Caucasian women. A trend of hypocortisolism was exhibited in both hypertensive ethnic groups. Systolic blood pressure (SBP) and a vascular marker arterial compliance (Cw), predicted hypertension in African women. Conversely, in Caucasian women, only SBP predicted hypertension.[50]

The findings suggest that stress-associated *alopecia areata* (AA) in some patients may be associated with stress. The ACTHR gene −2T > C variant may be one important factor that influences stress perception of patients with AA.[51]

Chronic breathlessness causes excessive stimulation of the HPA axis, resulting in dysfunctional regulation of the HPA axis and associated neuropsychological, metabolic, and immunological sequelae. A number of observations provide indirect support for this hypothesis. First, breathlessness and HPA axis are both associated with anxiety. Second, similar cortico–limbic system structures govern both breathlessness perception and HPA-axis regulation. Finally, breathlessness and HPA axis dysfunction are both independent predictors of survival.[52]

Compared with the healthy controls, the *oral lichen planus (OLP)* patients demonstrated a less effective coping ability, had higher scores in stress perception and loneliness, and had no significant variation in their anxiety and depressive symptoms. The OLP patients also showed dysregulation of the HPA axis activity with a significant reduction of diurnal salivary cortisol production, which was particularly significant in the morning hours. OLP subjects had a reduced capability of coping with stress events and presented a dysregulation of HPA-axis activity with hypocortisolism detected in the morning hours.[53]

2.2.12 Agents that Influence the HPA axis

The results suggested that *fish oil* (FO) attenuates the activation of the HPA axis induced by LPS challenge. The beneficial effects of FO on the HPA axis may be associated with decreasing the production of brain, or peripheral proinflammatory cytokines through inhibition of TLR4 and nucleotide-binding oligomerization domain protein signaling pathways.[54]

It is known that *heroin dependence* and withdrawal are associated with changes in the HPA axis. Both plasma ACTH and corticosterone (CORT) levels were elevated during acute withdrawal, and heroin challenge at 20 mg/kg (the last dose of chronic escalation) at this time point attenuated HPA hyperactivity. During chronic withdrawal, HPA hormonal levels returned to baseline, but heroin challenge at 5 mg/kg decreased ACTH levels. In contrast, this dose of heroin challenge stimulated the HPA axis in heroin naïve rats. In the anterior pituitary, pro-opiomelanocortin (POMC) mRNA levels were increased during acute withdrawal and retuned to control levels after chronic withdrawal. In the medial hypothalamus, however, the POMC mRNA levels were decreased during acute withdrawal, and increased after chronic withdrawal.[55]

Soluble *beta amyloid* (sBA), in transgenic AD model, resulted to be increased after chronic and acute stress and alterations in cortisol levels have been reported in AD. We found a significant decrease in plasma CORT concentrations in BA-treated rats.[56]

Stress-dependent and gender-specific neuroregulatory roles of the *apelin* receptor (APJ) in the HPA axis response to acute stress. We measured ACTH and CORT plasma levels in male and female mice lacking APJ (APJ knockout and APJ KO) and in WT controls, in response to a variety of acute stressors. Exposure to mild restraint, systemic injection of lipopolysaccharide (LPS), insulin-induced hypoglycemia, and forced swim (FS) stressors, and elevated plasma ACTH and CORT levels in WT mice. Acute mild restraint significantly increased plasma ACTH and CORT to a similar level in APJ KO mice as in WT mice. However, an intact APJ was required for a conventional ACTH, but not CORT, response to LPS administration in male mice and to insulin-induced hypoglycemia in male and female mice. In contrast, APJ KO mice displayed an impaired CORT response to acute FS stress, regardless of gender.[57]

Adiponectin is an adipocyte-derived hormone with antidiabetic and insulin-sensitizing properties. The i.c.v. injection of an adiponectin neutralizing antibody precipitates stress-induced depression-like behavior. Conversely, i.c.v. administration of exogenous adiponectin produces antidepressant-like behavioral effects in normal weighing mice and in diet-induced obese, diabetic mice.[58]

A fine-controlled interaction between plasma *corticosteron-binding globulin* (CBG) expression and stress-induced CORT release. Although plasma CBG levels, and therefore CBG-binding capacity, were higher in HR animals, CORT secretion overloaded the CBG buffering function in response to stressors, resulting in clearly higher free CORT levels in HR compared with IR and LR mice (HR > IR > LR; (*high-reactivity* (HR), *intermediate-reactivity* (IR), or *low-reactivity* (LR)), resembling the pattern of total CORT increase in all three lines. The stressors, *restraint, or FS, did not evoke fast CBG release from the liver into the bloodstream* and therefore CBG-binding capacity was not altered in our three-mouse lines.[59]

The neuropeptide *pituitary adenylate cyclase-activating polypeptide (PACAP)* regulates activation of the HPA axis and adrenal gland in response to various stressors. Defeat-exposed PACAP−/− mice showed a marked attenuation of stress-induced increases in serum CORT levels, cellular PVN ΔFosB immunostaining, and depression-like behaviors (social interaction and FS tests) compared to WT control mice. The PACAP−/− mice showed reduced PVN FosB-positive cell numbers, but relatively elevated cell counts in several forebrain areas including the medial prefrontal cortex (mPFC), after social stress. PACAP appears to be specific for mediating HPA activation *only in psychological stress* because marked elevations in plasma CORT after a systemic stressor (LPS administration) occurred regardless of genotype.[60]

2.2.13 Regulatory Mechanisms of HPA Axis

IMO stress-induced HPA activation, which could not be influenced by drugs regulating GCs or CRH-r1 receptors.[61]

Metabolic risk should be routinely assessed in depressed patients and also taken into account for therapeutic decisions. Alternative targets should be considered for innovative antidepressant agents, including cytokines and their receptors, intracellular inflammatory mediators, GRs, O&NS pathways, and peripheral mediators.[62] Prenatal ethanol exposure (PEE) induces enhanced susceptibility to metabolic syndrome in an adult offspring, fed a high-fat diet, and the underlying mechanism involves an *HPA axis-associated neuroendocrine metabolic programming alteration.*[63]

The major breakthrough in understanding the interactions between the CNS and *gut*, was the discovery of the enteric nervous system (ENS) in the nineteenth century. ENS (also called "little brain"), plays a crucial role in the regulation of the physiological gut functions. Furthermore, the identification of corticotropin-releasing factor (CRF) and the development of specific CRF receptor antagonists have permitted to characterize the neurochemical basis of the stress response. The neurobiological response to stress in mammals involves three key mechanisms, namely: (1) stress is perceived and processed by higher brain centers; (2) the brain mounts a neuroendocrine response by way of *the HPA axis and the autonomic nervous system (ANS);* and (3) the brain triggers feedback mechanisms by HPA and ANS stimulation to restore homeostasis. Various stressors such as anger, fear, painful stimuli, as well as life or social learning experiences affect both the individual's physiological and gastric function, revealing a two-way interaction between brain and stomach. There is overwhelming experimental and clinical evidence that stress influences gastric functions, thereby outlining the pathogenesis of gastric diseases such as functional dyspepsia, gastroesophageal reflux disease, and peptic ulcer disease.[64]

The actions of *testosterone* (T) in modulating *HPA axis* activity involve *5α-reductase* within the CNS.[65]

mPFC transection may enhance the *IMO stress*-induced *sympathoadrenal system (SAS)* activity without affecting the activity of the HPA axis. We found that the mPFC may exert an *inhibitory effect* on the SAS activity in the stressed animals.[66]

Gene expression of Ppid, a positive regulator of the GR, was predicted by plasma estradiol and 34% lower in control adolescent females than males, indicating that sex steroids may play a role in the sexually dimorphic response. After chronic adolescent stress, females displayed elevated hippocampal expression of Bag1 and Ppid genes that was not observed in males. Overall, the GR output to an acute stressor, illustrated by transcription of Nr3c1 (encoding the GR), Bag1, Fkbp5, Ppid, and Src1, was significantly upregulated and differed in a sex-specific and chronic stress-dependent manner. This study provides new evidence for sex differences during development and adaptation of the GR chaperone system.[67]

GCs bind to GRs in multiple body compartments, including the brain, and consequently have wide-reaching actions. For this reason, GCs serve a vital function in negative *feedback inhibition* of their own secretion. Negative feedback inhibition is mediated by a diverse collection of mechanisms, including fast, nongenomic feedback at the level of the PVN, stress shut-off at the level of the limbic system, and attenuation of ascending excitatory input through destabilization of mRNAs encoding neuropeptide drivers of the HPA axis. In addition, there is evidence that GCs participate in stress activation via feedforward mechanisms at the level of the amygdala. *Feedback deficits are associated with numerous disease* states, underscoring the necessity for adequate control of GC homeostasis. Thus, rather than having a single, defined feedback "switch," control of the stress response requires a wide-reaching *feedback "network"* that coordinates HPA activity to suit the overall needs of multiple body systems.[68]

Despite the male–female differences in HPA-axis activity, the effect of repeated stress exposure on *hippocampal cell differentiation* was not significantly different between sexes.[69]

3 DISCUSSION

Health is a condition when everything is all right with us. We are feeling good; we are happy, optimistic, and are ready for challenges. The French scientist Claude Bernard was the first to recognize that living organisms have their physiological parameters within physiological limits, which is designated as "normal" in *Medicine*. The scientific term for a healthy condition is *homeostasis,* which means our condition is stable, and at this stage only normal variability takes place.

Now *what about disease*? You may guess that there is disorganization, that is, various pathological conditions will take place, such as, infection, injury, and immune activation. All this can take place locally at the site of inflammation, like an abscess for instance.

This is a local process with little, if any systemic effect.

However, there is another systemic response where the entire organism is responding. Hans Selye worked with this reaction first and he called this the *stress response.*[70] The agents causing stress were called *stressors* and the reaction that was induced was the *stress response.*[71] Selye recognized that the stress response was a defense reaction, and he named it the general adaptation response (GAS).[8]

If we compare GAS with the APR as we call the acute immune/inflammatory response today, it is clear that we are talking about the same reaction (Figure 1.1).[72] So if APR is an acute defense response defending the host organism, then all kinds of diseases should activate this system. This is true for the stress reaction.

GAS must have the same function as APR. Clearly, the stress reaction is a defense response. For instance, think of acute febrile illness. How many times did you have fever in your lifetime? And as a rule, you recovered. So this is how well your GAS (APR) does the job!

If you examine thoroughly Figure 1.1, you notice that the INIM system is the first to react. INIM produces IL-beta, IL-6, TNF-α, and granulocyte macrophage colony stimulating factor (GM-CSF). These cytokines activate the HPA axis. The brain and the liver are activated and make acute phase proteins, and the bone marrow produces a lot of white cells. So there is *leukocytosis*. Fever and catabolism prevail. Sickness behavior takes place.

On the basis of this brief introduction, one could predict already that a number of diseases will activate the HPA axis because the host will defend it self, and the HPA axis coordinates the *neuroimmune host defense response*. If the HPA axis is not activated during systemic disease it is a lethal problem.

Stress may be beneficial. Some papers appeared recently, saying that stress is beneficial. It is claimed that stress would not harm you if you were healthy.[13,14] Recent evidence also shows that GR can *exert neuroprotective effects*.[15] These observations by others lend support to what we are saying.

Early life programming by GCs. If mothers are sick, then babies will be sick as a consequence. During sickness, the HPA axis is elevated and the adrenal gland produces excess GCs. It is imperative that the levels of GCs are right during embryonic development. Too much or too little GC is not correct and leads to a disease. The babies or baby animals cannot develop normally in such abnormal conditions, so they are born sick.

The fetal HPA axis is particularly susceptible to *long-term programming by GCs*. These effects will persist throughout the life of an organism. *Dysfunction of the HPA axis* as a result of fetal programming has been associated with impaired brain growth, altered behavior, and increased susceptibility to chronic disease (such as metabolic and cardiovascular diseases).

Evidence has emerged that early-life stress-induced *metabolic derangements*, for example, *hyperinsulinemia* and altered *insulin sensitivity*, on exposure to a high-energy diet later in life.[22]

Postnatal infections, necrotizing enterocolitis, and chronic lung disease were not significantly associated with the cortisol pattern at 18 months CA.[25]

Schizophrenia. Hormonal abnormality precedes disease![27]

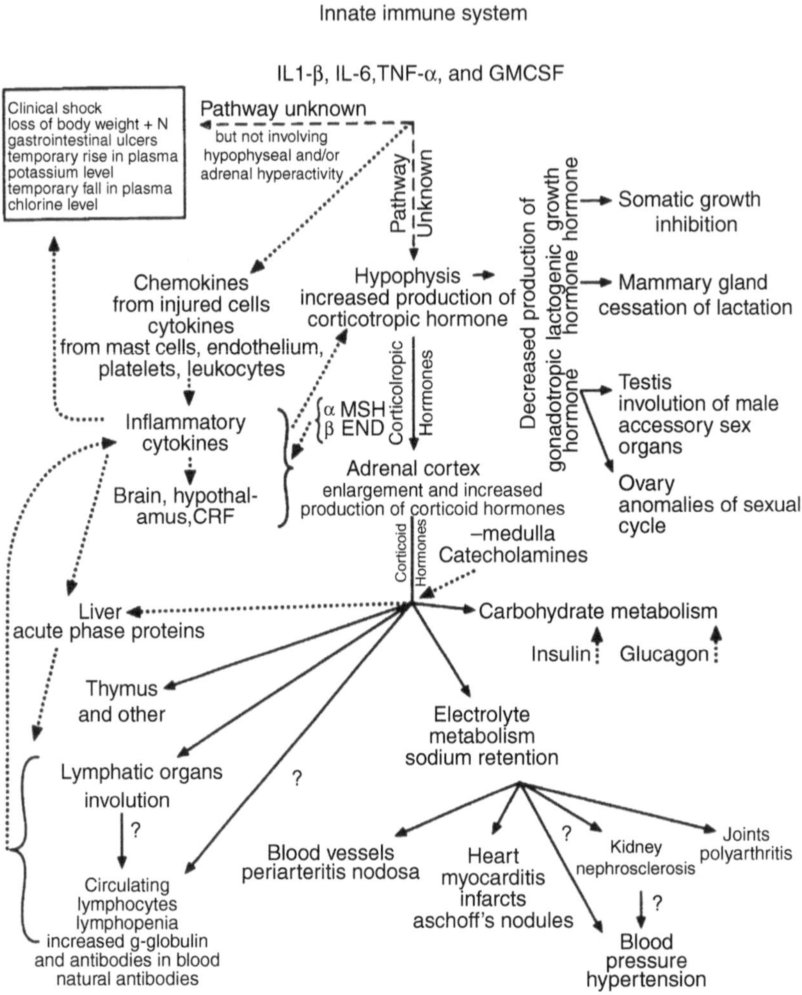

Figure 1.1 ***The APR.*** The solid lines and dashed lines are from a figure published by Hans Selye. Dotted lines indicate our recent understanding of the neuroendocrine response to stress or APR. There is complete overlapping on the neuroendocrine response to stress and of APR. The APR or febrile illness, are acute immune inflammatory reactions. Initially the INIM system senses the infection, and secretes cytokines, injury, noxious agents, and even mental abnormality is recorded by this system. The HPA axis stimulates GCs and catecholamines. SNS stimulates immunity whereas the parasympathetic nerve system is inhibitory for immune reactions. The liver is activated and produces acute phase proteins. The bone marrow produces excess number of leukocytes, which causes leukocytosis in the blood. The thymus and lymphoid organs undergo atrophy. Somatic growth and male sexual organs are inhibited. The ovary stops cycling and sex life is arrested. Fever and catabolism prevails. Most febrile diseases regress spontaneously so our stress system, or rather APR does an excellent job of healing and recovery. *(Source: Modified from Ref. [7].)*

4 CONCLUSIONS

Healthy organisms can cope with or even correct the effects of stress. This may explain the observation that stress epigenetic reprogramming is going on. Healthy animals would not get sick of stress but rather adapt to the changes induced by stress. These stress-induced changes will be inherited and will protect future generations from the same form of infections. This epigenetic programming makes evolution possible. It may be regarded as the prerequisite of Darwinian evolution.

Inflammation was quoted by several investigators as causing diseases affecting the nervous system. However, inflammation is only a symptom of the disease, and it cannot be a cure. Factors that cause inflammation must be detected. IS, T cells, phagocytic cells, monocyte – macrophages, cytokines, chemokines, some neurotransmitters, neuropeptides, and hormones that regulate immunity may be involved. Also some denatured antigen, foreign antigens, infectious diseases, trauma, noxious agents, and emotional factors may trigger immune reactions. *Inflammation is a symptom, not a disease!*

The *amyloid-β(25–35) peptide* given to rats will cause experimental AD. INIM activity has to be examined here: CTK, NK, NKT, MØ, and MOC should be studied. *In vitro* tests with the antigen, assay with CTK, or immune cells.

Cortisol abnormality is the first and later the HPA axis disturbances occur in AD. Environmental agents, stress, infection, and mental disorders can affect the HPA axis. INIM activation is possible for all these causes.

HPA and neural activation means that the *INIM system is also activated* because the nervous system, ES, and IS form an immune–neuroendocrine circuitry, and collaborate life for defending the host organism.[2] It provides immediate protection against dangerous infections. With immune reaction, there are neuroendocrine changes, so a more intense HPA axis response means better INIM and better protection of the host organism. Now, if this increased protection is inherited, it will function as the key to evolution. Create better protection, adaptation, and fix into the genome, so that the offspring inherits better genes for survival.

What is going on when stress has a harmful effect? Apparently, abnormal stress response means abnormal responding subjects, when animals or humans are involved. The HPA axis may not respond normally to stress stimuli, so this may be a problem. There are various forms of HPA abnormalities. However, the HPA axis is connected practically with the host defense system and *abnormality means disease.*

Epigenetic programming is going on during pregnancy. GCs are the major hormones regulating the imprinting process during gestation. If GCs are abnormal, small underdeveloped babies are born, with impaired growth of brain, altered behavior, and increased susceptibility to disease. Epigenetic programming occurs in man and has to do with type 2 diabetes, and also programs future maternal behavior. The PVN and stria terminalis are involved with regulating GC-sensitive genes in the PVN.

The i.c.v. injection of amyloid-β(25–35) peptide (Aβ(25–35)) in rat is a validated *acute model of AD*. Several investigators propose that inflammation is causing disease of the CNS. Inflammation is only a symptom. There must be causative agents for inflammation. Hormonal abnormality is often present prior to diagnosing disease. This could explain the disease. Sick people may react erratically to (stress) stimulus. The work with amyloid is interesting. This protein normally is present in brain tissue and has to do with regulating molecular folding. If amyloid denatures, protein structure will be abnormal and it stimulates autoimmune diseases. The INIM system is involved with stress and acute phase immune/inflammatory reactions. Inflammatory cells cause injury, which is raising cytokines by injured cells and the CNS and IS take notice of the problem.

REFERENCES

1. Bernard C, Lecons sur les phenomenes de la vie communs aux animaux et aux vegetaux. Paris: Balliere; 1878.
2. Berczi I. Integration and regulation of higher organisms by the neuroimmune supersystem. *Int J Integrative Biol* 2007;**1**(3):216–31.
3. Nowotny A. In search of active sites in endotoxins. In: Nowotny A, editor. *Beneficial effects of endotoxins*. New York: Plenum Press; 1983. p. 1–55.
4. Boivin A, Mesrobeanu J, Mesrobeanu L. Technique pour la preparation des polysaccharides microbiens specifiques. *C R Soc Biol* 1933;**113**:490.
5. Nagy Z, Berczi I, Bertok L. Experimental data on the pathogenesis of edema disease of swine. Clinical picture, gross and microscopic lesions related to endotoxin shock. *Zbl Vet Med Reihe B* 1968;**15**:504–11.
6. Bertók L. Radiodetoxified endotoxin as a potent stimulator of nonspecific resistance. *Persp Biol Med* 1980;**24**:61–6.
7. Berczi I. Neuroendocrine response to endotoxin. *Ann NY Acad Sci* 1998;**851**:411–5.
8. Selye H. The general adaptation syndrome and the diseases of adaptation. *J Clin Endocrinol* 1946;**6**:117.
9. Berczi I, Szentivanyi A. The acute phase response. In: Berczi I, Szentivanyi A, editors. *In neuroimmmune biology, volume 3: the immune-neuroendocrine circuitry. history and progress*. Amsterdam: Elsevier; 2003. p. 463–94.
10. Berczi I, Villalobos Hernández Egina C, Quintanar Stephano A, Campos R, Kovacs K. The healing power of vasopressin. *Adv Neuroimmune Biol* 2012;**2**:217–24.
11. Berczi I, Quintanar-Stephano A, Kovacs K. Neuroimmune regulation in immunocompetence, acute illness, and healing. *Ann NY Acad Sci* 2009;**1153**:220–39.

12. Filaretova L, Morozova O, Laszlo F, Morschl E, Zelena D. Does chronic stress enhance the risk of diseases? *Endocr Regul* 2013;**47**(4):177–88.
13. Brumby S, Chandrasekara A, Kremer P, Torres S, McCoombe S, Lewandowski P. The effect of physical activity on psychological distress, cortisol and obesity: results of the farming fit intervention program. *BMC Public Health* 2013;**13**:1018.
14. Vyas S, Maatouk L. Contribution of glucocorticoids and glucocorticoid receptors to the regulation of neurodegenerative processes. *CNS Neurol Disord Drug Targets* 2013;**12**: 1175–93.
15. Wolfram M, Bellingrath S, Feuerhahn N, Kudielka BM. Emotional exhaustion and over-commitment to work are differentially associated with hypothalamus-pituitary-adrenal (HPA) axis responses to a low-dose ACTH1-24 (Synacthen) and dexamethasone-CRH test in healthy school teachers. *Stress* 2013;**16**:54–64.
16. Uschold-Schmidt N, Nyuyki KD, Füchsl AM, Neumann ID, Reber SO. Chronic psychosocial stress results in sensitization of the HPA axis to acute heterotypic stressors despite a reduction of adrenal *in vitro* ACTH responsiveness. *Psychoneuroendocrinology* 2012;**37**:1676–87.
17. Moisiadis VG, Matthews SG. Glucocorticoids and fetal programming part 2: mechanisms. *Nat Rev Endocrinol* 2014;**10**:403–11.
18. Reynolds RM. Nick Hales award lecture 2011: glucocorticoids and early life programming of cardiometabolic disease. *J Dev Orig Health Dis* 2012;**3**:309–14.
19. Alexander N, Rosenlöcher F, Stalder T, Linke J, Distler W, Morgner J, Kirschbaum C. Impact of antenatal synthetic glucocorticoid exposure on endocrine stress reactivity in term-born children. *J Clin Endocrinol Metab* 2012;**97**:3538–44.
20. Maniam J, Antoniadis C. Morris MJ early-life stress, HPA axis adaptation, and mechanisms contributing to later health outcomes. *Front Endocrinol* 2014;**5**:73.
21. Brunton PJ. Effects of maternal exposure to social stress during pregnancy: consequences for mother and offspring. *Reproduction* 2013;**146**(5):R175–89.
22. Rodgers AB, Morgan CP, Bronson SL, Revello S, Bale TL. Paternal stress exposure alters sperm microRNA content and reprograms offspring HPA stress axis regulation. *J Neurosci* 2013;**33**:9003–12.
23. Gover A, Chau V, Miller SP, Brant R, McFadden DE, Poskitt KJ, Synnes A, Weinberg J, Grunau RE. Prenatal and postnatal inflammation in relation to cortisol levels in preterm infants at 18 months corrected age. *J Perinatol* 2013;**33**:647–51.
24. Sinclair D, Webster MJ, Fullerton JM, Weickert CS. Glucocorticoid receptor mRNA and protein isoform alterations in the orbitofrontal cortex in schizophrenia and bipolar disorder. *BMC Psychiatry* 2012;**12**:84.
25. Cullen AE, Zunszain PA, Dickson H, Roberts RE, Fisher HL, Pariante CM, Laurens KR. Cortisol awakening response and diurnal cortisol among children at elevated risk for schizophrenia: relationship to psychosocial stress and cognition. *Psychoneuroendocrinology* 2014;**46**:1–13.
26. Jacobson L. Hypothalamic-pituitary-adrenocortical axis: neuropsychiatric aspects. *Compr Physiol* 2014;**4**:715–38.
27. Notarianni E. Hypercortisolemia and glucocorticoid receptor-signaling insufficiency in Alzheimer's disease initiation and development. *Curr Alzheimer Res* 2013;**10**:714–31.
28. Brureau A, Zussy C, Delair B, Ogier C, Ixart G, Maurice T, Givalois L. Deregulation of hypothalamic-pituitary-adrenal axis functions in an Alzheimer's disease rat model. *Neurobiol Aging* 2013;**34**:1426–39.
29. Shirbin CA, Chua P, Churchyard A, Lowndes G, Hannan AJ, Pang TY, Chiu E, Stout JC. Cortisol and depression in pre-diagnosed and early stage Huntington's disease. *Psychoneuroendocrinology* 2013;**38**:2439–47.
30. Shirbin CA, Chua P, Churchyard A, Hannan AJ, Lowndes G, Stout JC. The relationship between cortisol and verbal memory in the early stages of Huntington's disease. *J Neurol* 2013;**260**:891–902.

31. Du X, Leang L, Mustafa T, Renoir T, Pang TY, Hannan AJ. Environmental enrichment rescues female-specific hyperactivity of the hypothalamic-pituitary-adrenal axis in a model of Huntington's disease. *Transl Psychiatry* 2012;**2**:e133.
32. Mo C, Pang TY, Ransome MI, Hill RA, Renoir T, Hannan AJ. High stress hormone levels accelerate the onset of memory deficits in male Huntington's disease mice. *Neurobiol Dis* 2014;**69**:248–62.
33. Kern S, Rohleder N, Eisenhofer G, Lange J, Ziemssen T. Time matters. Acute stress response and glucocorticoid sensitivity in early multiple sclerosis. *Brain Behav Immun* 2014; pii: S0889-1591(14)00119-6.
34. Kümpfel T, Schwan M, Weber F, Holsboer F, Trenkwalder C. Then Bergh F(2). Hypothalamo-pituitary-adrenal axis activity evolves differentially in untreated versus treated multiple sclerosis. *Psychoneuroendocrinology* 2014;**45**:87–95.
35. Melief J, de Wit SJ, van Eden CG, Teunissen C, Hamann J, Uitdehaag BM, Swaab D, Huitinga I. HPA axis activity in multiple sclerosis correlates with disease severity, lesion type and gene expression in normal-appearing white matter. *Acta Neuropathol* 2013;**126**:237–49.
36. Kern S, Krause I, Horntrich A, Thomas K, Aderhold J, Ziemssen T. Cortisol awakening response is linked to disease course and progression in multiple sclerosis. *PLoS ONE* 2013;**8**:e60647.
37. van Reedt Dortland AK, Vreeburg SA, Giltay EJ, Licht CM, Vogelzangs N, van Veen T, de Geus EJ, Penninx BW, Zitman FG. The impact of stress systems and lifestyle on dyslipidemia and obesity in anxiety and depression. *Psychoneuroendocrinology* 2013;**38**:209–18.
38. Vreeburg SA, Hoogendijk WJ, DeRijk RH, van Dyck R, Smit JH, Zitman FG, Penninx BW. Salivary cortisol levels and the 2-year course of depressive and anxiety disorders. *Psychoneuroendocrinology* 2013;**38**:1494–502.
39. Jarcho MR, Slavich GM, Tylova-Stein H, Wolkowitz OM, Burke HM. Dysregulated diurnal cortisol pattern is associated with glucocorticoid resistance in women with major depressive disorder. *Biol Psychol* 2013;**93**:150–8.
40. Bockting CL, Lok A, Visser I, Assies J, Koeter MW, Schene AH. DELTA study group. Lower cortisol levels predict recurrence in remitted patients with recurrent depression: a 5.5 year prospective study. *Psychiatry Res* 2012;**200**:281–7.
41. Farooq RK, Isingrini E, Tanti A, Le Guisquet AM, Arlicot N, Minier F, Leman S, Chalon S, Belzung C, Camus V. Is unpredictable chronic mild stress (UCMS) a reliable model to study depression-induced neuroinflammation? *Behav Brain Res* 2012;**231**(1):130–7.
42. Manenschijn L, Spijker AT, Koper JW, Jetten AM, Giltay EJ, Haffmans J, Hoencamp E, van Rossum EF. Long-term cortisol in bipolar disorder: associations with age of onset and psychiatric co-morbidity. *Psychoneuroendocrinology* 2012;**37**:1960–8.
43. Borges S, Gayer-Anderson C, Mondelli V. A systematic review of the activity of the hypothalamic-pituitary-adrenal axis in first episode psychosis. *Psychoneuroendocrinology* 2013;**38**:603–11.
44. Straub RH, Bijlsma JW, Masi A, Cutolo M. Role of neuroendocrine and neuroimmune mechanisms in chronic inflammatory rheumatic diseases – the 10-year update. *Semin Arthritis Rheum* 2013;**43**:392–404.
45. Keresztes M, Horváth T, Ocsovszki I, Földesi I, Serfőző G, Boda K, Ungi I. ACTH- and cortisol-associated neutrophil modulation in coronary artery disease patients undergoing stent implantation. *PLoS ONE* 2013;**8**:e71902.
46. Kim JE, Cho BK, Cho DH, Park HJ. Expression of hypothalamic-pituitary-adrenal axis in common skin diseases: evidence of its association with stress-related disease activity. *Acta Derm Venereol* 2013;**93**:387–93.
47. Patterson S, Moran P, Epel E, Sinclair E, Kemeny ME, Deeks SG, Bacchetti P, Acree M, Epling L, Kirschbaum C, Hecht FM. Cortisol patterns are associated with T cell activation in HIV. *PLoS ONE* 2013;**8**(7):e63429.

48. Prpić-Križevac I, Canecki-Varžić S, Bilić-Ćurčić I. Hyperactivity of the hypothalamic-pituitary-adrenal axis in patients with type 2 diabetes and relations with insulin resistance and chronic complications. *Wien Klin Wochenschr* 2012;**124**:403–11.
49. Lee SY, Hwang YK, Yun HS, Han JS. Decreased levels of nuclear glucocorticoid receptor protein in the hippocampus of aged Long-Evans rats with cognitive impairment. *Brain Res* 2012;**1478**:48–54.
50. Tolmay CM, Malan L, van Rooyen JM. The relationship between cortisol, C-reactive protein and hypertension in African and Causcasian women: the POWIRS study. *Cardiovasc J Afr* 2012;**23**:78–84.
51. Guo HW, Guo H, Li KS, Wu J, Yang SY, Liu BH, Hao F, Bai Y. The −2T/C polymorphism in the adrenocorticotropin receptor gene affects stress perception of patients with alopecia areata. *Int J Dermatol* 2013;**52**:441–5.
52. Ryan R, Spathis A, Clow A, Fallon M, Booth S. The biological impact of living with chronic breathlessness – a role for the hypothalamic-pituitary-adrenal axis? *Med Hypotheses* 2014;**83**:232–7.
53. Pippi R, Patini R, Ghiciuc CM, Sandu RB, Pasquali V, Scaccianoce S, Dima-Cozma LC, Patacchioli FR. Diurnal trajectories of salivary cortisol, salivary α-amylase and psychological profiles in oral lichen planus patients. *J Biol Regul Homeost Agents* 2014;**28**:147–54.
54. Liu Y, Chen F, Li Q, Odle J, Lin X, Zhu H, Pi D, Hou Y, Hong Y, Shi H. Fish oil alleviates activation of the hypothalamic-pituitary-adrenal axis associated with inhibition of TLR4 and NOD signaling pathways in weaned piglets after a lipopolysaccharide challenge. *J Nutr* 2013;**143**:1799–807.
55. Zhou Y, Leri F, Ho A, Kreek MJ. Suppression of hypothalamic-pituitary-adrenal axis by acute heroin challenge in rats during acute and chronic withdrawal from chronic heroin administration. *Neurochem Res* 2013;**38**(9):1850–60.
56. Morgese MG, Tucci P, Colaianna M, Zotti M, Cuomo V, Schiavone S, Trabace L. Modulatory activity of soluble beta amyloid on HPA axis function in rats. *Curr Pharm Des* 2014;**20**(15):2539–46.
57. Newson MJ, Pope GR, Roberts EM, Lolait SJ, O'Carroll AM. The neuropeptide apelin is expressed in hypothalamic paraventricular and supraoptic nuclei and mediates its effects via activation of the apelin receptor (APJ). *J Endocrinol* 2013;**216**:99–109.
58. Liu J, Guo M, Zhang D, Cheng SY, Liu M, Ding J, Scherer PE, Liu F, Lu XY. Adiponectin is critical in determining susceptibility to depressive behaviors and has antidepressant-like activity. *Proc Natl Acad Sci USA* 2012;**109**:12248.
59. Mattos GE, Heinzmann JM, Norkowski S, Helbling JC, Minni AM, Moisan MP, Touma C. Corticosteroid-binding globulin contributes to the neuroendocrine phenotype of mice selected for extremes in stress reactivity. *J Endocrinol* 2013;**219**:217–29.
60. Lehmann ML, Mustafa T, Eiden AM, Herkenham M, Eiden LE. PACAP-deficient mice show attenuated corticosterone secretion and fail to develop depressive behavior during chronic social defeat stress. *Psychoneuroendocrinology* 2013;**38**(5):702–15.
61. Belda X, Daviu N, Nadal R, Armario A. Acute stress-induced sensitization of the pituitary-adrenal response to heterotypic stressors: independence of glucocorticoid release and activation of CRH1 receptors. *Horm Behav* 2012;**62**:515–24.
62. Marazziti D, Rutigliano G, Baroni S, Landi P, Dell'osso L. Metabolic syndrome and major depression. *CNS Spectr* 2013;**8**:1–12.
63. Xia LP, Shen L, Kou H, Zhang BJ, Zhang L, Wu Y, Li XJ, Xiong J, Yu Y, Wang H. Prenatal ethanol exposure enhances the susceptibility to metabolic syndrome in offspring rats by HPA axis-associated neuroendocrine metabolic programming. *Toxicol Lett* 2014;**226**: 98–105.
64. Nardone G, Compare D. The psyche and gastric functions. *Dig Dis* 2014;**32**:206–12.
65. Handa RJ, Kudwa AE, Donner NC, McGivern RF, Brown R. Central 5-alpha reduction of testosterone is required for testosterone's inhibition of the hypothal-

amo-pituitary-adrenal axis response to restraint stress in adult male rats. *Brain Res.* 2013;**1529**:74–82.
66. Ondicova K, Kvetnansky R, Mravec B. Medial prefrontal cortex transection enhanced stress-induced activation of sympathoadrenal system in rats. *Endocr Regul* 2012;**46**: 129–36.
67. Bourke CH, Raees MQ, Malviya S, Bradburn CA, Binder EB, Neigh GN. Glucocorticoid sensitizers Bag1 and Ppid are regulated by adolescent stress in a sex-dependent manner. *Psychoneuroendocrinology* 2013;**38**:84–93.
68. Herman JP, McKlveen JM, Solomon MB, Carvalho-Netto E, Myers B. Neural regulation of the stress response: glucocorticoid feedback mechanisms. *Braz J Med Biol Res* 2012;**45**(4):292–8.
69. Hulshof HJ, Novati A, Luiten PG, den Boer JA, Meerlo P. Despite higher glucocorticoid levels and stress responses in female rats, both sexes exhibit similar stress-induced changes in hippocampal neurogenesis. *Behav Brain Res* 2012;**234**:357–64.
70. Selye H. A syndrome produced by diverse nocuous agents. *Nature* 1936;**138**:32.
71. Selye H. Thymus and adrenals in the response of the organism to injuries and Intoxication. *Brit J Exp Path* 1936;**17**:234.
72. Berczi I. The stress concept and neuroimmunoregulation in modern biology. *Ann NY Acad Sci* 1998;**851**:3–12.

CHAPTER 2

The Neuroimmune Supersystem

Istvan Berczi
Department of Immunology, Faculty of Medicine, The University of Manitoba, Winnipeg, Manitoba, Canada; Departamento de Fisiología y Farmacología, Centro de Ciencias Básicas, Universidad Autónoma de Aguascalientes, Aguascalientes, Estado Libre y Soberano de Aguascalientes, Mexico City, Mexico

1 INTRODUCTION

The fundamental role of the central nervous system (CNS) in the maintenance of vital bodily functions is already well recognized in primitive cultures. Interaction between the hypothalamus and the pituitary gland has been elucidated during the first half of the nineteenth century. These results prompted the term neuroendocrine system (NES),[1] rather than the CNS. These two systems interact continuously and form regulatory circuits for the numerous hormones that NES produces. The hypothalamus provides releasing factors toward the pituitary gland, which in turn regulates hormone secretion by the pituitary gland (e.g., thyrotropin-releasing hormone (TRH), thyroid-stimulating hormone (TSH), prolactin (PRL), and also growth hormone (GH); arginine vasopressin (AVP), stimulating adrenocorticotrophic hormone (ACTH), and PRL. Corticotropin-releasing hormone (CRH), for the release of (ACTH); gonadotropin-releasing hormone (GNRH), for the release of follicle-stimulating hormone (FSH) and luteinizing hormone (LH). More recently recognized factors are urocortin, which is a CRH analog and grehlin, which is a GH-releasing factor. These releasing hormones stimulate the release of specific hormones from the pituitary gland, involved in the regulation of growth, development, and adaptive immune functions (primarily GH and PRL). Thyroid hormones are considered to be metabolic regulators, however these hormones also affect immune and inflammatory responses. The regulation of hypothalamus–pituitary–adrenal (HPA) axis, which regulates metabolism, immune, and inflammatory reactions, and numerous other parameters in the body, such as glucose metabolism and blood pressure, for instance. The HPA axis plays a fundamental role in the stress response and in host defense during acute illness. The primary role of the GNRH–LH–FSH axis is the regulation of reproduction. However, other functions are also obvious, such as the modulation of immune/inflammatory reactions, which may be related to the primary function.[2–5]

Insights to Neuroimmune Biology
http://dx.doi.org/10.1016/B978-0-12-801770-8.00002-1

Traditionally ADIM was regarded as a self-regulated and intelligent system that has the capacity to recognize *foreign* microbes and material in general, as well as respond to it by antibody formation, by cell-mediated immunity, or by both kinds of responses. This system had the capacity to recognize *self*, and normally exhibited *self-tolerance*. Self–nonself recognition and the capacity to establish memory of immune stimulation, were thought to be fundamental to normal immune functions.[6] NATIM is an ancient form of immunity, which comprised of diverse mechanisms. Highly preserved homologous epitopes (homotopes) are recognized by germline-coded receptors, such as natural antibodies, toll-like receptors, endotoxin-binding protein, mannose-binding protein, C-reactive protein, and so on. Others introduced the concept of pathogen-associated recognition patterns, which are recognized by the cells of NATIM. This system is present at birth, it is never lost, and is ready to react instantaneously against microbes or against other infections of the body. This is in contrast with ADIM, which needs 5–7 days for the primary response and 4–5 days for the secondary response.[7–13]

Although the HPA axis was known to play a role in the stress response and in the regulation of inflammation, since the discovery of Hans Selye,[14] the interaction of other pituitary hormones with the immune system (IS) has been discovered more recently.[15] In this chapter, the development and current status of neuroimmune biology (NIB) is summarized. The significance of this newly emerging science is the demonstration that NISS coordinates host defense against infection, injury, mutations, and cancer. Indeed NISS is in control of the biology of higher organisms for their entire life cycle.[16–18]

In this book, an updated version of NISS has been presented. Much was discovered recently.

2 THE NEUROIMMUNE SUPERSYSTEM

Immune-derived cells and mediators are present in the CNS and are part of the neuroimmune regulatory equation. The immune cells in the CNS show enhanced activity after immunization, infection, or stress,[19–24] and in various pathological conditions, such as depression, and neurodegenerative disorders (e.g., multiple sclerosis (MS), Alzheimer's disease, Parkinson's disease, and stroke). These conditions are associated with many elements of inflammation and autoimmunity. Cytokines and chemokines initiate and propagate the inflammatory/immune response in these pathologies. In MS, there is a continuous realignment and redundancy in the inflammatory and immune responses.[25–31]

There is much more to these systems than simply sharing cytokines and other mediators. The neuroendocrine and ISs not only interact, but also rely on each other for mutual support both in health and disease. The thymus develops from the neural crest, which also gives rise to the CNS. Moreover, glial cells that represent roughly 50% of brain cells, are related to the monocyte–macrophage lineage and are bone marrow derived. A new and important information is that neurons themselves differentiate from bone marrow derived stem cells.[32–34] Thus the brain relies on the bone marrow for rejuvenation and healing (plasticity). Indeed, recent evidence indicates that inflammatory cells and cytokines exert a neuroprotective effect during traumatic brain injury.[6] For some time the brain shares adhesion molecules and numerous cell surface receptors with lymphocytes. It is now also clear that cell–cell and cell–matrix interactions play important roles in brain physiology and pathology.[9,36]

Until recently the IS has been considered as an autonomous system. Lymphocytes were known to be equipped with sophisticated receptors for the recognition of antigen and were capable of defending the host from pathogenic insults. It was also recognized that the IS was well organized and was regulated by internal regulatory pathways.[37] However, on the basis of common developmental origin, shared stem cells, receptors, mediators, and mutual interdependence, it was proposed that *the nervous, endocrine, and ISs are integrated parts of a united neuroimmune supersystem. This supersystem coordinates and regulates all the physiological and pathological processes in higher animals and humans for their entire life cycle.* Lymphocytes, such as neurons, are sensory cells with the capacity to recognize chemical structure and distinguish self from nonself. They store such information and are capable of memory responses. Lymphocytes are also capable of conveying information regarding chemical (antigenic) abnormalities in host tissues to the brain, via cytokine signals. Immune cells are essential for defending the body from foreign-invading pathogenic organisms as well as control mutation and eliminate aberrant cells from the host. It is now clear that immune cells are also involved in normal physiological regulation of host tissues and organs.[38–40]

In this chapter, the *extended version of NISS* have been discussed. Additional regulators are presented that tie into the GLH and HPA axis. AVP is a major immunoregulator, which will further be discussed in Chapter 5. In Clinical Medicine, it is observed that many diseases are associated with the HPA activation; the pathomechanism of this finding and the multifaceted regulation in biology have been discussed. The salient findings, such as role of neuropeptides in immunoregulation; acute stress and healing; chronic stress and disease; biologically controlled mutation which is correct in biology; microRNA in immunoregulation and neuroprotection have been pointed out here.

3 IMMUNOCOMPETENCE

3.1 Adaptive Immunocompetence

GH, PRL, and placental lactogen (PL) hormones represent the growth and lactogenic hormone family (GLH). These hormones regulate growth and development *in utero* and after birth. GH stimulates the production of insulin-like growth factor I (IGF-I) in the liver and in other tissues and organs, including the IS. IGF-I is a peptide hormone with structural relation to insulin and with cross-reactivity at the receptor level. Insulin and IGF-I do not only bind to each other's receptors, but also mutually regulate receptor levels, which is proportional to their ability to occupy that particular receptor. PRL also has the capacity to stimulate IGF-I in the liver and other tissues, which has been named by Nicoll and coworkers as synlactin. With the discovery of IGF, many investigators assumed that GH action is really mediated by IGF. This view is still held by a significant proportion of investigators.[41–46]

GH is necessary for target cell activation. One important aspect of GH action is the stimulation of IGF-I in responding cells. In turn, the locally-induced IGF-I will act as an autocrine and paracrine growth factor, completing the mitogenic cycle in GH targets.[47] There are other hormones/cytokines that are capable of inducing IGF-I in certain target cells. This implies, such hormones/cytokines also activate their targets similarly to that of GH, in a manner analogous to the competence–progression model of cell proliferation. Such hormones are interleukin-3 (IL-3), granulocyte macrophage colony-stimulating factor (GMCSF), and TGF-β1 in the bone marrow, and estradiol for the mammary gland.[1,48–50] The tropic effect of these hormones on their specific target organs is also based on two signals.

The PRL-dependent Nb2 rat T cell lymphoma cell line responds to serum factors (IGF-I) only after being primed by PRL. Most importantly, the magnitude of mitogenic response is directly proportional to the concentration of PRL used for priming.[51] *In vivo* observations in hypophysectomizsed (HYPOX) rats revealed that organ weight, DNA and RNA synthesis, and cell proliferation in the thymus, spleen and bone marrow, and immune reactivity require the presence of pituitary GH or PRL. These experiments show that lymphocyte proliferation in primary and secondary lymphoid tissue and immunocompetence are dependent on pituitary GLH. *Redundancy* was also demonstrated as either PRL or GH was sufficient to restore the IS. PL was also capable of immune restoration.[52]

Like GH, PRL is also a pleiotropic hormone with multiple targets. On the basis that GH is able to cause the proportional growth of the entire organism one may conclude that the entire body is a target for GH and PRL seems to affect most tissues except to stimulate body growth, although a limited growth promotion is present. Pituitary GLH induces IGF-I within the IS as observed by several investigators. Both GH and PRL were capable of inducing IGF-I within the thymus and spleen of HYPOX rats. This furnishes additional evidence for the ability of these hormones as growth promoters of the IS,[2,53–55] GLH are essential for the maintenance of vital bodily functions.[55,56] In mammals the fetus is exposed to high levels of placental GLH, as these hormones are present in the amniotic fluid in high concentrations.[57] During embryonic life, placental GLH plays a fundamental role in the growth and development of the fetus, the IS, and pituitary hormones are not required as fetuses with the congenital lack of the pituitary gland develop normally *in utero*.[58] During postnatal life, the pituitary gland assumes the role of growth control in the organism.[6] Bone marrow, thymus function, and the maintenance of immunocompetence during postnatal life depends entirely on pituitary GLH.[2] On this basis, it has been proposed that *GH and PRL are the hormones of adaptive immunocompetence*.[6,38]

Neuropeptides virtually regulate every area of immune function, which is not discussed here. A separate chapter presents this problem, which is a major area in today's immunology (Chapter 6).

3.2 Innate or Natural Immunocompetence

Higher animals and man are born with this innate immune system (INIM) and never loose it. The cells of the NATIM (natural killer cells, gamma–delta T lymphocytes, $CD5^+$ B lymphocytes, monocyte–macrophages, granulocytes, etc.) have genetically determined (germline coded) receptors for highly preserved *homologous epitopes* (*homotopes*) that are present in pathogenic microbes in injured/degenerated cells, and also in cancer cells.[11,12] Some examples of homotopes are, bacterial LPS, mannose residues in infectious agents, pathogen-associated recognition patterns in numerous pathogenic microbes,[10] and cancer antigens. NK cells kill targets that express tumor antigens and lack MHC antigens. Since natural immune cells are capable of responding instantaneously to the antigen they recognize, they provide the first line of defense for the host. However, in acute illness, natural immune mechanisms are amplified so the host organism relies on this system during the final struggle for survival. Since NATIM is hard wired for response to the antigen, if any regulation, little can take place at this level. Therefore, NATIM is regulated by cytokines, hormones, neurotransmitters, and neuropeptides.[12,13,59]

It is now clear that the genome undergoes epigenetic programming and it is changing to obtain adaptation to environmental impacts. So, if the genome is changing, INIM regulation, which depends on genes for antigen-receptor coding, will change accordingly. This is the correct way for biology. The INIM has to also adapt to the environment (Chapter 9).

4 GROWTH SIGNALING IN THE IS

4.1 Signaling by GLH

As already pointed out, GLH is considered as the first (competence) signal for lymphocyte proliferation. These hormones stimulate the production of mature T and B lymphocytes, which are required for the initiation of an immune response. Further, antigen-presenting cells (APCs) interact with helper T cells, and later on the helper T cells with B cells (antibody response) or immature antigen-sensitive T cells (cell-mediated immunity) stimulates the secretion of cytokines from helper T lymphocytes, which function as paracrine growth factors. The antigen signal is modified by additional "costimulatory" and inhibitory signals, which are also delivered by adhesion molecules.[38,41,50]

Cytokine signals complete the mitogenic stimulus and enable the immature cell to proliferate (clonal expansion). Multiple cytokines are available to deliver this signal, which varies according to the type and stage of the immune response.[12,41] It is known that lymphocytes are capable of producing GLH; however, the exact function of these hormones in immune reactions have not been elucidated in detail.

GLH shows redundancy in the maintenance of immunocompetence. Current evidence also indicates that IGF-I may be substituted in the IS by IGF-II or insulin. Clearly, immune function, as many other functions in the body, are maintained by multiple hormones and cytokines, which show overlapping and redundancy.[2,10]

Human GHs and other primate GHs are known to act on the PRL receptors and exert lactogenic activity in many species, which indicates functional overlap within GLH hormones. The major signal transduction pathway, which involves the Janus kinase (JAK) and signal transducers and activators of transcription (STAT) nuclear regulatory factors, is shared between cytokines and growth and lactogenic hormones. STAT knockout mice show severe developmental and immune deficiencies.[49]

This means signal functioning overlaps between the immunoregulatory function of GLH and cytokines that use the JAK/STAT signaling pathway. Sharing the JAK/STAT signal delivery system secures mutual substitution, should it be required. This mechanism creates important security for operating successfully. Other new mediators are *thyroid hormones, oxitocin, and grehlin* (Chapter 3).

Of the HPA axis, AVP is a major mediator (Chapter 5), which has been discovered in our laboratory. *Neuropeptide regulators are* urocortin, catecholamines, apelin, cholecystokinin, leptin, leukotrienes, natriuretic peptide, neuromedins, neuropeptides, nociceptin/orphanin, pituitary adenylate cyclase-activating polypeptide (PACAP), opioid peptides, orexin, relaxin, sauvagin/urotensin, stresscopin, and stresscopin-related peptides (urocortin and vasotocin) (Chapter 4).

4.2 Antigen Presentation

Adaptive immune reactions are mediated by lymphocyte clones that have specific receptors for the determinants (epitopes) of the antigen. Initially, an antigen was assumed to be the sole signal that instructed immunocytes to make antibodies by some sort of recognizing and copying mechanism.[59] Later it was realized that antibody specificity is genetically coded, the antigen has to be digested and presented by specialized APCs to T lymphocytes. Phagocytic mononuclear cells (monocyte–macrophages), B lymphocytes, and some specialized cells (e.g., dendritic cells, Langerhans cells, Kuppfer cells) are "professional" APCs that present antigens via major histocompatibility (MHC)-I and MHC-II. In the CNS, the macrophage-related microglia and to a lesser extent astrocytes, are involved in antigen presentation as well as in inflammatory responses. However, all nucleated cells are capable of antigen presentation by MHC-I, which is constitutively expressed by such cells. In addition, IFN is capable of inducing the expression of MHC-II in somatic cells, which enables them to present antigens by this pathway. MHC antigens bind peptides during their intracellular biosynthesis and carry them to the surface of the APC. In turn, the digested (processed) antigenic peptide–MHC complexes are recognized by T cells as "altered self."[2,6,38,59]

The T-cell receptor recognizes sequentially (linear recognition) the target peptide. The MHC selfpeptide complex is recognized as "self" and a complex with foreign peptide as "nonself" in this process. This self–nonself recognition has evolved to provide protection against autoimmune reactions, while foreign invaders are attacked. The requirement for T cells to recognize antigens in the context of MHC assures that infected and cancerous cells are specifically eliminated by killer cells. However, soluble antigen, which is not processed and not associated with MHC antigens, is not capable of triggering killer T lymphocytes. This mechanism prevents the exhaustion of T cells by viremia or by other antigen that is present in the circulation.[38] The immunoglobulin B cell antigen receptor is capable of recognition of the tertiary structure of native molecules, even microorganisms (conformational recognition). B cells engulf the antigen they recognize, digest (process) them, and present peptides from the antigen to helper T lymphocytes via their surface MHC antigens. Such helper cells adhere to the B cells and are activated for cytokine production (IL-4, -6, etc.), which will promote B cell proliferation and maturation to antibody producing plasma cells.[38,41]

5 IMMUNE NEUROENDOCRINE FEEDBACK SIGNALS

IL-1 was the first cytokine shown to stimulate the HPA axis.[61–63] Purified natural and recombinant human forms of IL-1 can stimulate ACTH and corticosterone output in mice and rats. Later, TNF-α, different types of interferons, IL-2, IL-6, IL-11, IL-12, leukemia inhibitory factor (LIF), GMCSF, oncostatin, and stem cell factor (SCF) were found to affect the HPA axis and release of other pituitary hormones.[61,64,65]

Cytokines are known to affect brain function directly, affect endocrine glands as well, and act on peripheral nerves, such as the vagus and sensory nerve fibers, which in turn transmit their signals to the hypothalamus. Eventually all forms of neural signaling to the brain act on the hypothalamus, on the paraventricular nucleus in particular, and also the supraoptic nucleus, which affect pituitary function, including the HPA axis, and GLH hormones. The brain may release immunostimulatory/proinflammatory signals (e.g., GH, PRL) and immunosuppressive/anti-inflammatory signals by HPA activation toward the adaptive IS.[40,61,66]

Catecholamines regulate cyclic AMP, GMP, and calcium influx. A number of other mediators, including opioid peptides, act through G-proptein linked adenylate cyclase receptors as well. Since Ca^{2+} is needed for cellular activation and phosphorylation, which is fundamental mechanisms of signal transduction by membrane-bound receptors, these mediators are designated as *signal modulators* in immune activation and in cellular activation in general. Catecholamines may play immunostimulatory (alpha-adrenergic) and immunosuppressive (beta adrenergic) roles, whereas acetylcholine is considered to be immunostimulatory. Substance P and calcitonin gene related peptide are proinflammatory and somatostatin are anti-inflammatory neuropeptides.[67,68]

The steroid hormones playing major roles in the neuroimmune regulatory network include glucocorticoids, aldosterone, estrogens, androgens, and vitamin D. These hormones have cytoplasmic or nuclear receptors, which are transcription factors that directly regulate gene expressions. Glucocorticoids are produced in the adrenal gland and are fundamental to immune function. Physiological levels are required for the normal development and functioning of the IS. Under pathophysiological conditions, the serum level of glucocorticoids is elevated, which plays an important role in the suppression of the adaptive immune response that depends on T lymphocytes. Glucocorticoids also exert a powerful anti-inflammatory effect. Elevated glucocorticoids during acute phase reactions augment natural immunity, the production of natural antibodies, and support the production of acute phase proteins (APPs) by the liver.[68] LH-releasing hormones and gonadotropins exert a direct regulatory influence on the IS, in addition to the regulation of sex steroid hormones. In turn immune-derived cytokines regulate the production of gonadotropins. These mechanisms insure the coordination of reproduction with health status and prevent inopportune conception.[68,73]

Estrogens regulate the thymus and suppress cell-mediated immune reactions. The antibody response and natural immunity (NK cytotoxicity, phagocytosis) are augmented by estradiol. Testosterone is immunosuppressive during trauma and shock. Many of the immunological effects of testosterone are due to its conversion to estradiol by aromatase in the thymus and in other lymphoid organs. The adrenal androgen, dehydroepyandrosterone stimulates immune reactions in experimental animals and in man. It antagonizes the immunosuppressive effect of glucocorticoids and its age-related decline may contribute to the immunodeficiency that develops in elderly individuals. Progesterone is a powerful immunosuppressive hormone. It plays a major role in the protection of the fetus, during mammalian reproduction. Progesterone also contributes to the generation of self tolerance and protects against excessive activation of the IS.[68]

1,25-Dihydroxyvitamin D3 (VD3) is a major immunoregulator with a powerful immunosuppressive potential. Deficiency ofVD3 plays an important role in the development of autoimmune disease. VD3 is also required for normal immune function and for proper defense against infectious disease and cancer.[68]

6 HOMEOSTASIS, THE IMMUNE–NEUROENDOCRINE CIRCUITRY

Under physiological conditions the *IS* provides continuous defense against infectious agents, injury, harmful mutations, and cancer, and it is part of the homeostatic neuroimmune regulatory circuit that coordinates the normal function of the entire organism. The IS provides local and mobile defense and regulation, and it has enormous capacity to deliver defense and regulatory molecules to sites that are in need. Every organ and tissue possesses stromal lymphoid elements that intervene locally to control autoimmune reactions, inflammation, and in general, participate in the physiological processes.[69]

Adaptive cell mediated and humoral immunity, and immunological memory are reactions exerted by T and B lymphocytes, in concert with members of the leukocyte series. *Innate* host defense relies on natural-immune mechanisms, on specialized immune cells, such as natural killer cells, T lymphocytes, and $CD5^+$ B cells. The complement system of T and B-lymphocytes activated by alternate pathways are also part of the natural IS. The natural IS relies on germline coded receptors that recognize evolutionarily highly preserved homologous epitopes (homotopes) on microbes and also on self components.[6,13]

Cell–cell interactions by the antigen receptors, MHC molecules, and other adhesion molecules are fundamental to immune activation, as well as to stromal regulation. Cytokines are an essential part of this regulatory system. In addition, the IS interacts with the neural elements and mediators, parenchymal and stromal cells as part of the local neuroimmune circuits that govern the organ/tissue under physiological circumstances.[12]

6.1 Immunoregulation by the HPA Axis

Glucocorticoid hormones exert a multitude of functions that virtually affect every cell in the body. The physiological significance of glucocorticoids is most remarkable at times of stress, when the HPA axis is fully mobilized. The same hypothalamic hormone that stimulates the HPA axis, CRF also mediates behavioral, autonomic, and neuroendocrine responses to stress in rodents and primates. Therefore, hyperfunction of CRF neurons, appears to underlie a variety of psychiatric, gastrointestinal, cardiovascular, metabolic, and reproductive illnesses attributable to stress.[61]

Clinical observations indicate that the HPA axis activation is often associated with disease. This problem has been discussed in Chapter 1.

7 ACUTE ILLNESS

7.1 Neurogenic Inflammation

The release of neuropeptides, including tachykinins, and calcitonin gene-related peptide, from sensory nerves, via an axon or local reflex may exert inflammatory effects in the airways. This neurogenic inflammation may be initiated by sensory nerves, which are activated by inflammatory mediators and irritants. The phenomenon of neurogenic inflammation is well known in rodents, and contributes to the inflammatory response to allergens, infections, and irritants in animal models. However, the role of neurogenic inflammation in airway inflammatory diseases, such as asthma and chronic obstructive pulmonary disease, is still uncertain. Still there is little direct evidence for the involvement of sensory neuropeptides in human airways. Initial clinical studies designed to block neurogenic inflammation have not been encouraging. Therefore, it is necessary to perform prolonged studies of severe forms of airway diseases in the future to explore the role of neurogenic inflammation.[70–72]

In Chapter 6, the immunoregulatory neuropeptides have been presented.

7.2 The Acute Phase Response (APR)

Mild infection or sublethal dose of endotoxin elicits a brief elevation of GH and PRL in the serum, which are proinflammatory and immunostimulatory. In severe trauma, sepsis, and shock, GH and PRL are suppressed, whereas glucocorticoids and catecholamines are elevated. Under these conditions, an APR is initiated by cytokines, produced by the INIM, primarily IL-1, IL-6, TNF-α, and GMCSF. These cytokines elicit a profound neuroendocrine and metabolic response. Fever and catabolism prevails, and the synthesis of APP in the liver, cell proliferation in the bone marrow, and protein synthesis by leukocytes are elevated. This is an emergency reaction to save the organism after the failure of the adaptive IS to protect the host. During sepsis and endotoxin shock, activation of the complement system and the release of leukocyte-derived enzymes, tissue-derived breakdown products, and highly toxic cytokines, seriously threaten survival. The HPA axis is activated. Glucocorticoids play a major role in the regulation of proinflammatory cytokine production and potentiate the secretion of APPs. Some APPs, such as C-reactive proteins, LPS-binding proteins, and mannose-binding proteins in the serum are designed to combine with microorganisms, and trigger their destruction by the activation of complement system and phagocytes. The increased production of some complement components also helps host resistance. The rise in serum fibrinogen promotes blood clotting. A number of enzyme inhibitors are produced as APP, which is supposedly serves to curb the nonspecific damage inflicted by enzymes. *Catecholamines* are also elevated, which serve to inhibit inflammatory responses, promote, and even initiate the APR. Serum *leptin* is also increased, which governs energy metabolism as well as it has an immunostimulatory effect. If the acute phase reaction fails to protect the host, shock will develop. Patients with subclinical *adrenal insufficiency* succumb to septic shock almost invariably, if glucocorticoid therapy is not given. However, glucocorticoid treatment of septic patients with normal adrenal function has not been helpful.[6,30,37,69,73–78]

During the APR T-cell regulated adaptive immune response is switched off and innate immune mechanisms are amplified several 100–1000 times within 24–48 h. This phenomenon has been designated as *immunoconversion*, which is initiated by immune-derived cytokines, and involves profound neuroendocrine and metabolic changes, all in the interest of host defense. Therefore, *innate immunity* is essential for a first and last line of defense and the NES is an important promoter and regulator of this fundamental form of immune defense.[11,12]

8 HEALING

Most people develop febrile illness on numerous occasions during a lifetime. These febrile episodes normally subside and are followed by healing, return to health, and normal adaptive immunocompetence. By now, we understand how the APR develops and what it is doing. However, we know little about the recovery phase. One would expect that in accord with the tight neuroendocrine regulation of APR, also the recovery phase would be regulated by neuroendocrine mechanisms. Some recent observations on the role of *vasopressin* in immune function and in APR, appears to provide indications of the mechanism of recovery, termed as *immunoreversion*.[13]

Now it is clear that when APR subsides, chronic inflammation will follow. This time the principal hypothalamic mediator will be AVP and not CRH. AVP will regulate chronic inflammation and brings it to recovery and healing. So AVP has healing power in biology.[1]

9 CONCLUSIONS: THE BIOLOGICAL SIGNIFICANCE OF NISS

The IS protects against infection, injury, cancer, and mutations (e.g., altered self) by self recognition and self preservation through host defense and by destroying foreign materials (antigens). Moreover, cytokines, chemokines, neurotransmitters, neuropeptides, and classical hormones are all shared by the IS and by the NES, and are required for the normal functioning of the CNS, ENS, IMS, and for the function of the *entire host organism*. At this point it is clear that the IS signals the CNS about infection, injury, mutation (altered self), and cancer by cytokine signals sent directly to the CNS and indirectly via sensory nerves and the vagus nerve. The CNS responds by nerve impulses, hormone and cytokine secretion, which are aimed at resolving the problem. Phagocytosis, neutralization by antibodies, cytotoxicity by antibody-complement and by killer cells, and an inflammatory response where defense molecules and effector cells are concentrated and exert complex interactions in the interest of host defense and of healing. It is now clear that the CNS has innate toll-like receptors so it is an immunocompetent organ, which directly senses infection, trauma, and injury, just like the INIM system. Most of the organs that had been tested expressed TLR, so it is likely that the increase of INIM reactions causes the entire body to be activated, such as by the APR.[78]

It is now known that inflammatory T cells do not only function as effector cells (e.g., killer or delayed hypersensitive type) but these cells may produce *nerve growth factor*, and other factors with similar effects, for protection rather than injure nerve cells during encephalomyelitis. Little is known about the factors that regulate destructive and constructive (healing) functions of inflammation. It is quite conceivable that the *neuroendocrine milieu* plays a decisive role regarding what the outcome would be in a local inflammatory response. Thus, in a CRH dominated *proinflammatory milieu* the destructive force of inflammation would be promoted systemically in order to eliminate the pathogen that caused the inflammatory response. However, later on VP takes over as a principal hypothalamic regulator and it is proposed to stimulate both the HPA axis and also GLH in a balanced fashion.

This would result in the restoration of *physiological (homeostatic) neuroendocrine milieu*, which can lead to healing and recovery. It is quite likely that the occurrence of healing by T lymphocytes is stimulated by the *homeostatic milieu*, which has emerged in response to the elevated VP levels.

Ultimately, the evidence reviewed earlier demonstrates that CNS has the capacity to sense foreign (infectious agents and other foreign compounds) materials, as well as altered self (mutations, cancer) by interacting with the IS, and that it is capable of directing and integrating the forces of host defense against these pathogenic situations. Here, an entirely new way of sensation by the CNS has been witnessed, which is mediated by *cytokines and neuropeptides.* The response to this sensation is mediated by hormones, neurotransmitters, neuropeptides, and cytokines.

REFERENCES

[1] Berczi I. Integration and regulation of higher organisms by the neuroimmune supersystem. *Int J Integr Biol* 2007;**1**(3):216–31.
[2] Berczi I, Szentivanyi A. Growth and lactogenic hormones, insulin-like growth factor and insulin. In: Berczi I, Szentivanyi A, series editors. *The immune-neuroendocrine circuitry. History and progress. Neuroimmune Biology*, vol. 3. Amsterdam: Elsevier; 2003. p. 129–153.
[3] Taub DD. Novel connections between the neuroendocrine and immune systems: the ghrelin immunoregulatory network. *Vitam Horm* 2008;**77**:325–46.
[4] Gravanis A, Margioris AN. The corticotropin-releasing factor (CRF) family of neuropeptides in inflammation: potential therapeutic applications. *Curr Med Chem* 2005;**12**(13):1503–12.
[5] Kronenberg HM, editor. *Williams textbook of endocrinology*. 11th ed. Elsevier; 2004.
[6] Paul WE. The immune system: an introduction. In: Paul WE, editor. *Fundamental immunology*. 4th ed. Philadelphia: Raven Press; 1999. p. 1–18.
[7] Berczi I. Neurohormonal host defence in endotoxin shock. *Ann NY Acad Sci* 1998;**840**:787–802.
[8] Berczi I, Chow DA, et al. Neuroimmunoregulation and cancer. *Int J Oncol* 1998;**13**:1049–51.
[9] Berczi I, Chow DA, et al. Neuroimmunoregulation and natural immunity. *Domestic Anim Endocrinol* 1998;**15**:273–81.
[10] Medzhitov R, Janeway Jr CA. Decoding the patterns of self and non-self by the innate immune system. *Science* 2002;**296**:298–300.
[11] Chow DA. Natural immune activation: stimulators/receptors. In: Bertok L, Chow DA, editors. *Natural immmunity*. Berczi I, Szentivanyi A, series editors. *Neuroimmune biology*, vol. 5. Amsterdam: Elsevier; 2005. p. 123–150.
[12] Berczi I. Neuroendocrine regulation of natural immunity. In: Chow D, Bertok L, editors. *Neuroimmune biology*, vol. 5. Berczi I, Szentivanyi A, series editors. Amsterdam: Elsevier; 2005. p. 215–262.
[13] Quan N, Banks WA. Brain-immune communication pathways. *Brain Behav Immun* 2007;**21**:727–35.
[14] Selye H. A syndrome produced by diverse nocuous agents. *Nature (Lond)* 1936;**138**:32.
[15] Wrona D. Neural–immune interactions: an integrative view of the bidirectional relationship between the brain and immune systems. *J Neuroimmunol* 2006;**172**:38–58.
[16] Webster JI, Tonelli L, et al. Neuroendocrine regulation of immunity. *Annu Rev Immunol* 2002;**20**:125–63.

[17] Penkowa M, Hidalgo J, et al. Immune and inflammatory responses in the CNS: modulation by astrocytes. In: Korneva HA, Phelps C, editors. *Neuroimmune biology*, vol. 6. Berczi I, Szentivanyi A, series editors. *Cytokines and the brain*. Amsterdam: Elsevier; 2007. p. 277–288.

[18] Suzumura A. Immune response in the brain: glial response and cytokine production. In: Korneva HA, Phelps C, editors. *Neuroimmune biology*, vol. 6. Berczi I, Szentivanyi A, series editors. *Cytokines and the brain*. Amsterdam: Elsevier; 2007. p. 289–303.

[19] Kannan Y, Moriyama M, et al. Lymphocytes and adrenergic sympathetic nerve system: the role of cytokines. In: Korneva HA, Phelps C, editors. *Neuroimmune biology*, vol. 6. Berczi I, Szentivanyi A, series editors. *Cytokines and the brain*. Amsterdam: Elsevier; 2007. p. 307–336.

[20] Goehler L. Cytokines in neural signaling to the brain. In: Korneva HA, Phelps C, editors. *Neuroimmune biology*, vol. 6. Berczi I, Szentivanyi A, series editors. *Cytokines and the brain*. Amsterdam: Elsevier; 2007. p. 337–352.

[21] Korneva EA, Kazakova TB. Interleukin-2 gene expression in the CNS cells after stress and antigen application. In: Korneva HA, Phelps C, editors. *Neuroimmune biology*, vol. 6. Berczi I, Szentivanyi A, series editors. *Cytokines and the brain*. Amsterdam: Elsevier; p. 353–372.

[22] Katafuchi T. Involvement of brain cytokines in stress-induced immunosuppression. In: Korneva HA, Phelps C, editors. *Neuroimmune biology*, vol. 6. Berczi I, Szentivanyi A, series editors. *Cytokines and the brain*. Amsterdam: Elsevier; p. 391–401.

[23] Phelps C, Chen L-T. Brain response to endotoxin. In: Korneva HA, Phelps C, editors. *Neuroimmune biology*, vol. 6. Berczi I, Szentivanyi A, series editors. *Cytokines and the brain*. Amsterdam: Elsevier; 2007. p. 435–454.

[24] Clarkson AN, Rahman R, et al. Inflammation and autoimmunity as a central theme in neurodegenerative disorders: fact or fiction? *Curr Opin Investig Drugs* 2004;**5**:706–13.

[25] Ketlinsky SA, Kalinina NM. Cytokines in demyelinating disease. In: Korneva HA, Phelps C, editors. *Neuroimmune biology*, vol. 6. Berczi I, Szentivanyi A, series editors. *Cytokines and the brain*. Amsterdam: Elsevier; 2007. p. 455–483.

[26] Dunn A. Cytokines and depression. In: Korneva HA, Phelps C, editors. *Neuroimmune biology*, vol. 6. Berczi I, Szentivanyi A, series editors. *Cytokines and the brain*. Amsterdam: Elsevier; 2007. p. 485–506.

[27] Summers WK. Clinical relevance: cytokines in Alzheimer's disease. In: Korneva HA, Phelps C, editors. *Neuroimmune biology*, vol. 6. Berczi I, Szentivanyi A, series editors. *Cytokines and the brain*. Amsterdam: Elsevier; 2007. p. 507–526.

[28] Aubert A, Renault J. Cytokines and immune-related behaviors. In: Korneva HA, Phelps C, editors. *Neuroimmune biology*, vol. 6. Berczi I, Szentivanyi A, series editors. *Cytokines and the brain*. Amsterdam: Elsevier; 2007. p. 529–547.

[29] Neveu PJ. The production and effects of cytokines depend on brain lateralization. In: Korneva HA, Phelps C, editors. *Neuroimmune biology*, vol. 6. Berczi I, Szentivanyi A, series editors. *Cytokines and the brain*. Amsterdam: Elsevier; 2007. p. 549–563.

[30] Szentivanyi A, Berczi I, et al. Some evolutionary morphoregulatory and functional aspects of the immune-neuroendocrine circuitry. In: Korneva HA, Phelps C, editors. *Neuroimmune biology*, vol. 6. Berczi I, Szentivanyi A, series editors. *Cytokines and the brain*. Amsterdam: Elsevier; 2007. p. 31–61.

[31] Gottfried-Blackmore A, Croft GF, et al. Sex hormones and cytokines in CNS pathology and repair. In: Korneva HA, Phelps C, editors. *Neuroimmune biology*, vol. 6. Berczi I, Szentivanyi A, series editors. *Cytokines and the brain*. Amsterdam: Elsevier; 2007. p. 373–390.

[32] Stewart R, Przyborski S. Non-neural adult stem cells: tools for brain repair? *Bioessays* 2002;**24**(8):708–13.

[33] Berczi I, Szentivanyi A. Adhesion molecules. In: Berczi I, Szentivanyi A, editors. *Neuroimmune biology*, vol. 3. The immune-neuroendocrine circuitry. history, progress. Amsterdam: Elsevier; 2003. p. 99–115.

[34] Wiranowska M, Plaas A. Cytokines and extracellular matrix remodeling in the central nervous system. In: Korneva HA, Phelps C, editors. *Neuroimmune biology*, vol. 6. Berczi I, Szentivanyi A, series editors. *Cytokines and the brain.* Amsterdam: Elsevier; 2007. p. 167–197.
[35] Blalock JE. The immune system as a sensory organ. *J Immunol* 1984;**132**:1067–70.
[36] Berczi I, Cosby H. Decreased bioactivity of circulating prolactin in patients with rheumatoid arthritis. *Br J Rheum* 1987;**26**:433–6.
[37] Conti B, et al. Cytokine receptors in the brain. In: Korneva HA, Phelps C, editors. *Neuroimmune biology*, vol. 6. Berczi I, Szentivanyi A, series editors. *Cytokines and the brain.* Amsterdam: Elsevier; 2007. p. 21–38.
[38] Berczi I, Szentivanyi A. Immunocompetence. In: Berczi I, Szentivanyi A, editors. *Neuroimmune biology, vol. 3: the immune-neuroendocrine circuitry. history and progress.* Amsterdam: Elsevier; 2003. p. 281–99.
[39] Phillips LS, Vassilopoulou-Sellin R. Somatomedins. *N Engl J Med* 1980;**302**:371–80 438–446.
[40] Blundell TL, Humbel RE. Hormone families: pancreatic hormones and homologous growth factors. *Nature (London)* 1980;**287**:781–7.
[41] Rosenfeld R, Hintz RL, et al. Insulin-induced loss of insulin-like growth factor-I receptors on IM-9 lymphocytes. *Diabetes* 1982;**31**:375–81.
[42] Mick CCW, Nicoll CS. Prolactin directly stimulates the liver in vitro to secrete a factor (synlactin) which acts synergistically with the hormone. *Endocrinology* 1985;**116**: 2049–53.
[43] Nicoll CW, Herbert NJ, et al. Lactogenic hormones stimulate the liver to secrete a factor that acts synergistically with prolactin to promote growth of the pigeon crop-sac mucosal epithelium in vivo. *Endocrinology* 1985;**116**:1449–53.
[44] Kelley KW, Arkins S, et al. Growth hormone, growth factors and hematopoiesis. *Horm Res* 1996;**45**(1-2):38–45.
[45] Kelley KW, Meier WA, et al. Insulin like growth factor-I inhibits apoptosis in hematopoietic progenitor cells. Implications in thymic aging. *Ann NY Acad Sci* 1998;**840**: 518–24.
[46] Fournier B, Gutzwiller S, et al. Estrogen receptor (ER)-alpha, but not ER-beta, mediates regulation of the insulin-like growth factor I gene by antiestrogens. *J Biol Chem* 2001;**276**:35444–9.
[47] Kveiborg M, Flyvbjerg A, et al. Transforming growth factor-beta1 stimulates the production of insulin-like growth factor-I and insulin-like growth factor-binding protein-3 in human bone marrow stromal osteoblast progenitors. *J Endocrinol* 2001;**169**:549–61.
[48] Berczi I, Nagy E. The effect of prolactin and growth hormone on hemolymphopoietic tissue and immune function. In: Berczi I, Kovacs K, editors. *Hormones and immunity.* Lancaster, UK: MTP Press; 1987. p. 145–71.
[49] Berczi I, Nagy E, et al. Pituitary hormones regulate c-myc and DNA synthesis in lymphoid tissue. *J Immunol* 1991;**146**:2201–6.
[50] Kelly PA, Djiane J, et al. The prolactin/growth hormone receptor family. *Endocr Rev* 1991;**12**:235–51.
[51] Nagy E, Berczi I. Pituitary dependence of bone marrow function. *Br J Haematol* 1989;**71**:457–62.
[52] Nagy E, Berczi I, et al. Endocrine control of the immunosuppressive activity of the submandibular gland. *Brain Behav Immun* 1992;**6**:418–28.
[53] Sinha YN, Vander Laan WP. Effect on growth of prolactin deficiency induced in infant mice. *Endocrinology* 1982;**110**:1871–8.
[54] Hill DJ. What is the role of growth hormone and related peptides in implantation and the development of the embryo and fetus. *Horm Res* 1992;**38**:28–34.

[55] Potter EL, Craig JM. . *Pathology of the fetus and the infant*. 3rd ed. Chicago: Year Book; 1975.
[56] Glasscock GF, Gelber SE, et al. Pituitary control of growth in the neonatal rat: effects of neonatal hypophysectomy on somatic and organ growth, serum insulin-like growth factors (IGF)-I and -II levels, and expression of IGF binding proteins. *Endocrinology* 1990;**127**:1792–803.
[57] Berczi I, Cosby H. Decreased bioactivity of circulating prolactin in patients with rheumatoid arthritis. *Br J Rheum* 1987;**26**:433–6.
[58] Berczi I. The role of the growth and lactogenic hormone family in immune function. *Neuroimmunomodulation* 1994;**1**:201–16.
[59] Berczi I, Szentivanyi A. Antigen presentation. In: Berczi I, Szentivanyi A, series editors. *The immune-neuroendocrine circuitry. History and progress. Neuroimmune biology*, vol. 3. Amsterdam: Elsevier; 2003. p. 301–313.
[60] Dorshkind K, Horseman ND. The roles of prolactin, growth hormone, insulin-like growth factor-I, and thyroid hormones in lymphocyte development and function. Insights from genetic models of hormone and hormone receptor deficiency. *Endocr Rev* 2000;**21**:292–312.
[61] Freeman ME, Kanyicska B, et al. Prolactin: structure, function, and regulation of secretion. *Physiol Rev* 2000;**80**:1523–631.
[62] Silverstein AM. The history of immunology. In: Paul WE, editor. *Fundamental immunology*. 4th ed. Philadelphia: Lippincott-Raven; 1999. p. 19–35.
[63] Germain RN. Fundamental immunology. In: Paul WE, editor. *Antigen processing and presentation*. 4th ed. Philadelphia: Lippincott-Raven; 1999. p. 287–340 CITY.
[64] Cresswell P. Assembly, transport and function of MHC class II molecules. *Ann Rev Immunol* 1994;**12**:259–93.
[65] York IA, Rock KL. Antigen processing and presentation by the class 1 major histocompatibility complex. *Ann Rev Immunol* 1996;**14**:369–96.
[66] Aloisi F. Immune function of microglia. *Glia* 2001;**36**:165–79.
[67] Gebicke-Haerter PJ. Microglia in neurodegeneration: molecular aspects. *Microsc Res Technol* 2001;**54**:47–58.
[68] Besedovsky HO, del Rey A, et al. Lymphokine-containing supernatants from Con A-stimulated cells increase corticosterone blood levels. *J Immunol* 1981;**126**:385–7.
[69] Besedovsky HO, del Rey A. Processing of cytokine signals at CNS levels: relevance for immune-HPA axis interactions. In: Korneva HA, Phelps C, editors. *Neuroimmune biology*, vol. 6. Berczi I, Szentivanyi A, series editors. *Cytokines and the brain*. Amsterdam: Elsevier; 2007. p. 227–240.
[70] Sapolsky R, Rivier C, et al. Interleukin-1 stimulates the secretion of hypothalamic corticotropin-releasing factor. *Science* 1987;**238**:522–4.
[71] Bernton EW, Beach JE, et al. Release of multiple hormones by a direct action of interleukin-1 on pituitary cells. *Science* 1987;**238**:519–21.
[72] McCann SM, Karanth S, et al. Induction by cytokines of the pattern of pituitary hormone secretion in infection. *Neuroimmunomodulation* 1994;**1**:2–13.
[73] Berczi I, Szentivanyi A. The acute phase response. In: Berczi I, Szentivanyi A, editors. *Neuroimmune biology*, vol. 3. *The immune-neuroendocrine circuitry. History, progress*. Amsterdam: Elsevier; 2003. p. 463–494.
[74] Berczi I, Szentivanyi A. Nerve growth factor, leptin and neuropeptides. In: Berczi I, Szentivanyi A, editors. *Neuroimmune biology*, vol. 3. *The immune-neuroendocrine circuitry. History, progress.* Amsterdam: Elsevier; 2003.p. 181–189.
[75] Berczi I, Nagy E. Steroid hormones. In: : Berczi I, Szentivanyi A, editors. *Neuroimmune biology*, vol. 3. *The immune-neuroendocrine circuitry. History, progress.* Amsterdam: Elsevier; 2003. p. 221–270.

[76] Berczi I, Szentivanyi A. The immune-neuroendocrine circuitry. In: Berczi I, Szentivanyi A, editors. *Neuroimmune biology*, vol. 3. *The immune-neuroendocrine circuitry. History, Progress*. Amsterdam: Elsevier; p. 561–592.
[77] Jancso N. Role of the nerve terminals in the mechanism of inflammatory reactions. *Bull Millard Fillmore Hosp* 1960;7–53.
[78] Barnes PJ. Neurogenic inflammation in the airways. In: Berczi I, Szentivanyi A, editors. *Neuroimmune biology*, vol. 3. *The immune-neuroendocrine circuitry. History, Progress*. Amsterdam: Elsevier; 2003. p. 437–449.

PART 2

Hypothalamic Immunoregulation

CHAPTER 3

Multifactorial Regulation of GLH Hormones

Istvan Berczi*,, Andrés Quintanar-Stephano****
*Department of Immunology, Faculty of Medicine, The University of Manitoba, Winnipeg, Manitoba, Canada
**Departamento de Fisiología y Farmacología, Centro de Ciencias Básicas, Universidad Autónoma de Aguascalientes, Aguascalientes, Estado Libre y Soberano de Aguascalientes, México City, Mexico City, México

1 IMMUNOREGULATION BY THE PITUITARY GLAND

The initial paper on pituitary function and immunity was published in 1978,[1] in which it was described that hypophysectomy (HYPOX) in young rats (100 g body weight) causes immunodeficiency. Treatment of HYPOX rats with replacement doses (20 μg/100 g rat, daily) of growth hormone (GH) or prolactin (PRL) restored the antibody response. The secondary antibody formation was partially inhibited and the antibody titer stayed at the level of primary response. After second immunization, it did not increase as it does normally during the secondary response.[2] It was interpreted that memory cells survive HYPOX, but they could not expand in the secondary response, mainly because of the lack of naive antigen-sensitive lymphocytes in HYPOX animals. This finding indicates that *GH and/or PRL are needed for the production of naive antigen-sensitive B cells.*

Human placental lactogen (PL) also restored immunocompetence in HYPOX rats.[3]

A form of cell-mediated immunity, contact sensitivity reactions to dinitrochlorobenzene (DNCB) were also suppressed by HYPOX in rats and could be restored by replacement doses (20 μg/100 g body weight, daily) of GH or PRL and also by human PL.[4,5]

Adjuvant-induced arthritis (AA) is an experimental autoimmune disease, which could be induced in rats by injection of complete Freund's adjuvant (mineral oil, detergent (arlacel), and 20 μg/mL *mycobacterial* antigen), 0.1 mL s.c. was injected on the base of the neck to each animal. AA was inhibited in rats by HYPOX and could be restored by 20 μg/100 g body weight, daily doses of PRL or GH. Adrenocorticotropic hormone (ACTH), inhibited all of the previously described reactions. Other pituitary hormones, such as,

Insights to Neuroimmune Biology
http://dx.doi.org/10.1016/B978-0-12-801770-8.00003-3

thyroid-stimulating hormone (TSH), luteinizing hormone (LH), follicle-stimulating hormone (FSH), and gonadotropic hormone (GNH), did not restore the immune function in HYPOX rats.[6]

Previously it has been established that the secretion of PRL is continuous through the pituitary gland and a neurotransmitter dopamine, exerts an inhibitory effect on PRL. We tested bromocriptine (BRC), a dopamine agonist drug, to know if it would influence antibody formation in rats. BRC was as immunosuppressive as was HYPOX in 100 g rats. Later, it was discovered that contact sensitivity responses and autoimmunity (AA) were similarly inhibited by BRC, as it was with HYPOX. Again the immunosuppressive effect of BRC could be antagonized by replacement doses of GH or PRL.[7] Long-surviving HYPOX rats were shown to depend on residual PRL in the blood (~50% that of normal) for survival.[8] Hypophysectomized rats depend on residual prolactin for survival.[9]

Bone marrow function was also stimulated by GH and PRL during homeostasis.[10] However, during acute illness, now called the acute phase response (APR), GH, and PRL are suppressed by suppressor/regulatory T cells (Tsr) and by glucocorticoids (GC) and catecholamines (CAT). During APR the bone marrow is likely to be stimulated by interleukin (IL)-6, granulocyte macrophage colony-stimulating factor (GMCSF), and possibly by some type I cytokines.[11] PRL and GH stimulate insulin-like growth factor (IGF-1) in thymus cells of HYPOX rats, and the growth factor gene promoter, *c-myc,* and DNA synthesis, were also stimulated by GH and PRL in lymphoid tissue.[12,13] Decreased bioactivity of circulating prolactin in sera of patients with rheumatoid arthritis was observed,[14] and PRL deficiency was demonstrated in rheumatoid arthritis.[15]

2 IMMUNOREGULATION BY GROWTH AND LACTOGENIC HORMONES

2.1 Immunoregulatory Pathways

2.1.1 Thymus

The thymus is a primary lymphoid organ producing T lymphocytes. It also has endocrine functions. Thymic physiology is under complex neuroendocrine control. It involutes during stress and is dramatically affected by the APR, a systemic reaction to tissue injury and/or infection accompanied by profound neuroendocrine and metabolic changes. The APR comprises of alterations in behavior, body temperature, and production and release of cytokines, particularly IL-1, IL-6, and TNF-α, and GCs. So innate immunity (INIM) is dramatically increased. The hypothalamus–pituitary–adrenal

(HPA) axis is activated, resulting in the suppression of adaptive immunity (ADIM), the immunostimulatory hormones (e.g., PRL, GH, IGF-1) are suppressed whereas the production of acute phase proteins (APPs) in the liver is stimulated by IL-6, CATs, and GCs. The most striking effect GCs on the immune system is the induction of apoptosis in the thymus. In concert with GCs, elevated levels of CATs also selectively suppress immune response mechanisms. APR may be regarded as an emergency response that represents a switch of the host defense from the ADIM response, which is slow in development. So it is commanded by the thymus and T-lymphocytes to perform the more rapid INIM reaction.[16]

2.1.2 Receptors for GLH in Various Cells and Tissues

The long prolactin receptor (lPRLR) internalized faster than the short (sPRLR), which would contribute to short-term down-regulation of lPRLR signaling at targets expressing both isoforms. Multiple motifs were required to mediate internalization of the lPRLR, including a phenylalanine (F290) plus a nearby d-leucine, and three d-leucines proximal to amino acid 272. The GH receptor requires only the phenylalanine-containing motif for endocytosis. Truncated lPRLR (cT272), which is the same length as the sPRLR and contained the proximal three dileucines, internalized at the same rate as the full-length lPRLR. Finally, the two dileucines shared by the sPRLR were able to mediate similar endocytic pathways as the lPRLR.[17]

PRL and GH are members of the immunological *cytokine/chemokine network* and share regulatory mechanisms with type I cytokines. These mediators are also regulated by the *suppressor of cytokine signaling* (*SOCS*) family of regulatory proteins.[18]

Human peripheral blood *granulocytes*, which express GH and PRL mRNA, contain high molecular-weight immunoreactive variants of GH and PRL (37 and 43 kDa, respectively), but not the pituitary-sized hormones. Secretion of these variants, or biologically active material as assessed by the Nb2 bioassay, was not detected.[19]

Cytokines IL-1β, IL-12p40, and IFN-γ were observed on the treatment of macrophages (MFs) with moderate doses of PRL or GH. However, higher doses of PRL (1000 ng/mL) induced the anti-inflammatory cytokine IL-10, with significant decrease of proinflammatory cytokines. PRL and GH induced the production of chemokines MIP-1α and RANTES. PRL but not GH significantly enhanced MCP-1 and IP-10. The p38 MAP kinase, STAT3, and NF-κB, which regulate PRL or GH, induced production of cytokines by MFs.[20]

Transmission of information from the immune system to the neuroendocrine system is mediated by cytokines, and by neural transmission. Efferent signals from the central nervous system (CNS) to the ADIM system are conveyed by both neuroendocrine and autonomic nervous systems. Communication is possible because the nervous and immune systems have common mediators (e.g., ligands and receptors, neurotransmitters, neuropeptides, growth factors, hormones, cytokines, and chemokines). This means that the brain participates in immune responses. Among protein hormones, immunoregulation has been most clearly documented for PRL, GH, IGF-I, and for TSH. Proinflammatory cytokines antagonize the function of a variety of hormones, which then causes endocrine resistance in both the periphery and in the brain.[21]

LPS-induced ACTH secretion, mediated by intrapituitary IL-6, may represent a pituitary-specific mechanism that stimulates the HPA axis during infection/inflammation.[22]

TNF-α and IL-1β, induce a state of IGF resistance. Proinflammatory cytokines induce resistance to IGF-I. IGF-I can induce resistance or reduce sensitivity to brain TNF-α, IL-1β, and IGF-I interact to regulate several aspects of behavior and cognition.[23]

2.1.3 Immune Reactions

After immunization of rats with sheep red blood cells (SRBC, a T-cell-dependent antigen), an increase of hypothalamic TRH mRNA was observed at 4–24 h postimmunization, and plasma PRL levels were increased, but TSH and GH levels did not change. The activation of the HPA suppressive response appeared 5–7 days after SRBC injection. In contrast, LPS stimulated a COS response and inhibited the release of other hormones. Intracerebroventricular injection of antisense oligonucleotide complementary to rat TRH mRNA in conscious rats immunized with SRBC, significantly inhibited antibody production and PRL release.[24]

2.1.4 Acute Febrile Illness (Acute phase response (APR))

During infection, bacterial and viral products, such as bacterial LPS, cause the release of cytokines from immune cells. These cytokines may reach the brain by several routes. Further, cytokines, such as IL-1, are induced in neurons within the brain by systemic injection of LPS. These cytokines determine the pattern of hypothalamic-pituitary secretion that characterizes infection. By the stimulation of *cholinergic neurons*, *IL-2* activates neural *nitric oxide synthase* (*NOS*). The nitric oxide (NO) released diffuses into

CRH-secreting neurons and *releases CRH*. IL-2 also acts in pituitary to stimulate *ACTH secretion*. On the other hand, *IL-1α blocks* the NO-induced release of *LHRH* from LHRH neurons, thereby blocking pulsatile LH but not FSH release, and also *inhibiting sex behavior* that is induced by LHRH. *IL-1α* and *granulocyte macrophage colony-stimulating factor* (*GMCSF*) blocks the response of the LHRH terminals to NO. GMCSF acts on its receptors on *γ-aminobutyric acid (GABA-)ergic* neurons to stimulate GABA release. GABA acts on GABAa receptors on the LHRH neuronal terminal to of LHRH release. IL-1α inhibits GH release by inhibiting GH-releasing hormone (GHRH) release, which is mediated by NO, and stimulating somatostatin (STA) release, also mediated by NO. IL-1α-induced stimulation of PRL release is also mediated by intrahypothalamic action of NO, which inhibits release of the PRL-inhibiting hormone *dopamine*. The actions of NO are brought about by its combined activation of guanylate cyclase-liberating cyclic guanosine monophosphate (cGMP), and activation of cyclooxygenase (COX) and lipoxygenase (LOX) with liberation of prostaglandin E2 and leukotrienes, respectively. Thus, NO plays a key role in inducing the changes in the release of hypothalamic peptides induced in infection by cytokines. Cytokines, such as IL-1β, also act in the anterior pituitary gland, at least in part via induction of inducible NOS. The NO produced inhibits the release of ACTH. The adipocyte hormone *leptin*, a member of the cytokine family, has largely opposite actions to those of the proinflammatory cytokines, stimulating the release of FSHRH and LHRH from the hypothalamus, and FSH and LH from the pituitary directly by NO.[25]

Adaptive immunocompetence is maintained by GH, PRL, and VP. Innate or *natural immunocompetence* depends on GCs, catecholamines, and VP. *The APR, or acute febrile illness* is an emergency defense reaction, wherein the adaptive, T-cell-dependent, immune reactions are suppressed and the innate immune functions are dramatically amplified. Infection and various forms of injury induce APR. *Cytokines* (primarily IL-1β, TNF-α, and IL-6) stimulate CRH and VP secretion and cause a "sympathetic outflow." Colony-stimulating factors activate leukocytes. CRH is a powerful activator of the pituitary–adrenocortical axis and elevates GC levels. Cytokines, GCs, and CATs play fundamental roles in the amplification of natural immune defense mechanisms. VP supports the APR at this stage. However, VP remains active and is elevated for a longer period than is CRH. VP, but not CRH, is elevated during chronic inflammatory diseases. VP controls adaptive immune function and stimulates ACTH and PRL secretion. PRL maintains the function of the thymus and the T-dependent ADIM system. The ACTH-adrenal axis stimulates natural immunity. VP also has a direct effect on lymphoid cells, the significance of which needs to be elucidated. It is suggested that VP regulates healing and recovery from acute illness.[10]

3 DOPAMINE IN NEUROIMMUNE BIOLOGY

3.1 Dopamine Receptors in Tissues and Cells

Dopamine transporters and receptors are expressed during *thymocyte development* (when CD4 and CD8 markers are differently expressed). D1-like receptors are expressed at high levels in *thymocytes, splenocytes, and peripheral lymphocytes.*[26] *Thymus and spleen expressed vesicular monoamine transporters (VMAT) such as, VMAT-1* and *VMAT-2* immunoreactivity, as well as *dopamine D1, D2, D3, D4,* and *D5 receptor* immunoreactivity.[27]

Human *PBL* synthesizes DA through the *tyrosine–hydroxylase/DOPA–decarboxylase pathway,* and expresses DA receptors and DA transporters (DAT) on their plasma membranes. The *feasibility of PBL as a cellular model for studying DA derangement in neuropsychiatric disorders, in particular Parkinsonian syndromes, schizophrenia, and drug- or alcohol-abuse has to be established.*[28]

The incubation of *MF* with 10 nM DOMP decreased oxidative burst, whereas the PMA-induced burst was decreased by DOMP 10 nM after 2 and 4 h. Treatment with PRL (10 and 100 nM) for 30 min decreased oxidative burst and rate of phagocytosis (10 nM). After 2 h of incubation, 10 nM PRL decreased oxidative burst and phagocytosis intensity, but increased the rate of phagocytosis. After 4 h, PRL 10 and 100 nM increased oxidative burst and the rate of phagocytosis, but decreased intensity of phagocytosis. Our data also indicated that both *PRL and dopamine exert their actions by acting directly on the peritoneal MFs.*[29]

Dendritic cells release dopamine to naive CD4+ T cells in Ag-specific cell–cell interaction, and also induce *Th17* differentiation through dopamine D1-like receptor (D1-like-R). D1-like-R antagonist *attenuates Th17*-mediated diseases such as *experimental autoimmune encephalomyelitis and autoimmune diabetes.* D1-like-R antagonist could suppress Th17-mediated *neutrophilic airway inflammation,* indicating that antagonizing D1-like-R may serve as a new strategy for treating neutrophil-dominant severe asthma.[30]

Dendritic cells are responsible for priming T-cells and for promoting their differentiation from naive T-cells into appropriate effector cells. Several studies have shown that *dopamine* modulates immune function.[31]

Dopamine released by DCs induces IL-6-Th17 axis and causes aggravation of synovial inflammation of RA. This is the first time that the pathological relevance of dopaminergic signaling was demonstrated for RA.[32]

Dendritic cells are a new source of *dopamine,* which function as a T(h)2-polarizing factor in DC-naive T-cell interface.[33]

Dopaminergic agonists can stimulate *keratinocytes* to produce *IL-6* and *IL-8*, which can cause *cutaneous inflammation*. Inflammation is mediated by *dopaminergic* and *β-adrenergic receptors* and by receptor-independent *oxidative* mechanisms.[34]

The *astrocytic dopamine D2 receptor* (*DRD2*) modulates *INIM* through *αB-crystallin* (*CRYAB*), which is known to *suppress neuroinflammation*. CRY-AB is critical for DRD2-mediated modulation of innate immune response in astrocytes.[35]

Five subtypes of DA receptors namely, D1R, D2R, D3R, D4R, and D5R, are expressed in *T lymphocytes and NK cells*. D1-like receptors enhanced *NK cell cytotoxicity*, but activation of D2-like receptors with *quinpirole* attenuated NK cells.[36]

Neurotransmitters (norepinephrine, dopamine, and substance P) are specific modulators of *CD8+ T lymphocytes.* The effect can be either stimulatory or suppressive depending on the activation status of the cells.[37]

3.2 Conserved Dopamine Neurotrophic Factor

Conserved dopamine neurotrophic factor (*CDNF*), a secretory neurotrophic factor, may provide therapy for *neuroinflammation* related to the *microglia.*[38]

3.3 BRC

BRC and other dopaminergic agents are being used or suggested for use in the management of *diabetes.*[39]

3.4 Neuropeptide

In 6-OHDA-treated rats, *NA nerve reinnervation was as normal in splenic compartments as was innervation by sympathetic nerves.*[40]

3.5 Interleukin

IFN-α is associated with decreased peripheral conversion of Phen to Tyr, which *reduced DA in the brain* as well as fatigue. These alterations may be related to oxidation of *tetrahydrobiopterin* (*BH4*) secondary to IFN-α-induced activation of a *CNS inflammatory response.*[41]

4 IMMUNOREGULATION BY DOPAMINE

Postsynaptic D(1)- and D(2)-receptors produced *immunostimulation*, while their blockade or activation of *presynaptic D(2)-receptors* mediated an *immunoinhibition*. Activation of *μ- and δ(1)-opioid receptors* also *increased* the immune reactions. Activation of *postsynaptic 5-HT(1A)-, 5-HT (2A)-receptors, δ(2)-, and κ-opioid* receptors resulted in *immunosuppression*, while the blockade of postsynaptic 5-HT(1A)-, 5-HT(2A)-receptors, or activation of somatodendritic 5-HT(1A)-autoreceptors resulted in the immune response stimulation. Immunomodulating effects of drugs depend on initial psychoemotional state of animals (aggression, submission, and depression).[42]

Activation of mu-opioid receptors (μ-OR) by the highly selective agonist DAGO (100 μg/kg) significantly increased the immune response in CBA mice. This was prevented by prior blockade of dopamine D(2) receptors with *haloperidol* (2 mg/kg).[43]

Activation of dopamine D(2) receptors by the agonist *quinpirole* lead to *immunostimulation* in control (no experience of confrontations) and in C57BL/6J mice subjected to psychoemotional tension (aggression and submission). Blockade of D(2) dopamine receptors by *haloperidol-immunosuppression* was seen only in controlled and aggressive animals, but were not seen in animals with *submissive* behavior.[44]

Intrastriatal 6-OHDA administration might cause neurodegeneration, increased sensitivity to the inflammation, which is caused by *impaired dopaminergic control of PRL and corticosterone (CORT) secretion*. The brain dopaminergic system is involved in peripheral immune regulation and central *dopaminergic hypoactivity*, which bears the risk of *excessive inflammation,* for example, during infection or tissue injury.[45]

5 OXYTOCIN AND GHRELIN IN IMMUNITY

OT receptors and expression. OT and neurophysin in thymus was in similar concentrations to that found in the hypothalamo-neurohypophysial system.[46] OT receptors bovine T lymphocytes are regulated by steroid hormones.[47] Rat uterine OT receptors are regulated by IL-6.[48,49] OT receptors from mounted female rats were significantly lower than of unmounted controls.[50,51]

OT in physiology. Pituitary OT *regulates PRL release*.[52] OT antiserum in female rats neutralized OT in pituitary portal blood and significantly *reduced the subsequent PRL surge*.[53] Mouse IL-2 caused a significant increase in VP and OT mRNA in the hypothalamus of nude mice.[54] CRH, VP, and OT mediate the IL-1β-induced *ACTH* secretion in the rat.[55] OT physiologically regulates ACTH secretion.[56]

Pharmacology. Cytokines were more reduced by *OT antagonists* in blood cells of females, compared to male donors.[57] OT significantly increased PBMC response to PHA, and induced CD25 and CD95.[58,59]

OT in pathology. OT protects against sepsis-induced oxidative damage, alleviates oxidative renal injury, inhibits sepsis, and the LPS response by inhibiting neutrophil infiltration as well as by *decreasing* the *proinflammatory* response and *oxidative stress.*[60–63]

Therapy. OT may provide a therapeutic benefit in diminishing burn-induced gastric inflammation.[64] OT may have stress buffering effects on women living with HIV reporting elevated stress. The addition of an oxytocin antagonist completely reversed the oxytocin activity.[65]

5.1 Ghrelin (GRL) and Immunity

Receptors and expression. Ghrelin stimulates phagocytosis in fish leukocytes.[66] Pit-1 mRNA levels are significantly lower in (*Ghsr−/−*) mice. This *Pit-1 loss* leads to *reduced GH* and *PRL production.*[67] GH and GHR contribute to *thymus* biology, also function in the *thyroid*. For the *ovary* GRL regulates energy metabolism and food intake, which are important for *ovarian function.*[68,69]

Ghrelin inhibits splenic *T cells* and suppressed Th(1) (IL-2 and IFN-γ) and Th(2) (IL-4 and IL-10) cytokines mRNA. Synthetic GRL inhibited the production of *IL-8 from TNF-α* or LPS-stimulated oral epithelial cells. TNF-α and ghrelin have opposite effects on hypothalamic regulation of eating behavior, the immune response, and the state of mental health.[70,71]

GRLs have anti-inflammatory effects in *MFs* and mediate the beneficial effects of exercise training. GRL reduced the increased *phagocytic activity* induced by acute cold-restraint stress (ACRS). GHR is important for MF phagocytosis. Ghrelin and obestatin stimulate *mast cells*. GHR correlates with the *nutritional status* of healthy children, but not in sick children.[72,73]

Ghrelin in immunopathology. The role of *GHR and other GH secretagogues* (GHS) are potent anti-inflammatory mediators. GHR also promotes lymphocyte development in the primary lymphoid organs (bone marrow and thymus) and ablates age-associated thymic involution.[74] Ghrelins regulate proinflammatory cytokines that contribute to the pathogenesis of neuroinflammatory and *mental diseases*. GHS-R and GRL are expressed in human T cells and monocytes, where GHR inhibits the expression of IL-1β, IL-6, and TNF-α. GRL inhibits leptin-induced cytokine expression.[75]

Ghrelin and its target receptors called growth hormone secretagogue receptors (GHS-Rs), have been localized to neutrophils, lymphocytes, and MFs, which suggests that ghrelin may be involved in immune modulation. Ghrelin has potent anti-inflammatory properties through modulation of secretion of both pro- and anti-inflammatory cytokines from LPS-stimulated MFs through distinct signaling cascades.[76] Ghrelin attenuates excessive inflammation and reduces organ injury after gut I/R through activation of the *cholinergic anti-inflammatory pathway*.[77] Elucidation of the roles of GH and ghrelin in the immune system may shed light on the treatment and prevention of *immunological disorders*.[78] Toll-like receptors and a plethora of neuroactive substances, such as NO, prostaglandins, cytokines, chemokines, growth factors, tryptases, and hormones are recognized today. Multidirectional signaling between the different components in the gastrointestinal wall, the spinal cord, and CNS impacts inflammation and its consequences.[79] GHR levels in *inflammatory bowel disease* (*IBD*) patients show a positive correlation with IGF-1. GHR level may be important for IBD patients and evaluation of nutritional status.[80]

In rats LPS decreased plasma *acyl ghrelin* (AG) and *desacyl ghrelin* (DG). *LPS* induces a long-lasting reduction of AG and DG levels that may decrease food intake.[81] Ghrelin attenuates the *LPS-induced* IL-6 secretion in mid-brain dopaminergic neurons.[82] LPS-induced fever in rats was attenuated by ghrelin.[59] Plasma ghrelin drops down with prolonged systemic inflammation after *esophagectomy*. The correction of low plasma *ghrelin* may help minimize excess inflammatory response in these patients.[83] The possible immunomodulatory roles of *gut hormones* in autoimmune inflammatory reactions deserves attention.[84]

Therapy. GRL improved colitis in most (but not all) animal models.[85] MKP-1 protein expression was significantly decreased 2 h after colonic ligation and puncture (*CLP*). *Ghrelin* treatment improved both mRNA and protein from vehicle group (0.58 ± 0.069 vs. 0.91 ± 0.16, $P < 0.05$; 0.14 ± 0.027 vs. 0.22 ± 0.017, $P = 0.013$), respectively.[86]

6 CONCLUSIONS

As we learnt about the *neuroimmune supersystem*, it is clear that an increasing number of regulatory factors tie into the *basic network of GLH and the HPA axis*. It was known for a long time that *dopamine* inhibited PRL synthesis. However, we thought that GH was also suppressed by DP as BRC suppressed the immune repose, as did HYPOX. Each of GH or PRL could restore immune function in HYPOX rats. So BRC had to suppress both hormones for immunosuppression. This is now proven by others as well that the HPA axis is also regulated by DP. Thyroxin also releases GH and PRL from pituitary glands transplanted under the kidney capsule of rats. Oxytocin and ghrelin also connect to GLH and the HPA axis.

REFERENCES

1. Nagy E, Berczi I. Immunodeficiency in hypophysectomized rats. *Acta Endocrinol* 1978;**89**:530–7.
2. Berczi I, Nagy E, Kovacs K, Horvath E. Regulation of humoral immunity in rats by pituitary hormones. *Acta Endocrinol* 1981;**98**:506–13.
3. Berczi I, Nagy E. Placental lactogen is a hemopoietic hormone. *Br J Haematol* 1991;**79**:355–8.
4. Nagy E, Berczi I, Friesen HG. Regulation of immunity in rats by lactogenic and growth hormones. *Acta Endocrinol* 1983;**102**:351–7.
5. Berczi I, Nagy E, Asa SL, Kovacs K. Pituitary hormones and contact sensitivity in rats. *Allergy* 1983;**38**:325–30.
6. Berczi I, Nagy E, Asa SL, Kovacs K. The influence of pituitary hormones on adjuvant arthritis. *Arthritis Rheum* 1984;**27**:682–8.
7. Nagy E, Berczi I, Wren GE, Asa SL, Kovacs K. Immunomodulation by bromocriptine. *Immunopharmacology* 1983;**6**:231–43.
8. Nagy E, Berczi I. Pituitary dependence of bone marrow function. *Br J Haematol* 1989;**71**:457–62.
9. Nagy E, Berczi I. Hypophysectomized rats depend on residual prolactin for survival. *Endocrinology* 1991;**128**:2776–84.
10. Berczi I, Szentivanyi A. The acute phase response. In: Berczi I, Szentivanyi A, editors. Neuroimmmune Biology. *The immune-neuroendocrine circuitry. History and progress*, vol. 3. Amsterdam: Elsevier; 2003. p. 463–494.
11. Berczi I, Quintanar-Stephano A, Kovacs K. Neuroimmune regulation in immunocompetence, acute illness, and healing. *Ann NY Acad Sci* 2009;**1153**:220–39.
12. Berczi I. Introduction to neuroimmune pathology. Berczi I, Gorczynski R, editors. *Neuroimmune biology: new foundation of biology*, vol. 1. Amsterdam: Elsevier; 2001. p. 349–350.
13. Berczi I, Nagy E, Matusik RJ, Friesen HG. Pituitary hormones regulate c-myc and DNA synthesis in lymphoid tissue. *J Immunol* 1991;**146**:2201–6.
14. Berczi I, Cosby H, Hunter T, Baragar F, McNeilly AS, Friesen HG. Decreased bioactivity of circulating prolactin in patients with rheumatoid arthritis. *Br J Rheum* 1987;**26**:433–6.
15. Nagy E, Chalmers IM, Baragar FD, Friesen HG, Berczi I. Prolactin deficiency in rheumatoid arthritis. *J Rheumatol* 1991;**18**:1662–8.
16. Haeryfar SM, Berczi I. The thymus and the acute phase response. *Cell Mol Biol* 2001;**47**:145–56.
17. Lu JC, Scott P, Strous GJ, Schuler LA. Multiple internalization motifs differentially used by prolactin receptor isoforms mediate similar endocytic pathways. *Mol Endocrinol* 2002;**16**:2515–27.
18. Redelman D, Welniak LA, Taub D, Murphy WJ. Neuroendocrine hormones such as growth hormone and prolactin are integral members of the immunological cytokine network. *Cell Immunol* 2008;**252**:111–21.
19. Kooijman R, Gerlo S, Coppens A, Hooghe-Peters EL. Growth hormone and prolactin expression in the immune system. *Ann NY Acad Sci* 2000;**917**:534–40.

20. Blanco EJ, Carretero-Hernández M, García-Barrado J, Iglesias-Osma MC, Carretero M, Herrero JJ, Rubio M, Riesco JM, Carretero J. The activity and proliferation of pituitary prolactin-positive cells and pituitary VIP-positive cells are regulated by interleukin 6. *Histol Histopathol* 2013;**28**:1595–604.
21. Sodhi A, Tripathi A. Prolactin and growth hormone induce differential cytokine and chemokine profile in murine peritoneal macrophages *in vitro*: involvement of p-38 MAP kinase, STAT3 and NF-kappaB. *Cytokine* 2008;**41**:162–73.
22. Kelley KW, Weigent DA, Kooijman R. Protein hormones and immunity. *Brain Behav Immun* 2007;**21**:384–92.
23. Gloddek J, Lohrer P, Stalla J, Arzt E, Stalla GK, Renner U. The intrapituitary stimulatory effect of lipopolysaccharide on ACTH secretion is mediated by paracrine-acting IL-6. *Exp Clin Endocrinol Diab* 2001;**109**:410–5.
24. Perez Castro C, Peñalva R, Páez Pereda M, Renner U, Reul JM, Stalla GK, Holsboer F, Arzt E. Early activation of thyrotropin-releasing-hormone and prolactin plays a critical role during a T cell-dependent immune response. *Endocrinology* 1999;**140**:690–7.
25. McCann SM, Kimura M, Karanth S, Yu WH, Mastronardi CA, Rettori V. The mechanism of action of cytokines to control the release of hypothalamic and pituitary hormones in infection. *Ann NY Acad Sci* 2000;**917**:4–18.
26. Mignini F, Sabbatini M, Capacchietti M, Amantini C, Bianchi E, Artico M, Tammaro A. T-cell subpopulations express a different pattern of dopaminergic markers in intra- and extra-thymic compartments. *J Biol Regul Homeost Agents* 2013;**27**:463–75.
27. Mignini F, Tomassoni D, Traini E, Amenta F. Dopamine, vesicular transporters and dopamine receptor expression and localization in rat thymus and spleen. *J Neuroimmunol* 2009;**206**:5–13.
28. Buttarelli FR, Fanciulli A, Pellicano C, Pontieri FE. The dopaminergic system in peripheral blood lymphocytes: from physiology to pharmacology and potential applications to neuropsychiatric. *Curr Neuropharmacol* 2011;**9**:278–88.
29. Carvalho-Freitas MI, Rodrigues-Costa EC, Nasello AG, Palermo-Neto J, Felicio LF. *In vitro* macrophage activity: biphasic effect of prolactin and indirect evidence of dopaminergic modulation. *Neuroimmunomodulation* 2008;**15**:131–9.
30. Nakano K, Yamaoka K, Hanami K, Saito K, Sasaguri Y, Yanagihara N, Tanaka S, Katsuki I, Matsushita S, Tanaka Y. Dopamine induces IL-6-dependent IL-17 production via D1-like receptor on CD4 naive T cells and D1-like receptor antagonist SCH-23390 inhibits cartilage destruction in a human rheumatoid arthritis/SCID mouse chimera model. *J Immunol* 2011;**186**:3745–52.
31. Pacheco R, Prado CE, Barrientos MJ, Bernales S. Role of dopamine in the physiology of T-cells and dendritic cells. *J Neuroimmunol* 2009;**216**:8–19.
32. Nakano K, Higashi T, Takagi R, Hashimoto K, Tanaka Y, Matsushita S. Dopamine released by dendritic cells polarizes Th2 differentiation. *Int Immunol* 2009;**21**:645–54.
33. Parrado AC, Canellada A, Gentile T, Rey-Roldán EB. Dopamine agonists upregulate IL-6 and IL-8 production in human keratinocytes. *Neuroimmunomodulation* 2012;**19**:359–66.
34. Shao W, Zhang SZ, Tang M, Zhang XH, Zhou Z, Yin YQ, Zhou QB, Huang YY, Liu YJ, Wawrousek E, Chen T, Li SB, Xu M, Zhou JN, Hu G, Zhou JW. Suppression of neuroinflammation by astrocytic dopamine D2 receptors via αB-crystallin. *Nature* 2013;**494**:90–4.
35. Zhao W, Huang Y, Liu Z, Cao BB, Peng YP, Qiu YH. Dopamine receptors modulate cytotoxicity of natural killer cells via cAMP-PKA-CREB signaling pathway. *PLoS ONE* 2013;**8**:e65860.
36. Strell C, Sievers A, Bastian P, Lang K, Niggemann B, Zänker KS, Entschladen F. Divergent effects of norepinephrine, dopamine and substance P on the activation, differentiation and effector functions of human cytotoxic T lymphocytes. *BMC Immunol* 2009;**10**:62.

37. Zhao H, Cheng L, Liu Y, Zhang W, Maharjan S, Cui Z, Wang X, Tang D, Nie L. Mechanisms of anti-inflammatory property of conserved dopamine neurotrophic factor: inhibition of JNK signaling in lipopolysaccharide-induced microglia for 24 h. *J Mol Neurosci* 2014;**52**:186–92.
38. Kalra S, Kalra B, Agrawal N, Kumar S. Dopamine: the forgotten felon in type 2 diabetes. *Recent Pat Endocr Metab Immune Drug Discov* 2011;**5**:61.
39. Lorton D, Lubahn C, Sweeney S, Major A, Lindquist CA, Schaller J, Washington C, Bellinger DL. Differences in the injury/sprouting response of splenic noradrenergic nerves in Lewis rats with adjuvant-induced arthritis compared with rats treated with 6-hydroxydopamine. *Brain Behav Immun* 2009;**23**:276–8.
40. Idova GV, Alperina EL, Cheido MA. Contribution of brain dopamine, serotonin and opioid receptors in the mechanisms of neuroimmunomodulation: evidence from pharmacological analysis. *Int Immunopharmacol* 2012;**12**:618–25.
41. Cheido MA, Idova GV. The differential contribution of dopamine D(1) and D(2) receptors to mu-opioidergic immunomodulation. *Neurosci Behav Physiol* 2007;**37**:721–4.
42. Idova GV, Cheido MA, Zhukova EN. Effects of activation and blockade of dopamine D(2) receptors on the immune response in mice with different types of behavior. *Neurosci Behav Physiol* 2007;**37**:725–9.
43. Engler H, Doenlen R, Riether C, Engler A, Niemi MB, Besedovsky HO, del Rey A, Pacheco-López G, Feldon J, Schedlowski M. Time-dependent alterations of peripheral immune parameters after nigrostriatal dopamine depletion in a rat model of Parkinson's disease. *Brain Behav Immun* 2009;**23**:518–26.
44. Matalka KZ, Attallah LJ, Qinna NA, Alhussainy T. Dopamine selectively modulates lipopolysaccharide-induced TNF-alpha, IFN-gamma and IL-10 within mice tissues. *Neuro Endocrinol Lett* 2011;**32**:176–86.
45. Tenk CM, Foley KA, Kavaliers M, Ossenkopp KP. Neonatal immune system activation with lipopolysaccharide enhances behavioural sensitization to the dopamine agonist, quinpirole, in adult female but not male rats. *Brain Behav Immun* 2007;**21**:935–45.
46. Assis MA, Pacchioni AM, Collino C, Paz MC, Sotomayor C, Basso AM, Cancela LM. A dopamine mechanism is implied in the acquisition and expression of amphetamine and stress-induced effects observed in the lymphocyte subpopulations. *Eur J Pharmacol* 2008;**584**:405–14.
47. Tomei G, Capozzella A, Ciarrocca M, Fiore P, Rosati MV, Fiaschetti M, Casale T, Anzelmo V, Tomei F, Monti C. Plasma dopamine in workers exposed to urban stressor. *Toxicol Ind Health* 2007;**23**:421–7.
48. Vuillermot S, Weber L, Feldon J, Meyer U. A longitudinal examination of the neurodevelopmental impact of prenatal immune activation in mice reveals primary defects in dopaminergic development relevant to schizophrenia. *J Neurosci* 2010;**30**: 1270–87.
49. Eyles D, Feldon J, Meyer U. Schizophrenia: do all roads lead to dopamine or is this where they start? Evidence from two epidemiologically informed developmental rodent models. *Transl Psychiatry* 2012;**2**:81.
50. Zager A, Mennecier G, Palermo-Neto J. Maternal immune activation in late gestation enhances locomotor response to acute but not chronic amphetamine treatment in male mice offspring: role of the D1 receptor. *Behav Brain Res* 2012;**232**:30–6.
51. Carvalho-Freitas MI, Anselmo-Franci JA, Teodorov E, Nasello AG, Palermo-Neto J, Felicio LF. Reproductive experience modifies dopaminergic function, serum levels of prolactin, and macrophage activity in female rats. *Life Sci* 2007;**81**:128–36.
52. Mori T, Kabashima K, Fukamachi S, Kuroda E, Sakabe J, Kobayashi M, Nakajima S, Nakano K, Tanaka Y, Matsushita S, Nakamura M, Tokura Y. D1-like dopamine receptors antagonist inhibits cutaneous immune reactions mediated by Th2 and mast cells. *J Dermatol Sci* 2013;**71**:37–44.

53. Mohammad SS, Sinclair K, Pillai S, Merheb V, Aumann TD, Gill D, Dale RC, Brilot F. Herpes simplex encephalitis relapse with chorea is associated with autoantibodies to *N*-methyl-D-aspartate receptor or dopamine-2 receptor. *Mov Disord* 2014;**29**:117–22.
54. Bałkowiec-Iskra E, Kurkowska-Jastrzebska I, Joniec I, Ciesielska A, Członkowska A, Członkowski A. Dopamine, serotonin and noradrenaline changes in the striatum of C57BL mice following myelin oligodendrocyte glycoprotein (MOG) 35-55 and complete freund adjuvant (CFA) administration. *Acta Neurobiol Exp (Wars)* 2007;**67**:379–88.
55. Bałkowiec-Iskra E, Kurkowska-Jastrzebska I, Joniec I, Ciesielska A, Muszynska A, Przybyłkowski A, Członkowska A, Członkowski A. MPTP-induced central dopamine depletion exacerbates experimental autoimmune encephalomyelitis (EAE) in C57BL mice. *Inflamm Res* 2007;**56**:311–7.
56. Block ML, Hong JS. Chronic microglial activation and progressive dopaminergic neurotoxicity. *Biochem Soc Trans* 2007;**35**(5):1127–32.
57. Sarkar C, Basu B, Chakroborty D, Dasgupta PS, Basu S. The immunoregulatory role of dopamine: an update. *Brain Behav Immun* 2010;**24**:525–8.
58. Brito-Melo GE, Nicolato R, de Oliveira AC, Menezes GB, Lélis FJ, Avelar RS, Sá J, Bauer ME, Souza BR, Teixeira AL, Reis HJ. Increase in dopaminergic, but not serotoninergic, receptors in T-cells as a marker for schizophrenia severity. *J Psychiatr Res* 2012;**46**:738–42.
59. Brunton PJ, Sabatier N, Leng G, Russell JA. Suppressed oxytocin neuron responses to immune challenge in late pregnant rats: a role for endogenous opioids. *Eur J Neurosci* 2006;**23**:1241–7.
60. Feleder C, Tseng KY, Calhoon GG, O'Donnell P. Neonatal intrahippocampal immune challenge alters dopamine modulation of prefrontal cortical interneurons in adult rats. *Biol Psychiatry* 2010;**67**:386–92.
61. Aguilar-Valles A, Flores C, Luheshi GN. Prenatal inflammation-induced hypoferremia alters dopamine function in the adult offspring in rat: relevance for schizophrenia. *PLoS ONE* 2010;**5**:10967.
62. Ferrari M, Termine C, Franciotta D, Castiglioni E, Pagani A, Lanzi G, Marino F, Lecchini S, Cosentino M, Balottin U. Dopaminergic receptor D5 mRNA expression is increased in circulating lymphocytes of Tourette syndrome patients. *J Psychiatr Res* 2008;**43**: 24–9.
63. Jafari M, Ahangari G, Saberi M, Samangoui S, Torabi R, Zouali M. Distorted expression of dopamine receptor genes in systemic lupus erythematosus. *Immunobiology* 2013;**218**:979–83.
64. Cosentino M, Marino F. Adrenergic and dopaminergic modulation of immunity in multiple sclerosis: teaching old drugs new tricks? *J Neuroimmune Pharmacol* 2013;**8**:163–79.
65. Ferreira TB, Kasahara TM, Barros PO, Vieira MM, Bittencourt VC, Hygino J, Andrade RM, Linhares UC, Andrade AF, Bento CA. Dopamine up-regulates Th17 phenotype from individuals with generalized anxiety disorder. *J Neuroimmunol* 2011;**238**:58–66.
66. Ferris MJ, Mactutus CF, Booze RM. Neurotoxic profiles of HIV, psychostimulant drugs of abuse, and their concerted effect on the brain: current status of dopamine system vulnerability in NeuroAIDS. *Neurosci Biobehav Rev* 2008;**32**:883–909.
67. Geenen V, Legros JJ, Franchimont P, Baudrihaye M, Defresne MP, Boniver J. The neuroendocrine thymus: coexistence of oxytocin and neurophysin in the human thymus. *Science* 1986;**232**(4749):508–11.
68. Diaye K, Poole DH, Pate JL. Expression and regulation of functional oxytocin receptors in bovine T lymphocytes. *Biol Reprod* 2008;**78**:786–93.
69. Işeri SO, Sener G, Saglam B, Gedik N, Ercan F, Yegen BC. Oxytocin protects against sepsis-induced multiple organ damage: role of neutrophils. *J Surg Res* 2005;**126**:73–81.
70. Fang X, Wong S, Mitchell BF. Effects of LPS and IL-6 on oxytocin receptor in non-pregnant and pregnant rat uterus. *Am J Reprod Immunol* 2000;**44**:65–72.

71. Caldwell JD, Walker CH, Pedersen CA, Mason GA. Sexual activity decreases oxytocin receptor densities in the thymus. *Life Sci* 1993;**52**:1781–6.
72. Sarkar DK. Immunoneutralization of oxytocin attenuates preovulatory prolactin secretion during proestrus in the rat. *Neuroendocrinology* 1988;**48**:214–6.
73. Pardy K, Murphy D, Carter D, Hui KM. The influence of interleukin-2 on vasopressin and oxytocin gene expression in the rodent hypothalamus. *J Neuroimmunol* 1993;**42**: 131–8.
74. Watanobe H, Sasaki S, Takebe K. Involvement of oxytocin and cholecystokinin-8 in interleukin-1 beta-induced adrenocorticotropin secretion in the rat. *Neuroimmunomodulation* 1995;**2**:88–91.
75. Gibbs DM. Immunoneutralization of oxytocin attenuates stress-induced corticotrophin secretion in the rat. *Regul Pept* 1985;**12**:273–7.
76. Geenen V, Martens H, Robert F, Vrindts-Gevaert Y, De Groote D, Franchimont P. Immunomodulatory properties of cyclic hexapeptide oxytocin antagonists. *Thymus* 1992;**20**:217–26.
77. Macciò A, Madeddu C, Chessa P, Panzone F, Lissoni P, Mantovani G. Oxytocin both increases proliferative response of peripheral blood lymphomonocytes to phytohemagglutinin and reverses immunosuppressive estrogen activity. *In Vivo* 2010;**24**:157–63.
78. Işeri SO, Sener G, Saglam B, Gedik N, Ercan F, Yegen BC. Oxytocin protects against sepsis-induced multiple organ damage: role of neutrophils. *J Surg Res* 2005;**126**:73–81.
79. Biyikli NK, Tuğtepe H, Sener G, Velioğlu-Oğünç A, Cetinel S, Midillioğlu S, Gedik N, Yeğen BC. Oxytocin alleviates oxidative renal injury in pyelonephritic rats via a neutrophil-dependent mechanism. *Peptides* 2006;**27**:2249–57.
80. Carnio EC, Moreto V, Giusti-Paiva A, Antunes-Rodrigues J. Neuro-immune-endocrine mechanisms during septic shock: role for nitric oxide in vasopressin and oxytocin release. *Endocr Metab Immune Disord Drug Targets* 2006;**6**:137–42.
81. Clodi M, Vila G, Geyeregger R, Riedl M, Stulnig TM, Struck J, Luger TA, Luger A. Oxytocin alleviates the neuroendocrine and cytokine response to bacterial endotoxin in healthy men. *Am J Physiol Endocrinol Metab* 2008;**295**:E686.
82. Oliveira-Pelegrin GR, Saia RS, Cárnio EC, Rocha MJ. Oxytocin affects nitric oxide and cytokine production by sepsis-sensitized macrophages. *Neuroimmunomodulation* 2013;**20**:65–71.
83. Işeri SO, Gedik IE, Erzik C, Uslu B, Arbak S, Gedik N, Yeğen BC. Oxytocin ameliorates skin damage and oxidant gastric injury in rats with thermal trauma. *Burns* 2008;**34**: 361–9.
84. Fekete EM, Antoni MH, Lopez C, Mendez AJ, Szeto A, Fletcher MA, Klimas N, Kumar M, Schneiderman N. Stress buffering effects of oxytocin on HIV status in low-income ethnic minority women. *Psychoneuroendocrinology* 2011;**36**:881–90.
85. Yada T, Kaiya H, Mutoh K, Azuma T, Hyodo S, Kangawa K. Ghrelin stimulates phagocytosis and superoxide production in fish leukocytes. *J Endocrinol* 2006;**189**:57–65.
86. Yang H, Dixit VD, Patel K, Vandanmagsar B, Collins G, Sun Y, Smith RG, Taub DD. Reduction in hypophyseal growth hormone and prolactin expression due to deficiency in ghrelin receptor signaling is associated with Pit-1 suppression: relevance to the immune system. *Brain Behav Immun* 2008;**22**:1138–45.

CHAPTER 4

An Update on Neural Regulators of the Hypothalamic–Pituitary–Adrenal Axis

Istvan Berczi*,, Andrés Quintanar-Stephano****
*Department of Immunology, Faculty of Medicine, The University of Manitoba, Winnipeg, Manitoba, Canada
**Departamento de Fisiología y Farmacología, Centro de Ciencias Básicas, Universidad Autónoma de Aguascalientes, Aguascalientes, Estado Libre y Soberano de Aguascalientes, México City, Mexico

The hypothalamic corticotropin-releasing hormone (CRH) is the major coordinator of acute inflammatory disease. Vasopressin (VP) has a secondary role in acute illness, but when the disease becomes chronic, VP becomes the hypothalamic mediator regulating healing and recovery. During homeostasis VP controls ADIM and INIM, and maintains homeostasis.

Proopiomelanocortin (*POMC*) is induced in the anterior pituitary (AP) by two independent pathways: corticotropin-releasing hormone (CRH) and interleukin (IL)-6.

Neuropeptide regulators of the hypothalamic–pituitary–adrenal (HPA) axis: urocortin, catecholamines (CAT), apelin, cholecystokinin (CCK), leptin, leukotrienes, natriuretic peptide, neuromedins, neuropeptides, nociceptin/orphanin, PACAP, opioid peptides, orexin, relaxin, sauvagin/urotensin, stresscopin (SCP), and stresscopin-related peptides (SRP), and vasotocin (VT) are involved.

CRH and the central nervous system (CNS): stimulates corticotropin-releasing hormone (CRH). The POMC gene is cytokine (CTK) regulated and the negative regulation is by suppressor of cytokine signaling SOCKS 3.

Parvocellular CRH neurons coexpress VP, a second "releasing factor," for ACTH and CRH. CRH and VP are affected by early life stress (Il-6 and LIF are involved). Here, regulation is complex and the HPA axis is the regulator of immune/inflammatory responses, whereas GC inhibits the HPA axis.

CRH-receptors 1 and 2 (CHR1, 2), and UCN receptors. CRH receptors recognize at least five different *G-proteins* (*Gs, Gi, Gq/11, Go, and Gz*), which stimulate at least two intracellular signaling cascades.

Peripheral, *immune CRH* from the spleen, thymus, and inflamed tissue has proinflammatory effects and act on CRH-R1 and CRH-R2 receptors on neighboring immune cells. Proinflammatory actions may be attributed

Insights to Neuroimmune Biology
http://dx.doi.org/10.1016/B978-0-12-801770-8.00004-5

to novel CRH-like peptides, *urocortin* acting on CRH-R2. *Human lymphocytes expressed urocortin but not CRH mRNA.* UCN is a CRH family peptide, it is involved in physiology, interacts with leptin, NPY, orexin, and with ACTH in the brain. It is also present in the heart, gut, and immune cells.

Neuroimmune biology deals with the functional interaction of the nervous, endocrine, and immune systems. Growth hormone (GH), prolactin (PRL), and placental lactogens are known as growth and lactogenic hormones (GLH) maintain the *adaptive immune system.* The HPA axis is inhibitory. In addition, CTKs, chemokines, neurotransmitters, neuropeptides, and stressful factors modify this system. Natural or innate immunity (INIM) is maintained by vasopressin (VP), glucocorticoids, and CATs. CTKs are important regulators of INIM.

VP regulates all immune functions during homeostasis. In case of acute illness, CRH is the primary regulator and the and VP is the secondary one. This *neuroimmune supersystem* regulates everything in the body from conception till death.

The submandibular salivary gland interacts with adaptive immune system. It stimulates the release of both GH and PRL, it also releases mediators from the superior cervical ganglion, stimulates the ovary, testis, thyroid gland, and lymph nodes, and the gastrointestinal (GI) tract. The GI tract will signal the immune system, which signals back to the salivary gland. All signals are stimulatory by growth factors, CTKs and immunoglobulins.

Finally, we discovered how to cure cancer by immunotherapy. The logical steps are as follows: (1) slow down the tumor by an inhibitory drug so that it could not grow and mutate; (2) treat with killer T, natural killer, or lymphokine-activated killer cells. Most of the tumors treated with this method will regress after a single treatment.

1 CORTICOTROPIN-RELEASING HORMONE

CRH is corticotropin-releasing hormone, which releases adrenocorticotropic hormone (ACTH) from the pituitary gland and in part, it mediates the CTK-induced ACTH release. CRH integrates the acute phase response (APR) in the CNS and acts centrally as an immunosuppressive agent, which is mediated by the stimulation of *sympathetic outflow.* CRH receptors are present in the immune system and immunocytes produce CRH.[1,2]

Immune CRF production gets increased during inflammatory responses and it is an important anti-inflammatory hormone. However, some forms of inflammation were enhanced by CRH.[3]

CRHR1-deficient (CRHR1-0) mice show marked impairment of the pituitary–adrenal axis. However, plasma ACTH concentrations of unstressed mutant mice are similar to those in wild-type animals. *VP* is a major ACTH secretagogue in CRHR1-0 mice under resting conditions. In null mutants, the administration of anti-VP serum, but not anti-CRF serum, markedly reduced (by 60%) resting plasma ACTH concentrations. The administration of turpentine into the hindlimb of CRHR1-0 mice produced a slightly (15–25%) smaller swelling of the limb, but a 10-fold greater rise in plasma IL-6 levels, compared to normal controls. Turpentine-induced local inflammation produced pronounced elevations in the plasma concentrations of both ACTH and corticosterone (CORT) in both CRHR1-0 and wild-type mice. Anti-CRH and anti-VP sera only inhibited ACTH secretion in wild type mice. These data indicate that resting ACTH secretion in CRHR1-0 mice is attributable in part to VP. Furthermore, mice deficient in CRHR1 are still able to mount a *pituitary–adrenal response* via mechanisms that do not depend critically on either CRHF or VP action.[4]

1.1 Proopiomelanocortin

IL-1α-induced activation of signal transducer and activator (STAT) 3 in the AP was suppressed in IL-6 knockout (KO) mice; thus, plasma IL-6 is first induced by IL-1α and IL-6 activates STAT 3 in AP, leading to the induction of POMC in concert with CRH. The induction of POMC in the AP may happen by two independent pathways namely; CRH and IL-6.[5]

2 NEUROPEPTIDE REGULATORS OF THE HPA AXIS

2.1 Urocortin

Urocortin (UCN) is a new mammalian member of the CRH family. UCN is a candidate endogenous ligand for type 2 CRH receptors. In adult male rats, UCN mRNA was abundant in the GI tract, thymus, and spleen. LPS induced a twofold time-dependent increase in thymic UCN mRNA levels within 6 h. Splenic UCN mRNA levels decreased after LPS administration. UCN mRNA in thymus, but not in the spleen, was significantly increased after ACTH injection. LPS did not influence UCN expression in the thymus of adrenalectomized rats that received CORT replacement, despite substantial increases in ACTH. CORT, given s.c., stimulated UCN mRNA similarly to that of LPS. These results indicate that UCN mRNA expression increases within the thymus after immune activation in a CORT-dependent manner, and that this is the consequence of HPA axis activation.[6]

Intracerebroventricular (i.c.v.) injection of UCN (1 ng) to rats produced a marked decrease in the proliferative response of splenocytes to a mitogen.

This suppressive effect of urocortin was abolished by pretreatment with a ganglionic blocking agent (chlorisondamine) or a β-adrenergic receptor antagonist (propranolol), but not by adrenalectomy. These results indicate that the immunosuppressive effect of UCN is mediated by the sympathetic nervous system.[7]

UCN 1 may activate the HPA axis by stimulation of the hypothalamic CRF production, and this process is mediated by CRFR1, and not by CRFR2. UCN 1 may stimulate the AVP production as well, but based on the results with AVPR3 antagonist, this effect is not involved in the regulation of the HPA axis.[8]

Members of the CRF family of neuropeptides, including *UCN 1, UCN 2, and UCN 3,* bind to the G protein-coupled receptors, CRF1 and CRF2 receptors. In addition, CRF-binding protein (CRFBP) binds both CRF and UCN 1, and can modulate their activities.[9]

Central administration of UCN 2 and UCN 3 time- and dose-dependently modulate the amygdalar and the hypothalamic CRF concentration; and it directly or indirectly alters the plasma CORT concentration. The present experiments suggest that the role of CRFR2 in the regulation of HPA axis can be inhibitory or stimulatory, depending on the actual concentration of their agonists.[10]

UCN is a new mammalian member of the CRH family. UCN is a candidate endogenous ligand for type 2 CRH receptors. In adult male rats, UCN mRNA was abundant in the GI tract, in the thymus, and spleen. LPS induced a twofold time-dependent increase in thymic UCN mRNA levels within 6 h. Splenic UCN mRNA levels decreased after LPS administration. UCN mRNA in thymus, but not in the spleen, was significantly increased after ACTH injection. LPS did not influence UCN expression in the thymus of adrenalectomized rats that received CORT replacement, despite substantial increases in ACTH. CORT, given s.c., stimulated UCN mRNA similarly to that of LPS. These results indicate that UCN mRNA expression increases within the thymus after immune activation in a CORT-dependent manner, and that this is the consequence of HPA axis activation.[11]

The i.c.v. injection of UCN (1 ng) to rats produced a marked decrease in the proliferative response of splenocytes to a mitogen. This suppressive effect of urocortin was abolished by pretreatment with a ganglionic blocking agent (chlorisondamine) or a β-adrenergic receptor antagonist (propranolol), but not by adrenalectomy. These results indicate that the *immunosuppressive effects of UCN* are mediated by the *sympathetic nervous system.*[12]

2.2 Catecholamines

It was demonstrated that: (1) the mechanism whereby LPS-induced hypoglycemia was independent of epinergic activity; (2) selective central inhibition of epinergic function reduced endotoxin-stimulated ACTH secretion, an effect that could mainly be due to a decrease in CRH-ergic activity; (3) the central epinergic system positively and negatively modulates CRH- and AVP-ergic functions, respectively; and (4) inhibition of peripheral PNMT activity reduced immune system function *in vivo* and *ex vivo*. It is further suggested that low peripheral levels of epinephrine (EPI) could be beneficial for the body's defense mechanisms during endotoxic shock.[13]

The catecholaminergic afferents of CRH and TRH neurons originate from both noradrenaline- and adrenaline-synthesizing cell groups located in the brainstem, and collectively represent one of the most well studied neural inputs of these neurons.[14]

2.3 Apelin

The neuropeptide apelin is expressed in the hypothalamic paraventricular and supraoptic nuclei. It mediates its effects via the activation of the apelin receptor (APJ). Evidence suggests a role of apelin and APJ in mediating the neuroendocrine response in stress.

These data indicate that APJ has a role in the regulation of the HPA axis response to some acute phase diseases and has a gender-specific function in peripheral immune activation of the HPA axis.[15]

2.4 Cholecystokinin

The expression of CCK mRNA in neuroendocrine CRH neurons of the hypothalamic paraventricular nucleus (PVN) of male rats was examined 8 h following an acute immune challenge by i.p. lipopolysaccharide (LPS, 250 μg/kg). Compared to controls, LPS-injected rats displayed an increased (185%) parvicellular CCK mRNA expression levels, occurring in a majority (70%) of CRH neurons.[16]

Sulfated CCK-8-induced hypothalamic cFos activation is strongly associated with A2 activation, whereas the relationship between mpPVN and glucagon-like peptide-1 (GLP-1) activation is less clear. Furthermore, activation of A2, GLP-1, and mpPVN neuron is significantly modulated by the feeding status, suggesting a mechanism through which food intake and metabolic state might impact hypothalamic neuroendocrine responses to homeostatic challenges.[17]

2.5 Leptin

Leptin suppresses peripheral lymphocyte functions indirectly through the activation of the CRH (UCN)-sympathetic nervous system.[18]

2.6 Leukotriens

LTB4 administered via icv, downregulates the airway contraction response and inflammation through activation of the HPA axis via its BLT1 receptor. This study expands our concept of the regulatory role of intracranial inflammatory mediators in inflammatory diseases including asthma. The favorable effects of LTB4 on the HPA axis may help to explain the phenomenon of self-relief after an asthmatic attack.[19]

2.7 Natriuretic Peptide

The relation of cortisol at 1 with NT-pro-BNP at 1 and 3 was demonstrated using logistic functions in all the participants together (for 1/cortisol at 1 $B = 63.40, 58.52$; $r = 0.41, 0.34$; $p = 0.003, 0.013$, respectively). ΔNT-pro-BNP2/1 linearly correlated with ΔACTH4/3 in WP and W ($r = -0.45, -0.48$; $p = 0.04, 0.04$, respectively). These results demonstrate for the first time that HPA axis and NT-pro-BNP interface in physical stress probably contributes to integrative regulation of cardiopulmonary performance.[20]

Chronic drug abuse profoundly alters stress-responsive systems like the HPA axis. Alterations of AVP, ANP, and cortisol appear to constitute an important component in the neurobiology of opiate-dependent patients.[21]

2.8 Neuromedins

Several lines of evidence suggest that within the HPA axis, NMU and NMS act at both hypothalamic and adrenal levels. Moreover, new data suggest that NMU and NMS are involved in the central and peripheral control of the *stress response*.[22]

In this study, we first demonstrate NMB expression in corticotrophs and its regulation by CRH and glucocorticoids. Furthermore, corticotrophs seemed to secrete NMB into the systemic circulation.[23]

2.9 Neuropeptides

NPSF augmented paraventricular CRH release and increased ACTH and CORT levels in the plasma. The release of CORT was successfully blocked by the pretreatment of the CRH antagonist α-helical CRH9-41.

Spontaneous and exploratory locomotor activity was also stimulated according to the telemetric and open-field studies. However, NPSF only tended to alter stereotyped behavior in the open-field experiments.[24]

The results show that single IN NPY can alter stress-triggered dysregulation of the HPA axis and activation of central noradrenergic activity. These findings provide proof of concept for the potential of IN NPY for noninvasive prophylactic treatment or early intervention in response to traumatic stress.[25]

2.10 Nociceptin/Orpanin

Antagonism of NOP receptors will produce antiobesity and antidepressant activities in humans. However, there are also contradictory data discussed. On the other hand, the current literature on the role of N/OFQ in anxiety and addiction points primarily to a role of agonist modulation being potentially therapeutic. Some drug-like molecules that function either as agonists or antagonists of NOP receptors have been optimized for human clinical study to test some of these hypotheses.[26]

Collectively, data in the literature support the notion that blockade of NOP receptor signaling could be a novel and interesting strategy for the development of innovative antidepressants.[27]

2.11 PACAP

The neuropeptide pituitary adenylate cyclase-activating polypeptide (PACAP) is a master regulator of central and peripheral stress responses required to restore and maintain homeostasis. PACAP modulates the HPA axis in response to *acute psychogenic*, but not systemic stressors, through the activation of CRH release to drive adrenal CORT output. During direct high-frequency stimulation of the splanchnic nerve that is designed to mimic stress, PACAP regulates adrenomedullary *CAT secretion*. In addition to the transmission, PACAP simultaneously facilitates the biosynthesis of adrenomedullary CATs through stimulus–secretion–synthesis coupling. During periods of chronic psychogenic stress, PACAP-mediated CORT elevation fails to desensitize and contributes to the development of *maladaptive behaviors* such as anxiety and depression. Based on these findings, PACAP regulates not only adaptive responses to stress but also maladaptive responses to sustained psychological stress.[28]

2.12 Opioid Peptides

β-Endorphin (BEP) exerts neuronal regulatory role in the control of CRH and NK cell cytolytic function, and identifies a possible novel therapy to treat stress hyperresponse and immune deficiency in fetal alcohol exposed subjects.[29]

Acutely CRH induced a significant rise of β-END in CSF both in subarachnoid hemorrhage (SAH) and sham SAH rats. However, in rats subjected to SAH, a single injection of CRH caused a prolonged rise of 5-END in CSF, which was also seen 2 days after SAH, during the chronic phase of vasospasm. The obtained results indicate that CRH increases neuroendocrine changes induced by SAH, probably by the activation of immune cells involved in the pathomechanism of chronic vasospasm.[30]

In a painful inflammation, exogenous, as well as endogenous *CRH can release opioid peptides* (mainly BEP) from various types of immune cells and produce antinociception by activating opioid receptors on peripheral sensory nerve endings. CRH mediates its central effects through two receptors, namely the CRH receptor 1 and 2. Both the selective CRH R1 antagonist CP-154,526 and the selective CRH R2 antagonist *astressin*-2B significantly attenuated peripheral antinociceptive effects of CRH indicating the involvement of both CRH receptor subtypes.[31]

2.13 Orexin

When orexin is administered in i.c.v. space, it activates the HPA axis, which is the main regulator of the stress response. The HPA axis is not the only player in the stress response evidence; this suggests that UCN 1, a member of the CRF neuropeptide family, also plays an important role in the stress response adaptation. UCN 1 is primarily synthesized in the centrally projecting Edinger–Westphal nucleus (EWcp). Taken together our data show that orexin in general, is not involved in the animal's acute stress response (plasma CORT) and it does not play a direct role in shaping the response of EWcp-UCN 1 neurons to acute stress either.[32]

These results illustrate that dual OXR antagonists, despite modulating stress-induced wakefulness, do not interfere with endocrine HPA axis function in the rat. They converge to suggest that endogenous orexin signaling plays a minor role in *stress* hormone release under basal conditions and under challenge.[33] Orexis has an immunoregulatory effect.[34]

2.14 Relaxin

Relaxin-3 is a member of the *insulin superfamily*. It is expressed in the nucleus incertus of the brainstem, which has projections to the hypothalamus. Relaxin-3 binds with high affinity to RXFP1 and RXFP3. RXFP3 is expressed within the *hypothalamic PVN*, an area central to the *stress response*.

The i.c.v. administration of *human relaxin-3 (H3)* (5 nmol) significantly increased plasma CORT at 30 min following injection compared with vehicle. Intra-PVN administration of H3 (1.8–1620 pmol) significantly increased plasma ACTH at 1620 pmol, H3 and CORT at 180–1620 pmol H3 at 30 min, following injection compared with vehicle. The stress hormone PRL was also significantly raised.[35]

2.15 Sauvagine and Urotensin

CRH and related peptides UCN (*sauvagine, urotensin*) play a central role in the coordination of autonomic, behavioral, cardiovascular, immune, and endocrine responses to *stressful stimuli*. Their actions are mediated through activation of two types of *G-protein-coupled receptors encoded by separate genes*.[36]

2.16 SCP and SRP

Human *SCP and SRP* were identified as specific ligands for the type 2 CRH receptor. The genes encoding these peptides were expressed in diverse peripheral tissues as well as in the CNS. *Treatment with SCP or SRP suppressed food intake, delayed gastric emptying, and decreased heat-induced edema.* Thus SCP and SRP might represent endogenous ligands for maintaining homeostasis after stress, and could allow the design of drugs to ameliorate stress-related diseases.[37]

2.17 Vasotocin

In the fourth transmembrane region (TM4), the vasotocin receptor (VT2R) may provide an important interface for effective receptor dimerization, suggesting that direct molecular interactions between VT2R and CRH-R1 receptors play a role in mediating an enhanced interaction between these two receptors. Their interaction at the AP level may potentiate the endocrine output of the avian HPA system.[38]

3 CRH AND THE CNS

The immune system programs the CNS. Early LPS challenge on postnatal (P) days 3 and 5-induced transient decrease of CRHR1 mRNA in the hippocampus, where GR mRNA increased in the CA1 region between P1 and P3. Hypothalamic CRH expression increased in LPS-mice on P28. Early life LPS challenge impacts the CNS stress neurocircuitry.[39]

During neonatal development, CTKs produce permanent alterations of HPA responses to stress. Overexpressing IL-6 or leukemia inhibitory factor (LIF) changes pituitary development and functions. Pituitary corticotroph (POMC) gene expression is regulated by CRH gp130 CTKs. The HPA axis regulates inflammatory disease. CTKs (IL-1, TNF, and gp130 CTK family) participates as mediators of an HPA axis response to stress and inflammation. Long exposure to proinflammatory CTKs increases the levels of glucocorticoid receptors, which are dominant negative regulators. Unresponsiveness of the HPA axis because of glucocorticoid negative feedback provides a defense from destructive effects of CTK excess. At the same time, gp130 CTKs stimulate pituitary suppressor of cytokine signaling (SOCS)-3, which represses CTK signaling and abrogates CTK-induced corticotrophic POMC gene transcription and ACTH secretion.[40]

Parvocellular CRH neurons coexpress *VP*, which is a second "releasing factor" for ACTH and CRH. The stress input to hypothalamic neurons, and other input signals are integrated to determine the output secretion of *CRH and VP.* Aminergic, cholinergic, GABAergic, glutamatergic, and a number of peptidergic inputs have all been implicated in the regulation of CRH/VP neurons. Glucocorticoids inhibit the HPA axis activity by negative feedback. IL-1 stimulates CRH and VP gene expression, which are involved in immune-neuroendocrine regulation. cAMP-response element-binding protein phosphorylation may mediate transcriptional activation of both CRH and VP genes, but the roles of AP-1 and other factors are not known. Expression profiles of CRH and VP genes are not uniform to stress exposure, and the VP gene is more sensitive to glucocorticoid suppression.[41]

CRH receptors preferentially activate the 45 kDa form of the Gs α-protein. CRH and UCNs activate two types of CRH receptors (CRH-R), *CRH-R1 and CRH-R2*, which are class-B1 family of G protein CRs (GPCRs). The CRH-Rs exhibit signaling promiscuity. CRH-R1 receptor in rat microglia is involved in NIM regulation. Cutaneous CRH/POMC responds to stressful stimulation.

4 CRH-RECEPTORS 1 AND 2 (CHR1, 2) AND UCN RECEPTORS

Photoaffinity experiments indicated that the *CRH receptors preferentially activate the 45 kDa form of the Gs α-protein.*[42]

CRH and UCNs exert their actions by activating two types of CRH receptors (CRH-R), *CRH-R1 and CRH-R2*, which belong to the class-B1 family of G protein CRs (GPCRs). The CRH-Rs exhibit signaling promiscuity as they couple to multiple G-proteins and regulate diverse intracellular networks that involve intracellular effectors such as cAMP and an array of PKs in an agonist and tissue-specific manner, a property that allows them to exert unique roles in the integration of homeostatic mechanisms.[43]

The functional expression of CRH-R1 receptor in rat microglia, indicating an important mechanism of interaction between immune and neuroendocrine systems in brain under physiological and pathological conditions.[44]

The cutaneous expression and action of CRH/POMC elements appear to be highly organized and entrained, representing general mechanism of cutaneous response to stressful stimuli. In such a CRH/POMC system, the CRH-Rs may be a central element.[45] CRH from the spleen, thymus, and inflamed tissue has proinflammatory function through peripheral CRHR1 and CRHR2. Proinflammatory action is not explained, as *urocortin* may be responsible. *Human lymphocytes expressed urocortin but not CRH mRNA*.

UCN is involved in physiology and interacts with leptin, NPY, orexin, and with ACTH in the brain. UCN is produced by the GI tract, heart, and *immune cells.*

UCN has blood concentrations ranging from 13 pg/mL to 152 pg/mL. Blood-borne UCN stimulates the cerebral endothelial cells composing the BBB and crosses the BBB by a unique transport system.[46]

The expression of urocortin mRNA in synovia of patients with rheumatoid arthritis was greater than that of patients with osteoarthritis. Also, UCN and CRH receptor immunoreactivity was detected in the synovial lining cell layer, subsynovial stromal cells, blood vessel endothelial cells, and mononuclear inflammatory cells from the joints of rheumatoid arthritis and osteoarthritis patients. The expression of *immunoreactive UCN* was significantly greater in rheumatoid arthritis than osteoarthritis ($p < 0.0001$) and correlated with the extent of inflammatory infiltrate. CRH receptor immunoreactivity was strong in mononuclear inflammatory cells of rheumatoid arthritis synovia. *Urocortin stimulated IL-1β and IL-6 secretion* by human peripheral blood mononuclear cells *in vitro*.[47]

Compared with wild type, *CRHR2-null mice* developed substantially reduced intestinal inflammation and had lower intestinal mRNA expression of the potent chemoattractants, keratinocyte, chemokine, and monocyte chemoattractant protein 1, following intraluminal exposure to *Clostridium difficile* toxin A, a potent enterotoxin that mediates antibiotic-associated diarrhea and colitis in humans.[48]

Unlike CRH-R1 inhibition in mice, *CRH-R2 inhibition attenuated stress-induced bacterial growth* and significantly prevented severe sepsis.[49]

5 HORMONAL IMMUNOREGULATION AND THE HPA AXIS

5.1 Immunoregulaton by the Pituitary Gland

It was discovered in my laboratory that hypophysectomized (HYPOX) rats are immunodeficient.[50] Subsequent experiments revealed that GLH (e.g, GH, PRL, and placental lactogens) stimulates adaptive immunity (ADIM) and the HPA axis is antagonistic and it suppresses *ADIM*.[51] It was also tested that placental lactogen is for immune regulation and gut positive results.[52] Therefore immunity in rats was regulated by lactogenic and GHs.[53] Bromocriptine was as immunosuppressive as was HYPOX. This drug inhibits dopamine, which is a PRL releasing factor. This is the discovery that dopamine is an immmunoregulator.[54] GLH hormones stimulate the development of adjuvant arthritis in rats. CRF inhibited the disease, which indicates the suppressing effect of glucocorticoids on inflammatory diseases.[55] Here we discussed the possibility if hyperprolactinemia in pregnancy would act upon autoimmune diseases. It has been observed that autoimmune reactions do regress during pregnancy.[56]

In patients with rheumatoid arthritis, PRL levels were normal but the mitogenic activity as detected by the Nb2 rat lymphoma cell line, was deficient. We concluded that PRL had decreased bioactivity in rheumatoid patients.[57]

While doing HYPOX, a fair number of animals die shortly. However, long surviving animals may live as long as 8 months. We were wondering what is going on with PRL in these animals. So we HYPOXed a group of rats and measured PPL values. Freshly operated animals have little PRL, but in long surviving animals PRL will regenerate in about 6–7 weeks, PRL will get to the level of 50% of normal. If we neutralized this PRL with specific antibodies they died quickly within 2 weeks. They were severely anemic when they died. The regenerated animals had normal immune functions and hematological parameters. So we concluded that PRL alone is capable to support long term survival of rats (perhaps up to 8 months, which was observed by others).[58]

The *c-myc* gene is regarded as a major house-keeping gene regulating important factors for cell survival. We found that *c-myc* was activated by GLH in rat lymphoid tissue. DNA synthesis was also activated at the same time.[59]

5.2 Endotoxin Shock

It was examined if the pituitary gland plays a role in the regulation of endotoxin shock. Adrenalectomized (ADX) mice were 500–700× more sensitive to the toxic moiety of LPS, Lipid-A, when compared to normal controls. This hypersensitivity was suppressed after treatment with dexamethasone, which is a glucocorticoid-agonist agent. Blood level of tumor necrosis factor (TNF)-alpha was elevated 60 times in ADX animals in comparison with controls. These experiments revealed for the first time that the pituitary glad responds to LPS by activating the HPA axis, which produce glucocorticoids to suppress the production of toxic CTKs, such as TNF. Immunosuppression was seen by glucocorticoids, which down regulated CTKs, such as TNF.[60]

5.3 Immunoregulation by the Submandibular Salivary Gland

A lack of funding led to cooperative work with Dr Edris Sabbadini. Work on the submandibular salivary gland was done, which produced immunoregulatory substances. More and more involvement in the work amazingly led to the fact that a different aspect of pituitary immunoregulation was being studied,[61] which was an unknown territory of science. The submandibular salivary glad (SMG) releases pituitary GH and PRL by stimulation with epidermal growth factor (EGF) and nerve growth factor (NGF). By this effect, it stimulates ADIM. The same messengers act on the superior cervical ganglion, which releases α- and β-adrenergic CATs, substance-P (SP), and vasoactive intestinal peptide (VIP); SMG acts on the thyroid gland, ovaries, testes, and adrenal glands (via EGF). SMG secretes into the intestine antimicrobial products (AMP), C-reactive protein (CRP), EGF, glandular kallikrein (GK), NGF, and secretory IgA (sIgA) (Figure 4.1). These are major regulators for the GI tract. Also, there is no feedback loop here. Instead, the stomach stimulates (possibly by using EGF and NGF) the immune system (thymus, lymph nodes, and spleen) and the immune system provides feedback to the SMG (by using EGF, GK, and NGF).[62]

5.4 Combination Immunotherapy of Cancer

Another side project was with Edward Baral, who was a clinician at the Cancer Foundation in Winnipeg, Canada. Edward observed that breast tumors react to the antiestrogen tamoxifen (TX), whether they are estrogen receptor positive or negative. He suspected that TX interacts with the immune system. This is how receptor negative tumors (only one receptor was known at that time) get affected. We examined TX and another antiestrogen, toremifene (TO). Both drugs stimulated cytotoxic immune reactions. When we examined the effect on estrogen receptor positive and receptor

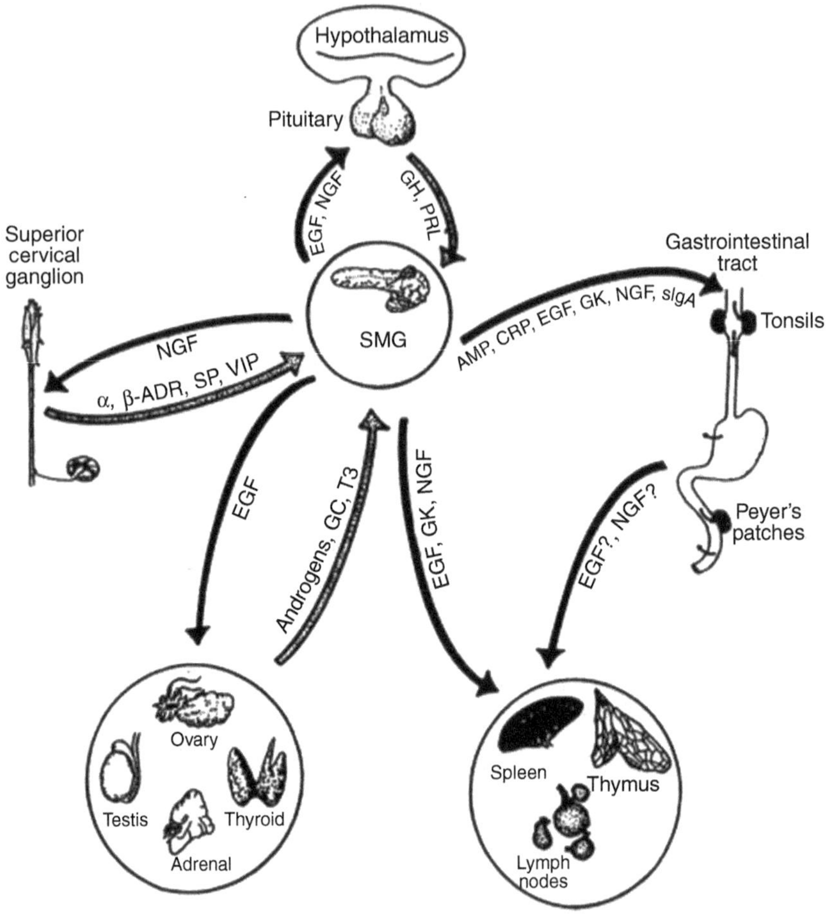

Figure 4.1 ***The Hypothalamus Pituitary Submandibular Axis.*** The submandibular gland (SMG) secretes EGF and NGF and with this it stimulates the pituitary gland to secrete GH and PRL; interacts with the sympathetic nerves system (superior cervical ganglion (NGF) and stimulates α- and β-adrenergic CATs, SP, and VIP; acts on the thyroid gland, ovaries, treatises and adrenal glands (EGF); secretes into the intestine AMP, CRP, EGF, GK, NGF and sIgA. These are major regulators for the GI tract. And there is no feedback loop here instead the stomach stimulates (possibly by using EGF and NGF) the immune system (thymus, lymph nodes, and spleen). The immune system also interacts with SMG (by using EGF, GK, and NGF). *Modified from Ref. [58].*

negative tumors both kind of tumors were inhibited by TX or TO. It was concluded that, a second estrogen receptor is responsible for this cytotoxic activity.[63,64] There was no funding to find the receptor, so within 2 years it had been isolated.[65]

Table 4.1 VP and immune function

Neurointermediate pituitary lobectomy (NIL) inhibits serums IgG and IgM, antibody
to T-dependent, and T-independent antigens.[76]
to intestinal IgA antibody formation.[77]
to the Arthus reaction.[74]
to contact sensitivity reaction (CMI to DNCB).[73]
to in EAE clinical score (desmopressin restores disease activity).[74]
to EAE: *ACTH, CORT, and PRL response* to inflammation depend on VP.[75]
VP receptor antagonist drugs alter immune function.[79]
VP *regulates* the *HPA axis* and *PRL* secretion.[76]
VP stimulates CTKs of INIM.[79]
VP regulates healing and recovery from disease.[79]
During homeostasis VP regulates adaptive immunocompetence as it stimulates both PRL and HPA axis. VP also regulates healing and recovery from acute illness.[79]

Inbred mice were treated with transplantable tumors with TX or TO and they gave various kind of killer cells (e.g., natural killer (NK); lymphokine activated killer (LAK); and cytotoxic T lymphocytes (CTL)). Each of these combinations was therapeutic.[66–69] Human breast carcinomas and lung tumors were examined and cells from tumor tissue were isolated by digestion with collagenase. Then the lymphocytes and cancer cells were separated on Ficoll density gradients. The autologous killer cells were stimulated for cytotoxicity by TO and TX against the autologous tumor cells. This finding indicates that our results are valid for human cancer.[70]

So the principles of *combination immunotherapy of cancer* were discovered: (1) inhibit tumor growth by a suitable medication. (2) Attack the tumor by cytotoxic lymphocytes. Since the tumor is not capable to mutate out of this situation, it will regress. So cure is achieved. This is the principle of *Combination Immunotherapy of Cancer.*[71,72]

5.5 VP as an Immunoregulator

Another form of immunity with which we are born, is called natural, or INIM. It turned out that VP functions as the hypothalamic regulator of both ADIM and INIM[73–77] as it regulates PRL, the HPA axis, and INIM-derived CTKs. Also, VP is the hypothalamic regulator for chronic inflammatory disease and healing.[78–80] Other factors stimulating INIM are glucocorticoids and CATs (Table 4.1).[78]

During acute febrile illness, which is now known as the APR, CRH commands the acute inflammatory response and VP has a secondary role. However, when the illness subsides, and a chronic inflammation follows, it is VP and not CRH, which controls immunity. VP brings the disease to conclusion in most cases, and healing will follow. VP will serve as the hypothalamic coordinator of immune function (Table 4.1).

Thus, VP is the hypothalamic hormone, which promotes healing and recovery.[78]

6 DISCUSSION

The terms, acute febrile illness, stress response, and APR refer to the same condition in pathology.[79] The mediators of APR are acute phase proteins.

The HPA axis is built by certain hierarchy: VP is the hypothalamic regulator of the entire immune system during homeostasis. VP also regulates chronic disease and recovery and *supports healing.* CRH regulates the HPA axis during APR. It also controls the POMC gene, which secures its superiority. VP supports the APR, but here its role is secondary.

What we see here is a functional division. CRF has only one function and here it controls everything including gene activation. VP can substitute for CRH if necessary. In fact, VP is the most important acute phase protein, so it can do whatever is necessary during APR. This is *total substitution.*

Some years ago, it turned out that there is another hormone, urocortin, which is associated with the HPA axis. This is interesting if there is another pathway of regulating this axis. Actually it is turning out that it is not complete substitution rather sharing some of the workload of the HPA axis.

Now, in this chapter, 20 neuropeptides have been described, which are associated with the HPA axis. There is limited information about what most of these peptides do.

Apellin and its receptor APJ regulate the HPA axis and mediate the neuroendocrine response to stress. *CCK* is produced in the PVN. Sulfated CCK-8-induced hypothalamic cFos activation is strongly associated with A2 activation. *Leptin* suppresses peripheral lymphocyte functions indirectly through the activation of the *CRH (UCN)-sympathetic nervous system.* LTB4 administered via icv downregulates the airway contraction response and inflammation through activation of the HPA axis via its BLT1 receptor. The HPA axis and *natriuretic peptide* (NT-pro-BNP) interface in physical *stress* probably contribute to integrative regulation of cardiopulmonary performance. AVP, ANP, and cortisol appear to constitute important components in the neurobiology of opiate-dependent patients.

Neuromedins: Within the HPA axis NMU and NMS act at both hypothalamic and adrenal levels. Also NMU and NMS are involved in the control of *stress response*.

Neuropeptides: NPSF-augmented paraventricular CRH release and increased ACTH and CORT. NPSF only tended to alter stereotyped behavior in the open-field experiments.

It appears from this brief and incomplete survey that neuropeptides, which are associated with the HPA axis, are involved with less function than what the HPA axis has. Stress and APR are very important problems, physiological regulation, food intake, antinociception, behavior, and fortifying some regulatory axes are vitally important, so it makes sense to secure these functions by more than one mediators. This means that for every major area of HPA function, there would be one or several spear mediators to perform the same function. This makes a lot more sense than having duplication of the entire regulatory gene, which have full power to regulate all functions of the HPA axis. This would be a terrible waste. But if this system can substitute for total or partial failure of the HPA axis, this makes a lot of sense. Important areas like APR are covered by multiple mediators. Mother Nature worked out a perfect solution for the safest regulation possible. One of which is insurmountable in its perfection!

GLH hormones are also associated with some 18 or so CTKs and neuropeptides. We know that type I CTKs share signal transmission pathway (Jak-STAT) with GLH, so this means that type I CTKs can substitute fully for GLH functions. The other mediators may work by *partial substitution*. This is to be examined.[78]

For the regulation of an entire organism, so that the tissues and organs can talk to each other, a common language of mediators is necessary. As an example, how the submandibular salivary gland talks to the nerves, the endocrine system, the GI tract, and the immune system, has been shown. This is an important circuitry in the hypothalamus–pituitary–submandibular gland axis.

Living organisms were thought to evolve by gene duplication and modification. Now we can add *gene splicing*. Big genes would handle big problems, little genes handle little problems. It is likely that neuropeptides originate from splicing up a big gene to little genes that handle parts of the original problem.

REFERENCES

1. Besedovsky HO, del Rey A, Klusman I, Furukawa H, Monge Arditi G, Kabiersch A. Cytokines as modulators of the hypothalamus-pituitary-adrenal axis. *J Steroid Biochem Mol Biol* 1991;**40**:613–8.
2. Tsagarakis S, Grossman A. Corticotropin-releasing hormone: interactions with the immune system. *Neuroimmunomodulation* 1994;**1**:329–34.
3. Karalis K, Sano H, Redwine J. Autocrine or paracrine inflammatory actions of corticotropin-releasing hormone *in vivo*. *Science* 1991;**254**:421–3.
4. Turnbull AV, Smith GW, Lee S, Vale WW, Lee KF, Rivier C. CRF type I receptor-deficient mice exhibit a pronounced pituitary-adrenal response to local inflammation. *Endocrinology* 1999;**140**:1013–7.
5. Chida D, Imaki T, Suda T, Iwakura Y. Involvement of corticotropin-releasing hormone- and interleukin (IL)-6-dependent proopiomelanocortin induction in the anterior pituitary during hypothalamic-pituitary-adrenal axis activation by IL-1alpha. *Endocrinology* 2005;**146**(12):5496–502.
6. Kageyama K, Bradbury MJ, Zhao L, Blount AL, Vale WW. Urocortin messenger ribonucleic acid: tissue distribution in the rat and regulation in thymus by lipopolysaccharide and glucocorticoids. *Endocrinology* 1999;**140**:5651–8.
7. Okamoto S, Ishikawa I, Kimura K, Saito M. Potent suppressive effects of urocortin on splenic lymphocyte activity in rats. *Neuroreport* 1998;**9**:4035–9.
8. Bagosi Z, Csabafi K, Palotai M, Jászberényi M, Földesi I, Gardi J, Szabó G, Telegdy G. The effect of *urocortin I* on the hypothalamic ACTH secretagogues and its impact on the hypothalamic-pituitary-adrenal axis. *Neuropeptides* 2014;**48**:15–20.
9. Fox JH, Lowry CA. Corticotropin-releasing factor-related peptides, serotonergic systems, and emotional behavior. *Front Neurosci* 2013;**7**:169.
10. Bagosi Z, Csabafi K, Palotai M, Jászberényi M, Földesi I, Gardi J, Szabó G, Telegdy G. The interaction of Urocortin II and Urocortin III with amygdalar and hypothalamic corticotropin-releasing factor (CRF) – reflections on the regulation of the hypothalamic-pituitary-adrenal (HPA) axis. *Neuropeptides* 2013;**47**(5):333–8.
11. Kageyama K, Bradbury MJ, Zhao L, Blount AL, Vale WW. Urocortin messenger ribonucleic acid: tissue distribution in the rat and regulation in thymus by lipopolysaccharide and glucocorticoids. *Endocrinology* 1999;**140**:5651–8.
12. Okamoto S, Ishikawa I, Kimura K, Saito M. Potent suppressive effects of urocortin on splenic lymphocyte activity in rats. *Neuroreport* 1998;**9**:4035–9.
13. Giovambattista A, Chisari AN, Gaillard RC, Spinedi E. Modulatory role of the epinergic system in the neuroendocrine-immune system function. *Neuroimmunomodulation* 2000;**8**(2):98–106.
14. Wittmann G. Regulation of hypophysiotrophic corticotrophin-releasing hormone- and thyrotrophin-releasing hormone-synthesising selective ligands for the type 2 corticotropin-releasing hormone receptor. *Nat Med* 2001;**7**(5):605–11.
15. Juaneda C, Lafon-Dubourg P, Ciofi P, Sarrieau A, Wenger T, Tramu G, Corio M. CCK mRNA expression in neuroendocrine CRH neurons is increased in rats subjected to an immune challenge. *Brain Res* 2001; 901(1–2): 277–280.
16. Maniscalco JW, Rinaman L. Overnight food deprivation markedly attenuates hindbrain noradrenergic, glucagon-like peptide-1, and hypothalamic neural responses to exogenous cholecystokinin in male rats. *Physiol Behav* 2013;**121**:35–42.
17. Okamoto S, Irie Y, Ishikawa I, Kimura K, Saito M. Central leptin suppresses splenic lymphocyte functions through activation of the corticotropin-releasing hormone-sympathetic nervous system. *Brain Res* 2000;**855**(1):192–7.
18. Zhang SJ, Deng YM, Zhu YL, Dong XW, Jiang JX, Xie QM. Intracerebroventricular injection of *leukotriene B4* attenuates antigen-induced asthmatic response via BLT1 receptor stimulating HPA-axis in sensitized rats. *Respir Res* 2010;**11**:39.

19. Popovic D, Popovic B, Plecas-Solarovic B, Pešic´ V, Markovic V, Stojiljkovic S, Vukcevic V, Petrovic I, Banovic M, Petrovic M, Vujisic-Tesic B, Ostojic MC, Ristic A, Damjanovic SS. The interface of hypothalamic-pituitary-adrenocortical axis and circulating brain *natriuretic peptide* in prediction of cardiopulmonary performance during physical stress. *Peptides* 2013;**47**:85–93.
20. Glahn A, Heberlein A, Dürsteler-MacFarland KM, Lenz B, Frieling H, Gröschl M, Wiesbeck GA, Kornhuber J, Bönsch D, Bleich S, Hillemacher T. Atrial *natriuretic peptide, arginine vasopressin* peptide and *cortisol* serum levels in opiate-dependent patients. *Neuropsychobiology* 2013;**67**(2):111–5.
21. Malendowicz LK, Ziolkowska A, Rucinski M. Neuromedins U and S involvement in the regulation of the hypothalamo-pituitary-adrenal axis. *Front Endocrinol* 2012;**3**:156.
22. Kameda H, Miyoshi H, Shimizu C, Nagai S, Nakamura A, Kondo T, Chida D, Atsumi T. Expression and regulation of *neuromedin B* in pituitary corticotrophs of male melanocortin 2 receptor-deficient mice. *Endocrinology* 2014;**155**:2492–9.
23. Jászberényi M, Bagosi Z, Csabafi K, Palotai M, Telegdy G. The actions of *neuropeptide SF* on the hypothalamic-pituitary-adrenal axis and behavior in rats. *Regul Pept* 2014;**188**: 46–51.
24. Serova LI, Tillinger A, Alaluf LG, Laukova M, Keegan K, Sabban EL. Single intranasal *neuropeptide Y* infusion attenuates development of PTSD-like symptoms to traumatic stress in rats. *Neuroscience* 2013;**236**:298–312.
25. Witkin JM, Statnick MA, Rorick-Kehn LM, Pintar JE, Ansonoff M, Chen Y, Tucker RC, Ciccocioppo R. The biology of Nociceptin/Orphanin FQ (N/OFQ) related to obesity, stress, anxiety, mood, and drug dependence. *Pharmacol Ther* 2014;**141**(3):283–99.
26. Gavioli EC, Calo' G. Nociceptin/orphanin FQ receptor antagonists as innovative antidepressant drugs. *Pharmacol Ther* 2013;**140**:10–25.
27. Mustafa T. Pituitary adenylate cyclase-activating polypeptide (PACAP): a master regulator in central and peripheral stress responses. *Adv Pharmacol* 2013;**68**:445–57.
28. Boyadjieva NI, Ortigüela M, Arjona A, Cheng X, Sarkar DK. Beta-endorphin neuronal cell transplant reduces corticotropin releasing hormone hyperresponse to lipopolysaccharide and eliminates natural killer cell functional deficiencies in fetal alcohol exposed rats. *Alcohol Clin Exp Res* 2009;**33**(5):931–7.
29. Henryk S, Jośko J, Jedrzejowska-Szypułka H, Jarzab B, Döhler KD. Corticotropin releasing hormone (CRH) increases beta-endorphin (beta-end like) concentration in cerebrospinal fluid of rats with vasospasm following subarachnoid hemorrhage. *J Physiol Pharmacol* 1999;**50**(3):419–28.
30. Mousa SA, Bopaiah CP, Stein C, Schäfer M. Involvement of corticotropin-releasing hormone receptor subtypes 1 and 2 in peripheral opioid-mediated inhibition of inflammatory pain. *Pain* 2003;**106**(3):297–307.
31. Emmerzaal TL, vd Doelen RH, Roubos EW, Kozicz T. *Orexinergic* innervation of urocortin1 and cocaine and amphetamine regulated transcript neurons in the midbrain centrally projecting Edinger–Westphal nucleus. *J Chem Neuroanat* 2013;**54**:34–41.
32. Steiner MA, Sciarretta C, Brisbare-Roch C, Strasser DS, Studer R, Jenck F. Examining the role of endogenous orexins in hypothalamus-pituitary-adrenal axis endocrine function using transient dual orexin receptor antagonism in the rat. *Psychoneuroendocrinology* 2013;**38**:560–71.
33. Perekrest SV, Abramova TV, Novikova NS, Loskutov YV, Rogers VJ, Korneva EA. Changes in immunoreactivity of orexin-A-positive neurons after intravenous lipopolysaccharide injection. *Med Sci Monit* 2008;**14**(7):BR127–33.
34. Gowan BM, Minnion JS, Murphy KG, Roy D, Stanley SA, Dhillo WS, Gardiner JV, Ghatei MA, Bloom SR. Relaxin-3 stimulates the neuro-endocrine stress axis via corticotropin-releasing hormone. *J Endocrinol* 2014;**221**(2):337–46.
35. Hillhouse EW, Randeva H, Ladds G, Grammatopoulos D. Corticotropin-releasing hormone receptors. *Biochem Soc Trans* 2002;**30**(4):428–32.

36. Hsu SY, Hsueh AJ. Human stresscopin and stresscopin-related peptide are selective ligands for the type 2 corticotropin-releasing hormone receptor. *Nat Med* 2001;**7**:605–11.
37. Cornett LE, Kang SW, Kuenzel WJ. A possible mechanism contributing to the synergistic action of vasotocin (VT) and corticotropin-releasing hormone (CRH) receptors on corticosterone release in birds. *Gen Comp Endocrinol* 2013;**188**:46–53.
38. Amath A, Foster JA, Sidor MM. Developmental alterations in CNS stress-related gene expression following postnatal immune activation. *Neuroscience* 2012;**220**:90.
39. Chesnokova V, Melmed S. Minireview: neuro-immuno-endocrine modulation of the hypothalamic-pituitary-adrenal (HPA) axis by gp130 signaling molecules. *Endocrinology* 2002;**143**(5):1571–4.
40. Itoi K, Jiang YQ, Iwasaki Y, Watson SJ. Regulatory mechanisms of corticotropin-releasing hormone and vasopressin gene expression in the hypothalamus. *J Neuroendocrinol* 2004;**16**(4):348–55.
41. Grammatopoulos DK, Randeva HS, Levine MA, Kanellopoulou KA, Hillhouse EW. Rat cerebral cortex corticotropin-releasing hormone receptors: evidence for receptor coupling to multiple G-proteins. *J Neurochem* 2001;**76**(2):509–19.
42. Grammatopoulos DK. Insights into mechanisms of corticotropin-releasing hormone receptor signal transduction. *Br J Pharmacol* 2012;**166**(1):85–97.
43. Wang W, Ji P, Riopelle RJ, Dow KE. Functional expression of corticotropin-releasing hormone (CRH) receptor 1 in cultured rat microglia. *J Neurochem* 2002;**80**(2):287–94.
44. Slominski A, Wortsman J, Pisarchik A, Zbytek B, Linton EA, Mazurkiewicz JE, Wei ET. Cutaneous expression of corticotropin-releasing hormone (CRH), urocortin, and CRH receptors. *FASEB J* 2001;**15**(10):1678–93.
45. Pan W, Kastin AJ. Urocortin and the brain. *Prog Neurobiol* 2008;**84**(2):148–56.
46. Kohno M, Kawahito Y, Tsubouchi Y, Hashiramoto A, Yamada R, Inoue KI, Kusaka Y, Kubo T, Elenkov IJ, Chrousos GP, Kondo M, Sano H. Urocortin expression in synovium of patients with rheumatoid arthritis and osteoarthritis: relation to inflammatory activity. *J Clin Endocrinol Metab* 2001;**86**(9):4344–52.
47. Kokkotou E, Espinoza DO, Torres D, Karagiannides I, Kosteletos S, Savidge T, O'Brien M, Pothoulakis C. Melanin-concentrating hormone (MCH) modulates C difficile toxin A-mediated enteritis in mice. *Gut* 2009;**58**:34–40.
48. Kim BJ, Kayembe K, Simecka JW, Pulse M, Jones HP. Corticotropin-releasing hormone receptor-1 and 2 activity produces divergent resistance against stress-induced pulmonary Streptococcus pneumoniae infection. *J Neuroimmunol* 2011;**237**:57–65.
49. Nagy E, Berczi I. Immunodeficiency in hypophysectomized rats. *Acta Endocrinol* 1978;**89**:530–7.
50. Berczi I, Nagy E, Kovacs K, Horvath E. Regulation of humoral immunity in rats by pituitary hormones. *Acta Endocrinol* 1981;**98**:506–13.
51. Berczi I, Nagy E. Placental lactogen is a hemopoietic hormone. *Br J Haematol* 1991;**79**:355–8.
52. Berczi I, Nagy E, Berczi I, Friesen HG. Regulation of immunity in rats by lactogenic and growth hormones. *Acta Endocrinol* 1983;**102**:351–7.
53. Nagy E, Berczi I, Wren GE, Asa SL, Kovacs K. Immunomodulation by bromocriptine. *Immunopharmacology* 1983;**6**:231–43.
54. Berczi I, Nagy E, Asa SL, Kovacs K. The influence of pituitary hormones on adjuvant arthritis. *Arthritis Rheum* 1984;**27**:682–8.
55. Berczi I. Editorial: prolactin, pregnancy and autoimmune disease. *J Rheumatol* 1993;**20**:1095–100.
56. Nagy E, Chalmers IM, Baragar FD, Friesen HG, Berczi I. Prolactin deficiency in rheumatoid arthritis. *J Rheumatol* 1991;**18**:1662–8.
57. Nagy E, Berczi I. Hypophysectomized rats depend on residual prolactin for survival. *Endocrinology* 1991;**128**:2776–84.
58. Berczi I, Nagy E, Matusik RJ, Friesen HG. Pituitary hormones regulate c-myc and DNA synthesis in lymphoid tissue. *J Immunol* 1991;**146**:2201–6.

59. Ramachandra RN, Sehon AH, Berczi I. Neuro-hormonal host defence in endotoxin shock. *Brain Behav Immun* 1992;**6**:157–69.
60. Nagy E, Berczi I, Sabbadini S. Endocrine control of the immunosuppressive activity of the submandibular gland. *Brain Behav Immun* 1992;418–28.
61. Sabbadini E, Berczi I. The submandibular gland: a key organ in the neuro-immunoregulatory network? *Neuroimmunomodulation* 1995;**2**:184–202.
62. Baral E, Kwok S, Berczi I. Suppression of lymphocyte mitogenesis by tamoxifen. *Immunopharmacology* 1989;**18**:57–62.
63. Nagy E, Berczi I, Sabbadini S. Endocrine control of the immunosuppressive activity of the submandibular gland. *Brain Behav Immun* 1992;418–28.
64. Baral E, Nagy E, Berczi I. Modulation of natural killer cell mediated cytotoxicity by tamoxifen and estradiol. *Cancer* 1995;**75**:591–9.
65. Baral E, Nagy E, Berczi I. Modulation of lymphokine-activated killer cell-mediated cytotoxicity by estradiol and tamoxifen. *Int J Cancer* 1996;**66**:214–8.
66. Kuiper GG, Shughrue PJ, Merchenthaler I, Gustafsson JA. The estrogen receptor beta subtype: a novel mediator of estrogen action in neuroendocrine systems. *Front Neuroendocrinol* 1998;**19**:253–86.
67. Baral E, Nagy E, Kangas L, Berczi I. Anti-estrogens enhance the therapeutic effect of lymphokine activated killer cells on the P815 murine mastocytoma. *Int J Cancer* 1996;**67**:580–5.
68. Nagy E, Baral E, Kangas L, Berczi I. Anti-estrogens potentiate the immunotherapy of the P815 murine mastocytoma by cytotoxic T lymphocytes. *Anticancer Res* 1997;**17**:1083–8.
69. Baral E, Nagy E, Kangas L, Berczi I. Immunotherapy of the SL2-5 murine lymphoma with natural killer cells and tamoxifen or toremifene. *Anticancer Res* 1997;**17**:77–84.
70. Baral E, Nagy E, Krepart GE, Lotocki RJ, Unruh HW, Berczi I. Antiestrogens sensitize human ovarian and lung carcinomas for lysis by autologous killer cells. *Anticancer Res* 2000;**20**:2027–31.
71. Haeryfar SMM, Nagy E, Baral E, Krepart GE, Lotocki RJ, Berczi I. Antiestrogens affect both pathways of killer cell-mediated oncolysis. *Anticancer Res* 2000;**20**:1849–53.
72. Nagy E, Berczi I, Baral E. Combination immunotherapy of cancer. In: Berczi I, Gorczynski R, editors. *Neuroimmune biology*, vol. 1. New foundation of biology. UK: Elsevier; 2001. p. 417–432.
73. Quintanar-Stephano A, Kovacs K, Berczi I. Effects of neurointermediate pituitary lobectomy on humoral and cell-mediated immune responses in the rat. *Neuroimmunomodulation* 2004;**11**:233–40.
74. Quintanar-Stephano A, Chavira-Ramírez R, Kovacs K, Berczi I. Neurointermediate pituitary lobectomy decreases the incidence and severity of experimental autoimmune encephalomyelitis in Lewis rats. *J Endocrinol* 2005;**184**:51–8.
75. Berczi I, Quintanar-Stephano A, Kovacs K. Neuroimmune regulation in immunocompetence, acute illness, and healing. *Ann NY Acad Sci* 2009;**1153**:220–39.
76. Quintanar-Stephano A, Abarca-Rojano E, Jarillo-Luna RA, Rivera-Aguilar V, Ventura-Juárez J, Berczi I, Kovacs K, Campos-Rodríguez R. Hypophysectomy and neurointermediate pituitary lobectomy decrease humoral immune responses to T-independnent and T-dependnent antigens. *J Physiol Biochem* 2010;**66**:7–13.
77. Campos-Rodríguez R, Stephano AQ, Drago-Serrano ME, Abarca-Rojano E, Berczi I, Ventura-Juárez J, Kormanovski A. The central nervous system modulates the immune response to *Salmonella*. InTech – Open Access Publisher. Available from: http://www.intechweb.org/
78. Berczi I, Villalobos Hernández Egina C, Quintanar Stephano A, Campos R, Kovacs K. The healing power of vasopressin. *Adv Neuroimmune Biol* 2012;**2**:217–24.
79. Berczi I, Quinatrar-Stephano A, Campos R, Kovacs K. Vasopressin and oxytocin in immunoregulation. *Adv Nerimm Biol* 2012;**4**(3):313–46.
80. Berczi I. The stress concept and neuroimmunoregulation in modern biology. *Ann NY Acad Sci* 1998;**30**:3–12.

CHAPTER 5

Arginine Vasopressin: An Immunoregulatory Hormone

Andrés Quintanar-Stephano*, Alejandro Organista-Esparza*, Roberto Chavira-Ramírez, Rafael Campos-Rodríguez†, Istvan Berczi*,‡, Kalman Kovacs§**

*Departamento de Fisiología y Farmacología, Centro de Ciencias Básicas, Universidad Autónoma de Aguascalientes, Aguascalientes, Estado Libre y Soberano de Aguascalientes, Mexico City, México
**Departamento de Hormonas esteroideas, ICMyN "SZ", Mexico City, México
†Departamento Bioquímica, ESM, IPN, México
‡Department of Immunology, Faculty of Medicine, The University of Manitoba, Winnipeg, Manitoba, Canada
§Department of Pathology and Laboratory Medicine, St. Michael's Hospital, University of Toronto, Toronto, Ontario, Canada

1 INTRODUCTION

Numerous studies over the last 35 years demonstrated regulatory interactions between the immune, nervous, and endocrine systems under normal conditions, and during immune/inflammatory reactions.[1] The transmission of information from the immune system to the nervous and endocrine systems is initiated by cytokines released at the site of the immune insult, and it reaches the central nervous system (CNS) or pituitary gland via the circulatory system or activation of cytokine neural receptors in the sensorial branch of the vagus nerves or cytokine receptors in the peripheral nervous system. Feedback by neural transmission via sympathetic and afferent vagus nerves are also documented. It is also known that the nervous system influences the immune system through endocrine and autonomic nervous signals. Communication is possible because the nervous, endocrine, and immune systems have a common biochemical language involving shared ligands and receptors, including neurotransmitters, neuropeptides, growth factors, neuroendocrine hormones, and cytokines.[1,2]

1.1 Pituitary Gland

In rats, the pituitary gland is composed of three lobes namely: the anterior or adenohypophysial lobe, which synthesizes and releases growth hormone (GH), prolactin (PRL), adrenocorticotropin (ACTH), thyrotropin (TSH), follicle-stimulating hormone (FSH), and luteinizing hormone (LH); the intermediate lobe, which synthesizes melanocyte-stimulating hormone (α-MSH)

Insights to Neuroimmune Biology
http://dx.doi.org/10.1016/B978-0-12-801770-8.00005-7

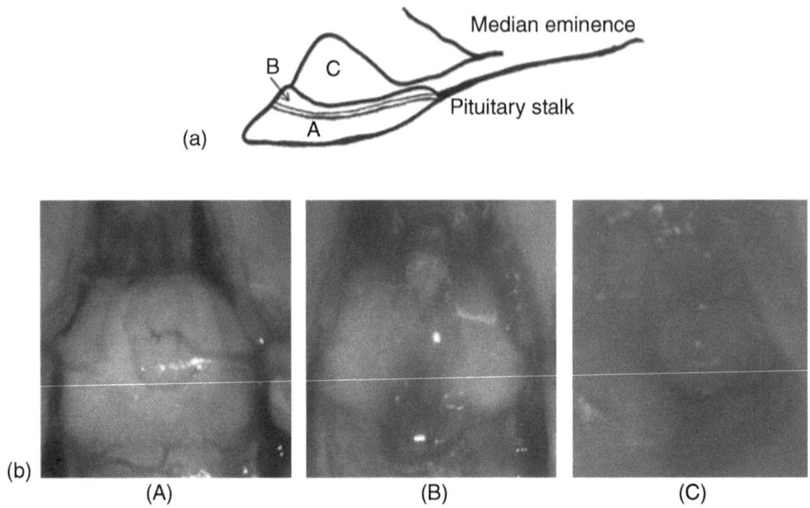

Figure 5.1 (a) Sagittal view of the rat pituitary showing the several lobes, which were selectively removed: adenohypophyseal lobe (A), neurointermediate pituitary lobe (A + B), and entire pituitary (A + B + C lobes). (b) Pituitaries from SHAM, NIL and AL lobectomized rats. Photographs were taken 4 weeks after surgeries. Original magnification: 6×. *(Modified with permission from Ref. [3].)*

and the posterior or neural lobe, which stores and releases the hypothalamic hormones arginine vasopressin (AVP) and oxytocin (OXT). Anatomically, the anterior lobe is loosely connected to the intermediate lobe, whereas the intermediate and neural lobes are tightly attached to each other and together form the neurointermediate lobe (Figure 5.1a). The anatomical disposition of several lobes, the pituitary stalk, and the median eminence in the rat, facilitate the removal of the entire hypophysectomy (HYPOX), the selective excision of the neurointermediate lobe (NIL) (Figure 5.1b) or the anterior lobe (AL) without damaging the neural and other neighboring structures (Figure 5.1b).[3]

The role of the anterior pituitary hormones in the regulation of the adaptive immune system (ADIM) has been widely documented. The PRL and GH, are stimulatory,[1,4–13] while ACTH, an important component of the hypothalamic–pituitary–adrenocortical (HPA) axis, modulates/suppresses ADIM.[2,14–16] Thyroxin (T4) regulates pituitary GH and PRL, which in turn regulate ADIM.[17] T4 also modulates innate immune responses (INIM),[18,19] whereas FSH and LH via estrogens, are stimulatory and through testosterone, suppress ADIM.[20] Although a direct effect of neurohypophysial hormones AVP and OXT on the immune system was reported,[21–25] their role during the immune responses and/or inflammation has been only partially studied. Here, we review the role of AVP in immunoregulation and discuss the possible use of AVP receptor antagonist drugs as a therapeutic agents for some autoimmune diseases.

1.2 Arginine Vasopressin

AVP is a nonapeptide hormone synthesized primarily in the magnocellular neurons of the hypothalamic paraventricular and supraoptic nuclei, which are axonally transported via the pituitary stalk to the posterior pituitary lobe and released to the general circulation. In addition, the parvocellular neurons of the paraventricular nuclei coexpress AVP and corticotropin-releasing hormone (CRH), which are transported to the median eminence and released into the hypothalamic–pituitary portal vessels. In the pituitary CRH regulates the synthesis and release of ACTH. CRH and AVP together control the secretion of ACTH. CRH and AVP play a crucial role in the coordination of the HPA axis during stress and autoimmune/inflammatory diseases.[26–30] Thus, through the control of corticosterone secretion (in rats), AVP indirectly participates in the regulation of the immune system. AVP also affects the contraction of vascular smooth muscle cells, vasodilatation through endothelial cells, urinary output (for this effect, AVP is also called antidiuretic hormone (ADH)), homeostasis, hypertrophy of cardiomyocytes, stimulation of liver glycogenolysis, and modulation of memory, anxiety, and depression.[31–39]

1.3 AVP Receptors

AVP interacts with its target tissues through three specific cell membrane receptors, namely: V1a, V1b (also called V3), and V2;[34,40–43] these are all G protein-coupled. The V1a and V1b receptors act through phosphatidylinositol hydrolysis to increase intracellular Ca^{2+} concentration, whereas V2 receptors act via Gs proteins to increase cAMP levels.[44] AVP receptors have been described in the brain (V1b), pituitary corticotrophs (V1b), heart (V1a), platelets and endothelial cells (V1b and V2), liver (V1a), vascular smooth muscle cells (V1a), renal tubules (V1a, V2), and cells of the immune system (V1a and V2).[39,45–51]

1.4 AVP Receptors in the Immune Cells

The relationship between AVP and immune system comes from the initial report of Dicker and Rozengurt.[52] These authors found a synergistic effect between several mitogenic drugs and the potent tumor promoter

12-O-tetradecanoylphorbol-13-acetate (TPA) resulting in increased DNA synthesis in 3T3 cells. AVP has a mitogenic effect on 3T3 cells, but fails to act on 3T3-TPA treated cells to stimulate DNA synthesis, ornithine decarboxylase activity or 2-deoxyglucose uptakes. It was concluded that TPA acts via a pathway which converges with that used by AVP, and at least some of its biological effects occur through mechanisms normally utilized by specific hormones.[52] Johnson et al.[21] had demonstrated that AVP was able to replace interleukin 2 (IL-2) requirement for γ-interferon (IFNγ) production by Lyt-2' cells from C57BL/6 mouse spleen cells. AVP replacement did not stimulate DNA synthesis in the target lymphocytes. It was suggested that AVP was capable of replacing IL-2 without the stimulation of cell proliferation or DNA synthesis. OXT, which is structurally related to AVP, was also capable of replacing IL-2 requirement for IFNγ production. These data showed that AVP and OXT possess a lymphokine-like activity and indicate a functional and/or structural relationship between the immune and neuroendocrine systems.[21] In turn, Johnson and Torres[22] demonstrated that AVP and OXT can replace IL-2 for T cell mitogen induction by INFγ in mouse spleen cell cultures. The structural basis for the helper signal by these hormones reside in the six N-terminal amino acids of AVP based on the relative ability of AVP, OXT, vasotocin, and pressinoic acid to help IFNγ induction. It should be mentioned that OXT and vasotocin were 10-fold less effective than AVP. Since the AVP competitive antagonist of vasopressor activity ([1-(*p*-mercapto-P,P-cyclopentamethylene propionic acid), 2-(O-methyl) tyrosine] AVP) blocks the AVP helper signal for production of INFγ with no effect on IL-2, it is possible that the AVP helper signal is operated via binding to an AVP vasopressor-type receptor on lymphocytes.[22]

The search for AVP receptors in immune cells was undertaken by binding studies with radiolabeled ligands. By using ^{125}I-8-L-AVP, Block et al. [45] described the binding of AVP to human mononuclear phagocytes. In addition, the specificity of vasopressin-sensitive sites on human mononuclear phagocytes was demonstrated by using several vasopressin analogues with various degrees of antidiuretic potency. Another important finding was the increased level of cyclic AMP in the cytoplasm of mononuclear phagocytes, indicating that AVP may initiate an intracellular signaling pathway. It was also verified that OXT and DP ([deamino-Cys 1,D-Arg8]-vasopressin acetate hydrate), a synthetic analog agonist for V2 receptors, stimulates the cyclic AMP intracellular signaling pathway in several immune cells.[45,48,50,51] The presence of AVP receptors was reported in diverse cells of the immune system, namely: V1a receptors in blood monocytes, macrophages, splenic lymphocytes, and B cells from peripheral blood,[45–48,53] V2 receptors in peripheral blood cell cultures[50,51] and V1b receptors in thymus and spleen cells.[49,54] It was also demonstrated that in splenic and blood lymphocytes DP, the synthetic agonist for V2 receptors, is able to stimulate the cyclic AMP intracellular pathway through V2 receptors.[51] Thus, evidence clearly indicates that acting directly on immune cell AVP may play a role in the regulation of immune responses.

1.5 Effects of AVP on Cells of the Immune System; *In Vitro* Observations

Locher et al.[55] described increased synthesis of prostaglandin E2 (PGE2) by human mononuclear phagocytes in response of AVP. These authors also showed that although less effective than AVP, the AVP analog DP also increases the synthesis of PGE2. Macrophages pretreated with indomethacin enhanced AVP-mediated accumulation of intracellular cyclic AMP, whereas the addition of PGE2 to cultured macrophages did not modify its capability to bind [^{3}H]AVP.[55] Bell et al.,[46,56] demonstrated that AVP may modulate cell immunity by stimulating the autologous mixed lymphocyte response, an effect that occurs over a wide dose range of AVP. Enhanced proliferation appears to be a specific response that is influenced by arginine residues in position 8 of this nonapeptide. The increased lymphocyte proliferation, and the presence of AVP receptors in peripheral blood mononuclear cells and spleen cells, may indicate a direct immunostimulatory effect of AVP. The direct communication between neuroendocrine and immune systems is also supported by the blocking effects of AVP V1a, V1b, and V2 receptor peptide-antagonists, which inhibit binding of the labeled AVP with several AVP receptors.[22,23,46,56] In summation, results from *in vitro* studies clearly prove the immunostimulating role of AVP.

1.6 Effects of AVP on the Immune System; *In Vivo* Observations

Fernández-Repollet et al.[25] described that in the vasopressin-deficient Brattleboro rats, a strain with central diabetes insipidus (DI) induced by a mutation in the gene that regulates the production of hypothalamic AVP,[57] the peritoneal macrophages phagocytic activity is decreased compared to normal Long–Evans rats. The phagocytic activity in Brattleboro and Long–Evans rats were significantly increased by AVP administration. These authors concluded that endogenous AVP may play a modulatory role in the phagocytic activity of the reticuloendothelial system (mononuclear phagocytic system).

Immune disorders were studied in DI rats by Khegai et al.[58] They observed that during ontogenetic development, DI animals showed a permanent decrease in the number of blood lymphocytes, increase in neutrophils, reduced activity of macrophages, early involution of the thymus and spleen, as well as suppression of antibody production. These changes were accompanied for an impaired general resistance. These results support the view that AVP is needed to maintain immunocompetence. In Lewis rats, a highly susceptible strain to develop experimental autoimmune diseases, neutralization of AVP with specific antibodies decreased the inflammatory response to carrageenan injection.[59] These authors suggested that in the rats with strain-specific higher AVP plasma levels (Lewis strain), the hormone plays a role in the increased susceptibility to inflammatory/autoimmune diseases, and concluded that AVP may mediate the inflammatory response.[59,60]

In our experiments the role of AVP in the regulation of immune responses was documented in animals subjected to several lobe-specific pituitary surgeries. A brief description of the various surgical procedures is included (for anatomical details refer Figure 5.1). The surgical procedures have been previously described by Alvarez-Buylla et al.;[61] Ben-Jonathan and Peters;[62] Mena et al.;[63] and Berczi et al.[3]

The surgical approach to the pituitary is through the parapharyngeal route under a stereoscopic dissecting microscope. A hole is drilled into the ventral aspect of the occipital–sphenoid joint until the pituitary capsule is exposed. In sham operated (SHAM) rats, surgery is terminated when the pituitary capsule and the pituitary became directly visible. For neurointermediate pituitary lobectomy (NIL), to remove the neural lobe, the posterior border of the pituitary capsule is cut and the caudal border of the gland lifted. After a direct view of the neural lobe it is carefully aspirated with a special adapted curved needle. In the case of anterior pituitary lobectomy (AL), the pituitary capsule is widely opened, and the adenohypophysial lobe is cut into two halves; each half is then carefully aspirated with a special adapted curved needle, thus leaving the posterior lobe, the infundibular stalk, and the hypothalamic median eminence untouched. For total hypophysectomy (HYPOX), the entire gland is aspirated.

1.7 Effects of Pituitary Lobe-Specific Surgeries on Endocrine Activity

HYPOX causes a significant diminution of all pituitary hormones (panhypopituitarism). In NIL rats, adenohypophyseal hormone levels remain within the normal range, whereas AVP decreases by 80% as compared to intact

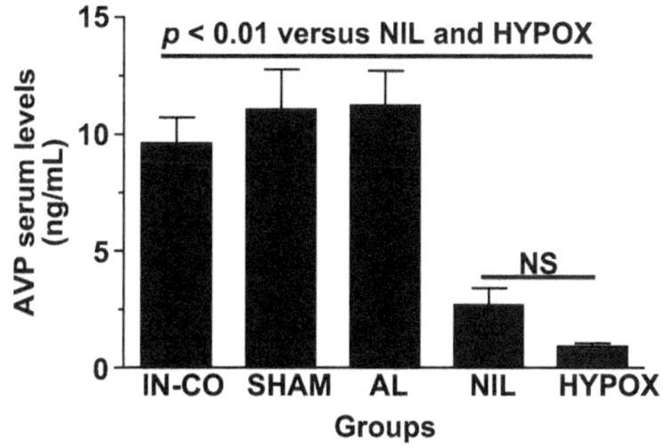

Figure 5.2 ***Effects of SHAM, AL, NIL and Total HYPOX on AVP Serum Levels Three Weeks After Surgery.*** For reference the intact control group (IN-CO) was included. NS, no statistical difference. Here, $n = 6–10$ animals/group. *(From with permission from Ref. [3].)*

controls (Figure 5.2).[3] Also, NIL animals develop central DI (increase of urine output and water intake). In AL rats, all adenohypophysial hormones are decreased to similar levels of that seen in HYPOX animals, whereas AVP blood levels remain in normal range (Figure 5.2).[3] In addition to changes in blood pituitary hormone levels, other important consequences of the different pituitary surgeries are evident in the immune responses, body weights, hormone secretion, histology, and weights of the pituitary target glands (adrenals, thyroid, gonads, and GH-dependent organs such as; lymphoid tissue, liver, kidney, and gastrointestinal tract).

1.8 NIL and Humoral Immune Responses

In order to evaluate the capability of NIL animals (AVP and OXT deficient) to respond to immune challenges, serum hemagglutinin levels and footpad inflammatory responses were assessed in NIL rats immunized with sheep red blood cells (SRBC). Results showed a significant reduction in the humoral immune response; NIL rats had lower hemagglutinin titers and diminished inflammatory response to SRBC footpad injection.[64] These results in the AVP–OXT deficient NIL animals suggest that AVP or OXT or both hormones are required to maintain immunocompetence.[64] The effects of neurohypophysial hormone deficiency on the immune response were also demonstrated in NIL animals that were orally infected with sublethal doses of *Salmonella typhimurium* (STY).[65,66] Results showed that STY was eliminated from the intestinal lumen at the same rate than those observed in SHAM and HYPOX animals, whereas STY colonization of Peyer's patches and spleen was significantly higher in NIL and HYPOX rats. Serums IgG, IgM, and intestinal secretory IgA against surface proteins of STY, were also significantly lower in NIL and HYPOX rats. It was noteworthy that HYPOX rats showed a significantly higher Peyer's patch and spleen bacterial colonization and decreased IgG, IgM, and secretory IgA than the NIL rats. Altogether, these results indicate that adenohypophysial and neurohypophysial hormones play a role in the regulation of the intestinal immune response. This view is supported by the results of Edwards et al.[67–70] These authors found that in HYPOX rats intraperitoneally infected with STY, GH, and PRL administration, restored the ability of the animals to overcome the infection. Since an increased severity of the STY infection occurred in the NIL animals than the controls, it is suggested that AVP is required to overcome the disease.

In order to obtain a deeper insight into the differential role played by adenohypophysial and neurohypophysial hormones on humoral immune regulation, SHAM, HYPOX, and NIL animals were immunized with the T cell-independent (TI) type 1 antigen DNP-LPS, TI type 2 antigen DNP-FICOLL, T cell-dependent (TD) antigens of ovalbumin (OVA), and bovine serum albumin (BSA). In comparison to the SHAM rats, both HYPOX and NIL induced a similar and significant decrease in IgM responses to TI-1 antigens, NIL but HYPOX did not induce a decreased IgM response to TI-2 antigens, and both HYPOX and NIL induced similar and significant decrease in IgG responses to TI-2 antigens. As compared with the SHAM rats, no changes were observed in IgM responses to TD antigens in HYPOX and NIL animals, whereas the IgG responses to OVA and BSA were significantly decreased in these groups, supporting the view that hormones from the anterior and posterior pituitary lobes play a different role in the regulation of humoral immune responses. These results also indicate that anterior and posterior pituitary hormones act in a complementary or redundant fashion in the control of immune responses.[71]

1.9 NIL and Cell-Mediated Immunity

Experimental autoimmune encephalomyelitis (EAE) is an autoimmune/inflammatory disease of the CNS caused by activated T lymphocytes, which induce axonal demyelination, paralysis, and sensorial deficits. EAE is widely used as experimental model for human multiple sclerosis.[72] The Lewis rats are very susceptible to the induction of EAE and other forms of autoimmune disease. Acute EAE may be induced by injecting guinea pig brain–spinal cord homogenate or purified myelin basic protein, myelin oligodendrocyte glycoprotein, or proteolipidic peptide suspended in complete Freund adjuvant. The clinical symptoms get manifested during the second week of immunization, reaching the peak at around 15–17 days postimmunization (limping, tail, and hind leg paralysis), followed by partial or complete recovery within 30 days.[73] The histopathological findings in the CNS depend on severity of the disease. In acute form of the disease, histology demonstrates inflammatory reactions almost invariably localized in the lumbosacral segment of the spinal cord. In addition, patchy inflammatory infiltrates can be observed along the thoracic and cervical spinal cord, brainstem, and cerebellum, and in more severe cases, in the cerebral hemispheres. The inflammatory infiltration has a predilection for perivascular areas (perivascular cuffs) in the white matter. The typical infiltrate consists of T lymphocytes, plasma cells, macrophages, polymorphonuclear neutrophils, and eosinophil leukocytes. Tissue demyelination is almost completely absent in the acute form of the disease.[74]

In order to correlate the role of different pituitary hormones on EAE, the effects of NIL, AL, HYPOX, and DP administration on the clinical course, CNS histological changes (perivascular cuffs), and the activation of the HPA axis (ACTH and corticosterone [CO] serum levels) at 15 days post EAE immunization were studied. SHAM animals showed the clinical signs of a full-blown disease, increased number of CNS perivascular cuffs, and the activation of HPA axis; HYPOX animals developed a rapid, severe, and a lethal form of the disease. NIL animals had only mild clinical and pathological symptoms of EAE with no changes in the HPA axis activity. In NIL + DP treated animals, the susceptibility to EAE was restored, showing full-blown clinical and histological signs of EAE and an activated HPA axis (Figures 5.3–5.5).[75,76] The decreased and increased HPA axis responses in NIL and NIL + DP animals could be due to AVP as an immunoregulatory hormone, which acts on AVP receptors in immune cells, stimulates/modulates the cascade of activation of the immune response, including the secretion of proinflammatory ILs that activate the HPA axis.[75–77] The role of AVP in the regulation of the immune responses is now proven. The involvement of OXT on the immune system requires further studies.

Previous experiments strongly suggested that AVP exerts its immunoregulatory effects through AVP receptors on the immune cells, suggesting that the use of specific antagonists for AVP V1a and V2 receptors may help control the immune responses. To test this hypothesis, effects of three peptide-derived AVP V1a *receptor antagonists*: [deamino-Pen1,O-Me-Try2,Arg8]-vasopressin (V1880), [β-mercapto-β,β-cyclopentamethylene propionyl^1,O-me-Try^2,Arg^8]-vasopressin (V2255), and [phenylacetyl1,

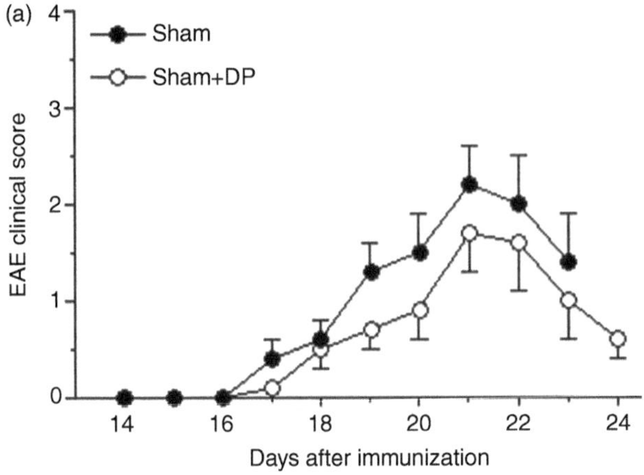

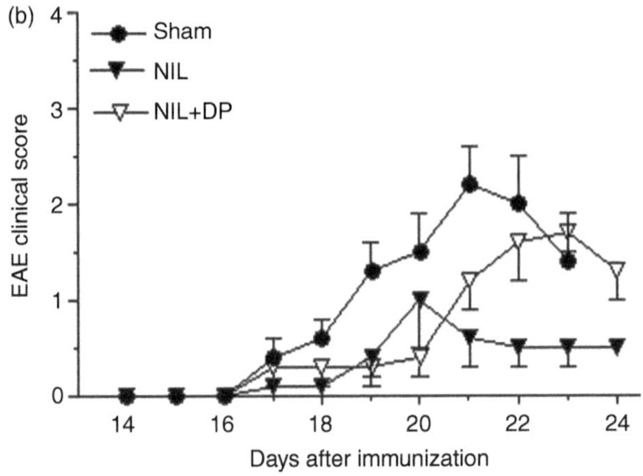

Figure 5.3 ***Clinical Development of EAE in SHAM and SHAM + DP Treated Rats (a) and SHAM, NIL and NIL+ DP Treated Rats (b).*** Points represent the means ± SEM of the daily scores of EAE. No significant differences were observed between SHAM and SHAM + DP treated groups (a). In (b), Significant differences between SHAM and NIL animals occurred at days 19, 21, 22, and 23, post immunization ($p < 0.01$). Significant differences between NIL and NIL + DP treated animals occurred at days 21, 22, 23, and 24 ($p < 0.01$). No significant differences were observed between SHAM and DP treated NIL animals at days 22 and 23, postimmunization. *(Reproduced from Karger and Wiley.)*

O-Me-D-Try2,Arg6,8,Lys9]-vasopressin amide (V2138) were studied on the development of clinical signs of EAE. Results showed that all three antagonists induced a similar and significant reduction in the severity of the disease (Figure 5.6).[3,78] However, in some experiments, no effects were noted on the clinical course of the disease, and in others worsening of EAE

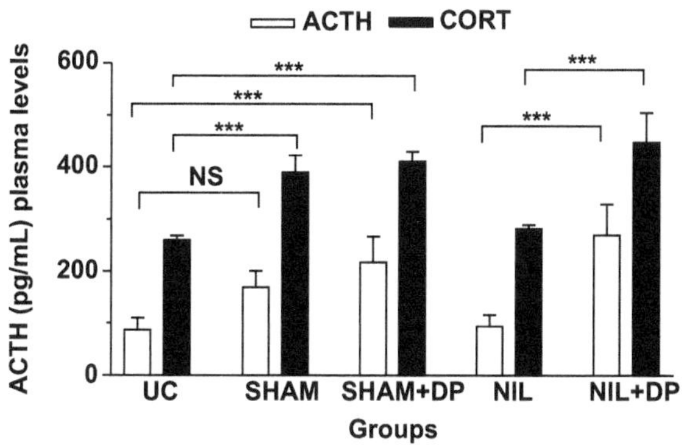

Figure 5.4 ***ACTH and CORT Serum Levels in SHAM and NIL Rats Treated or Untreated with DP.*** Blood samples were collected 15 days after immunization (peak of the clinical signs of EAE). Results from an untreated control (UC) group are included. NS, nonsignificant; ***, $p < 0.001$. *(Reproduced from Karger and Wiley.)*

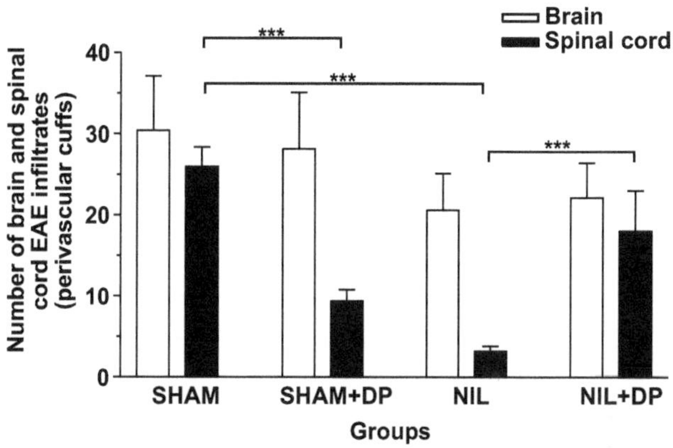

Figure 5.5 ***Number of Inflammatory Infiltrates (Perivascular Cuffs) in SHAM and NIL Rats Immunized for EAE and Treated or Untreated with DP (0.75 μg/s.c. Each 12 h).*** No significant differences were apparent in the number of cerebral lesions in various groups. DP administration started once a week before EAE immunization. The figure shows the diminution of lesions in the NIL group and the increased number of lesions in the NIL + DP treated group. ***, $p < 0.001$. *(Reproduced from Karger and Wiley.)*

occurred. The inconsistent results may be explained by the fact that these peptide-derived AVP antagonists have a weak AVP intrinsic agonist activity, which after repeated administration may increase its agonistic effect.[79–81] The recently described nonpeptide drugs called vaptans, with specific antagonistic effects on V1a, V1b, and V2 AVP receptors,[43,79,82–88] may be also used as modulators of the immune system.

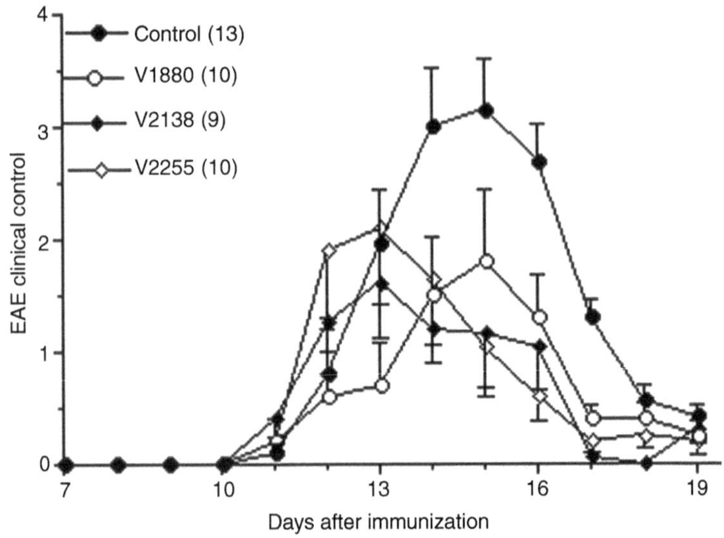

Figure 5.6 ***The Effects of V1a AVP Receptor Antagonists (Deamino-Pen1, O-Me-Try2, Arg8)-Vasopressin (V1880), (β-Mercapto-β, β-Cyclopentamethylene Propionyl1, O-Me-Try2, Arg8)-Vasopressin (V2255), and (Phenylacetyl1, O-Me-D-Try2, Arg6,8, Lys9]-Vasopressin Amide (V2138) on Development of EAE.*** Points represent the means ± SEM of the daily scores of EAE. Significant differences were observed between control and V1a AVP receptor antagonist treated groups on days 14–17 postimmunization ($p < 0.05$). In parenthesis the number of animals is given for each group. *(From Ref. [3] with permission.)*

Adjuvant-induced arthritis (AIA), an experimental model for human rheumatoid arthritis (RA), is a T cell-dependent autoimmune disease. It may be induced in Lewis rats by subcutaneous injection of heat-killed *Mycobacterium tuberculosis* in Freund adjuvants. Inflammatory signs in the joints appear during the second week postimmunization, reach the peak at around 21 days, and subside subsequently. Histologic findings include inflammatory infiltration of the joints containing macrophages, lymphocytes, plasma cells, and polymorphonuclear leukocytes, pannus formation, periostitis, as well as erosions of the cartilage and bone.[89,90] In AIA, the subcutaneously injected arthritogen antigens are taken up by antigen presenting cells, transported via blood and lymphatic vessels to lymph nodes and spleen, where the afferent arm of the immune response is mounted.[91] In these organs the immune response includes changes in ILs mRNA expression, release of pro- and anti-inflammatory cytokines, activation of CD4 and CD8 lymphocytes and attraction of circulating monocytes, which are transformed to macrophages.[91]

Cytokines regulate a broad range of inflammatory processes that are implicated in the pathogenesis of RA.[92] Thus, although a relationship between

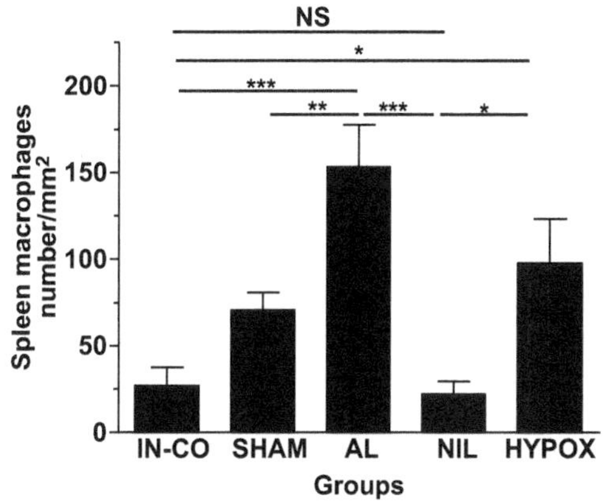

Figure 5.7 Effects of SHAM, NIL, AL, and HYPOX on spleen macrophage numbers. As reference, an intact control (IN-CO) group was included. CD68 was used as macrophage marker. Except the IN-CO the remaining groups were immunized for AIA 15 days before sacrifice. Pituitary surgeries were performed 3 weeks before immunization. $N = 3–4$ animals/group. $^*p < 0.05$, $^{**}p < 0.01$, and $^{***}p < 0.001$. *(From Ref. [3] with permission.)*

PRL and AVP in the AIA and human RA has been established,[8,10,33,93–95] it remains unclear how cytokines are organized within a hierarchical regulatory network during the induction of a disease.[92] The role of AVP in the regulation of the spleen ILs mRNA expression during the AIA is also undefined.

In order to obtain a deeper insight on the role of AVP in spleen cytokine expression, we correlated the effects of different pituitary surgeries on AVP serum levels, spleen macrophage numbers, and spleen mRNA expressions of ILs 1, 2, 6, 10, 12, INF-γ, and TNF-α in the AIA.[3] AVP serum levels before AIA immunizations (3 weeks after pituitary surgeries) were similar to that showed in Figure 5.2. Whereas AVP serum levels were similar in nonimmunized intact controls, SHAM, and AL groups in NIL and HYPOX animals, and the surgeries induced a significant diminution in AVP.[3] A strong correlation between AVP serum levels and clinical symptoms of AIA was apparent. In SHAM and AL animals, full-blown joint inflammation developed, whereas in NIL and HYPOX rats no symptoms of AIA were apparent. These results suggest that AVP is required to stimulate/modulate the immune response. This view is supported by the increased spleen macrophage number and ILs, in SHAM and AL animals having normal AVP serum levels, and unchanged spleen macrophage number and ILs in the AVP-deficient NIL animals (Figures 5.7 and 5.8). These results indicate that the low serum AVP levels were not enough to stimulate monocyte–macrophage recruitment

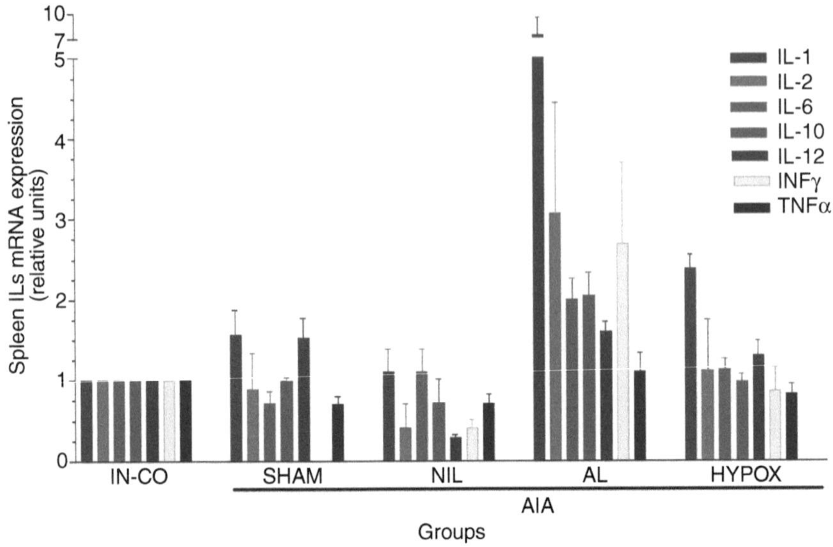

Figure 5.8 Effects of SHAM, NIL, AL, and HYPOX on spleen ILs mRNA expression 15 days after immunization for AIA. The intact control (IN-CO) group served as basal ILs expression. Rats were operated 3 weeks before immunization. $N = 3–4$ animals/group. *(Modified with permission from Ref. [3].)*

and ILs expression. It is noteworthy that in the HYPOX animals, while clinical symptoms of AIA were not apparent, the spleen macrophage number and ILs expression increased (Figures 5.7 and 5.8). These findings suggest that AVP maintains INIM, which in turn produces ILs for immune defense.

Altogether, these results indicate that immune responses are dependent on AVP serum levels. This view is further supported by increased macrophage phagocytosis stimulated by DP in Brattleboro and Long–Evans rats[25] and by the restored susceptibility to EAE in the NIL DP-treated animals.[76,78]

The presence of OXT receptors and some immunomodulatory effects of OXT on cells of the immune system have been demonstrated in the thymus, peripheral blood mononuclear cells, T lymphocytes, and spleen cells.[22,24,96–98] However, experiments are required to establish the importance of OXT in the control of EAE and AIA. It is unknown whether the development of these diseases may be affected by OXT receptor antagonists. It is possible that both AVP and OXT are involved in a coordinate fashion in the regulation of the immune system. Experiments are needed to test these possibilities.

1.10 Regulation of Intestinal Morphology and Gut-Associated Lymphoid Tissue (GALT) by Neurophypophysial Hormones

GH plays an important role in the maintenance of intestinal function and morphology, as well as in gut associated immune mechanisms.[67–70] It is also an important hormone for the survival of small gut grafts and treatment of severe trauma.[99–101] The role of the neurohypophysial hormones in the modulation of immune system is indicated by the decreased immune response to STY infection in NIL rats.[65,66] The function and morphology of the small intestine mucosa is affected by the immune system (GALT) in rats and are influenced by several pituitary lobe-specific surgeries.[103] The results indicate that AVP and OXT also participate in the maintenance and regulation of intestinal mucosa and GALT, as compared with the SHAM group, the height of the mucosa villi decreased significantly in AL, NIL, and HYPOX groups (25, 22, and 34%, respectively), whereas the number of goblet cells decreased significantly in AL (38%) and HYPOX (27%) animals. In the lamina propria, CD4+ cells were significantly less in NIL than in AL group, whereas CD8+ cells decreased significantly in the HYPOX group. Intraepithelial CD8+ cells were significantly decreased in NIL animals. No changes were observed in Paneth cells of the various groups, whereas IgA and IgM cells and the secretory IgA were significantly decreased in NIL, AL, and HYPOX animals. As compared with the AL group, in NIL animals a more significant loss of IgA cells and secretory IgA were noted (Table 5.1). From these observations, it is clear that both, neurohypophysial and adenohypophysial hormones exert lobe-specific regulatory effects on intestinal mucosa cells and GALT. It appears that in the lamina propria, CD8+ cells, and goblet cells are more dependent on the anterior pituitary hormones, mainly GH and PRL, as supported by observations of Shulman et al.;[98]

Table 5.1 Effects of SHAM, NIL, AL, and HYPOX on ileal paneth, IgM, and IgA cell numbers and on intestinal IgA secretion (μg/mL)

Groups	Paneth cells	IgM cells	IgA cells	IgA secretion (μg/mL)
SHAM	2.6 ± 0.2(a)	18.5 ± 1.1(a)	21.9 ± 2.0(a)	38.6 ± 0.1(a)
NIL	2.2 ± 0.1(a)	9.2 ± 0.4(b)	11.2 ± 0.7(b)	23.7 ± 0.5(b)
AL	2.0 ± 0.05(a)	13.6 ± 0.08(c)	15.3 ± 1.0(b)	31.8 ± 0.5(c)
HYPOX	2.2 ± 0.05(a)	12.1 ± 0.7(b,c)	14.7 ± 1.2(b)	32.5 ± 0.8(c)

Columns with same letters mean nonsignificant differences and different letters mean significant differences. Values are mean ± SEM in each group.

Source: Reproduced from Karger and Wiley.

Jeschke et al.;[99] Zhang et al.[100], whereas the size of villi, CD4+ cells, IgA, IgM cells, and secretory IgA are more dependent on neurohypophysial hormones.[102] It can also be concluded that hormones from these two pituitary lobes act in a complementary fashion and participate in the maintenance of gut morphology, function, and immune responses. Further experiments are required to reveal the specific role of AVP and OXT on gut homeostasis.

2 CONCLUSIONS

The decreased humoral and cell-mediated immune responses in NIL rats and the increased severity of EAE and AIA in AL animals indicates that neurohypophysial hormones, mainly AVP, and to a lesser extent OXT, play important roles in the maintenance of immunocompetence. The restored susceptibility of NIL animals to EAE by DP administration supports the view that AVP, through AVP V2 receptors, directly participates in the function and modulation of the immune system.

The decreased severity of EAE by V1a AVP receptor antagonists raises the possibility that AVP-derived drugs and the nonpeptide-derived AVP receptor antagonists (vaptans), may be used in the treatment of autoimmune diseases.

This review emphasizes that adenohypophysial and neurohypophysial hormones have their own roles in regulating the immune system. However, they work complementarily to restore homeostasis during immune responses. More work is needed to better understand the role of AVP and OXT in the regulation of the immune system.

2.1 Clinical Relevance and Future Directions

AVP is a hormone with multiple functions. From the present review, it is clear that substantial interactions exist between the immune system and AVP. The discovery of AVP receptors in the immune cells and the use of specific AVP antagonists to modulate the immune response promises further discoveries and may provide the opportunity to treat autoimmune diseases including those as EAE and AIA, multiple sclerosis, RA, allograft rejection, and lupus erythematous in animals and humans. The presence of OXT receptors in cells of the immune system suggests that OXT may be another important hormone in the regulation of the immune system. More experiments are necessary to answer the questions emerging from this chapter and other works.

ACKNOWLEDGMENTS

Authors wish to thank Manuel Tinajero-Ruelas, Roberto Ramírez-Ruíz, Jesús Martínez-Hernández, José Luis Ponce-Márquez, Adriana Rodríguez-Peralta, Juan Esparza-Valenzuela, Eleazar Luévano-Delgadillo, Norma Adela Carrasco-Esparza, and Verónica Viñuela-Berni for their skillful technical assistance. This work was supported by UAA-PIBB-11-5 and CONACYT 221262 grants, México (AQS) and SIP-IPN-20110764 grants, México (RCR).

DISCLOSURES

The authors declare that there is no conflict of interest that would interfere with the impartiality of this work.

REFERENCES

1. Kelley KW, Weigent DA, Kooijmand R. Protein hormones and immunity. *Brain Behav Immun* 2007;**21**:384–92.
2. Sternberg EM. Neuroendocrine regulation of autoimmune/inflammatory disease. *J Endocrinol* 2001;**169**:429–35.
3. Berczi I, Quintanar-Stephano A, Villalobos-Hernández EC, Campos-Rodriguez R, Kovacs K. The healing power of vasopressin. *Adv Neuroimm Biol* 2012;**3**:217–24.
4. Nagy E, Berczi I. Prolactin and contact sensitivity. *Allergy* 1981;**36**:429–31.
5. Nagy E, Berczi I. Hypophysectomized rats depend on residual prolactin for survival. *Endocrinology* 1991;**128**:2776–84.
6. Berczi I, Nagy E, Asa SL, Kovacs K. Pituitary hormones and contact sensitivity in rats. *Allergy* 1983;**38**:325–30.
7. Berczi I, Nagy E, Asa SL, Kovacs K. The influence of pituitary hormones on adjuvant arthritis. *Arthritis Rheum* 1984;**27**:682–8.
8. Nagy E, Berczi I, Friesen HG. Regulation of immunity in rats by lactogenic and growth hormones. *Acta Endocrinol (Copenhagen)* 1983;**102**:351–7.
9. Gala RR. Prolactin and growth hormone in the regulation of the immune system. *Proc Soc Exp Biol Med* 1991;**198**:513–27.
10. Neidhart M, Fluckiger EW. Hyperprolactinaemia in hypophysectomized or intact male rats and the development of adjuvant arthritis. *Immunology* 1992;**77**:449–55.
11. Berczi I, Szentivanyi A. The immune-neuroendocrine circuitry. History and progress. Berczi I, Szentivangyi A, editors. *Neuroimmune Biology*, vol. 3. Amsterdam, The Netherlands: Elsevier; 2003. p. 495–536.
12. Redelman D, Welniak LA, Taub D, Murphy WJ. Neuroendocrine hormones such as growth hormone and prolactin are integral members of the immunological cytokine network. *Cell Immunol* 2008;**252**:111–21.
13. Díaz L, Díaz-Muñoz M, González L, Lira-Albarrán S, Larrea F, Méndez I. Prolactin in the immune system. In: Nagy GM, Toth BE, editors. *Prolactin*. InTech Open Science, p. 53–81.
14. Chrousos GP. The hypothalamic-pituitary-adrenal axis and immunemediated inflammation. *N Engl J Med* 1995;**332**:1351–62.
15. Webster JI, Tonelli L, Sternberg EM. Neuroendocrine regulation of the immunity. *Ann Rev Immunol* 2002;**20**:125–63.
16. Webster JI, Sternberg EM. Role of hypothalamic-pituitary adrenal axis, glucocorticoids and glucocorticoid receptors in toxic sequelae of exposure to bacterial and viral products. *J Endocrinol* 2004;**181**:207–21.

17. Berczi I, Quintanar-Stephano A, Campos-Rodríguez R, Kovacs K. New aspects of immunoregulation by growth and lactogenic hormones. *Adv Neuroimm Biol* 2014;**5**: 43–60.
18. De Vito P, Incerpi S, Pedersen JZ, Luly P, Davis FB, Davis PJ. Thyroid hormones as modulators of immune activities at the cellular level. *Thyroid* 2011;**21**:879–90.
19. De Vito P, Balducci V, Leone S, Percario Z, Mangino G, Davis PJ, Davis FB, Affabris E, Luly P, Pedersen JZ, Incerpi S. Nongenomic effects of thyroid hormones on the immune system cells: new targets, old players. *Steroids* 2012;**77**:988–95.
20. Eskandari F, Webster JI, Sternberg EM. Neural immune pathways and their connection to inflammatory diseases. *Arthritis Res Ther* 2003;**5**:251–65.
21. Johnson HM, Farrar LM, Torres BA. Vasopressin replacement of interleukin 2 requirement in gamma interferon production: lymphokine activity of a neuroendocrine hormone. *J Immunol* 1982;**129**:983–6.
22. Johnson HM, Torres BA. Regulation of lymphokine production by arginine vasopressin and oxytocin: modulation of lymphocyte function by neurohypophysial hormones. *J Immunol* 1985;**135**:773s–5s.
23. Torres BA, Johnson HM. Arginine vasopressin (AVP) replacement of helper cell requirement in IFN-gamma production. Evidence for a novel AVP receptor on mouse lymphocytes. *J Immunol* 1988;**140**:2179–83.
24. Torres BA, Johnson HM. Neuroendocrine peptide hormone regulation of immunity. In: Blalock JE, editor. Neuroimmunoendocrinology. *Chem Immunol* 69, 155–184.
25. Fernandez-Repollet E, Opava-Stitzer S, Tiffany S, Schwartz A. Effects of endogenous antidiuretic hormone (ADH) on macrophage phagocytosis. *J Histochem Cytochem* 1983;**31**:956–9.
26. Antoni FA. Vasopressinergic control of pituitary aderenocorticotropin secretion comes of age. *Front Neuroendocrinol* 1993;**14**:76–122.
27. Buckingham JC, Cowell AM, Gilles GE, Herbison AE, Steel JH. The neuroendocrine system: anatomy, physiology and responses to stress. In: Buckingham JC, Gillies GE, Cowell A-M, editors. *Stress, hormones and the immune system*. Chichester, UK: Wiley & Sons; 1997. p. 9–47.
28. Harbuz MS, Conde GL, Marti O, Lightman SL, Jessop DS. The hypothalamic-pituitary adrenal axis in autoimmunity. *Ann NY Acad Sci* 1997;**823**:214–24.
29. Harbuz MS, Richards LJ, Chover-Gonzalez AJ, Marti-Sistac O, Jessop DS. Stress in autoimmune disease models. *Ann NY Acad Sci* 2006;**1069**:51–61.
30. Goncharova ND. Stress responsiveness of the hypothalamic-pituitary-adrenal axis: age-related features of the vasopressinergic regulation. *Front Endocrinol (Lausanne)* 2013;**12**:4–26.
31. Kjaer A, Knigge U, Bach FW, Warberg J. Histamine-and stress-induced secretion of ACTH and beta-endorphin: involvement of corticotropin-releasing hormone and vasopressin. *Neuroendocrinology* 1992;**56**:419–28.
32. De Wied D, Diamant M, Fodor M. Central nervous system effects of the neurohypophysial hormones and related peptides. *Front Neuroendocrinol* 1993;**14**:251–302.
33. Chikanza IC, Petrou P, Chrousos G. Perturbations of arginine vasopressin secretion during inflammatory stress. Pathophysiologic implications. *Ann NY Acad Sci* 2000;**917**: 825–34.
34. Ferguson JW, Therapondos G, Newby DE. Therapeutic role of vasopressin receptor antagonism in patients with liver cirrhosis. *Clin Sci (London)* 2003;**105**:1–8.
35. Berczi I, Quintanar-Stephano A, Kovacs K. Immunoconversion in the acute phase response. In: Plotnikoff NP, Faith RE, Murgo AJ, Good RA, editors. *Cytokines, stress and immunity*. Boca Raton: CRC; 2007. p. 215–54.
36. Laycock J. Vasopressin and its renal effects. In: Laycock JF, editor. *Perspectives on vasopressin*. London: Imperial College Press; 2010. p. 133–66.

37. Laycock J. Vasopressin and the cardiovascular system. In: Laycock JF, editor. *Perspectives on vasopressin*. London: Imperial College Press; 2010. p. 176–95.
38. Leng G, Meddle S. Vasopressin: the central systems. In: Laycock JF, editor. *Perspectives on vasopressin*. London: Imperial College Press; 2010. p. 257–87.
39. Koshimizu TA, Nakamura K, Egashira N, Hiroyama M, Nonoguchi H, Tanoue A. Vasopressin V1a and V1b receptors: from molecules to physiological systems. *Physiol Rev* 2012;**92**:1813–64.
40. Derick S, Cheng LL, Voirol MJ, Stoev S, Giacomini M, Wo NC, Szeto HH, Ben-Mimoun M, Andres M, Gaillard RC, Guillon G, Manning M. [1-deamino-4-cyclohexylalanine] Arginine vasopressin: a potent and specific agonist for vasopressin V1b receptors. *Endocrinology* 2002;**143**:4655–64.
41. Jard S. Vasopressin receptors. A historical survey. *Adv Exp Med Biol* 1998;**449**:1–13.
42. Arai Y, Fujimori A, Sasamata M, Miyata KJ. New topics in vasopressin receptors and approach to novel drugs: research and development of conivaptan hydrochloride (YM087), a drug for the treatment of hyponatremia. *Pharmacol Sci* 2009;**109**:53–9.
43. Manning M, Misicka A, Olma A, Bankowski K, Stoev S, Chini B, Durroux T, Mouillac B, Corbani M, Guillon G. Oxytocin and vasopressin agonists and antagonists as research tools and potential therapeutics. *J Endocrinol* 2012;**24**:609–28.
44. Barret KE, Barman SM, Boitano S, Brooks HL. *Hypothalamic regulation of hormonal functions, in Ganong's review of medical physiology*. New York: Lange Medical Books/McGraw-Hill; 2010. p. 232–55.
45. Block LH, Locher R, Tenschert W, Siegenthaler W, Hofmann T, Mettler E, Vetter W. ^{125}I-8-L-Arginine vasopressin binding to human mononuclear phagocytes. *J Clin Invest* 1981;**68**:374–81.
46. Bell J, Adler MW, Greenstein JI, Liu-Chen LY. Identification and characterization of [125I] arginine vasopressin binding sites on human peripheral blood mononuclear cells. *Life Sci* 1993;**52**:95–105.
47. Elands J, Resink A, De Kloet R. Neurohypophysial hormone receptors in the rat thymus, spleen, and lymphocytes. *Endocrinology* 1990;**126**:2703–10.
48. Wickramasinghe SN, Valdimarsson H, Williams B. Interaction between a synthetic analogue of vasopressin and human peripheral blood lymphocytes. *J Clin Lab Immunol* 1982;**7**:67–70.
49. Lolait SJ, O'Carroll AM, Mahan LC, Felder CC, Button DC, Young III WS, Mezey E, Brownstein MJ. Extrapituitary expression of the rat V1b vasopressin receptor gene. *Proc Natl Acad Sci* 1995;**15**:6783–7.
50. Pereira A, del Valle Onorato M, Sanz C. DDAVP enhances the ability of blood monocytes to form rosettes with activated platelets by increasing the expression of P-selectin sialylated ligands on the monocyte surface. *Br J Haematol* 2003;**120**:814–20.
51. Yamaguchi Y, Yamada K, Suzuki T, Wu YP, Kita K, Takahashi S, Ichinose M, Suzuki N. Induction of uPA release in human peripheral blood lymphocytes by [deamino-Cysl,D-Arg8]-vasopressin (dDAVP). *Am J Physiol Endocrinol Metab* 2004;**287**:E970–6.
52. Dicker P, Rozengurt E. Phorbol esters and vasopressin stimulate DNA synthesis by a common mechanism. *Nature* 1980;**287**(5783):607–12.
53. Thibonnier M, Conarty DM, Preston JA, Plesnicher CL, Dweik RA, Erzurum SC. Human vascular endothelial cells express oxytocin receptors. *Endocrinology* 1999;**140**:1301–9.
54. Hansenne I, Rasier G, Charlet-Renard Ch, DeFresne MP, Greimers R, Breton C, Legros JJ, Geenen V, Martens H. Neurohypophysial receptor gene expression by thymic T cell subsets and thymic T cell lymphoma cell lines. *Clin Dev Immunol* 2004;**11**:45–51.
55. Locher R, Vetter W, Block LH. Interactions between 8-L-arginine vasopressin and prostaglandin E2 in human mononuclear phagocytes. *J Clin Invest* 1983;**71**:884–91.

56. Bell J, Adler MW, Greenstein JI. The effect of arginine vasopressin on the autologous mixed lymphocyte reaction. *Int J Immunopharmacol* 1992;**14**:93–103.
57. Majzoub JA, Carrazana EJ, Shulman JS, Baker KJ, Emanuel RL. Defective regulation of vasopressin gene expression in Brattleboro rats. *Am J Physiol* 1987;**252**:E637–42.
58. Khegai II, Gulyaeva MA, Popova NA, Zakharova LA, Ivanova LN. Immune system in vasopressin-deficient rats during ontogeny. *Bull Exp Biol Med* 2003;**136**:448–50.
59. Patchev VK, Mastorakos G, Brady LS, Redwine J, Wilder RL, Chrousos GP. Increased arginine vasopressin secretion may participate in the enhanced susceptibility of Lewis rats to inflammatory disease. *Neuroendocrinology* 1993;**58**:106–10.
60. Patchev UVK, Kalogeras KT, Zelazowski P, Wilder RL, Chrousos GP. Increased plasma concentrations, hypothalamic content, and *in vitro* release of arginine vasopressin in inflammatory disease-prone, hypothalamic corticotropin-releasing hormone-deficient Lewis rats. *Endocrinology* 1992;**131**:1453–7.
61. Alvarez-Buylla R, Quintanar Stephano A, Quintanar Stephano JL, De Alvarez-Buylla E. Technical note: removal of the unfragmented pituitary gland (hypophysectomy) in the rat. *Bol Estud Med Biol (Mexico)* 1991;**39**:33–8.
62. Ben-Jonathan N, Peters LL. Posterior pituitary lobectomy: differential elevation of plasma prolactin and luteinizing hormone in estrous and lactating rats. *Endocrinology* 1982;**110**:1861–5.
63. Mena F, Aguayo D, Vigueras M, Quintanar-Stephano A, Perera G, Morales T. Effect of posterior pituitary lobectomy on *in vivo* and *in vitro* secretion of prolactin in lactating rats. *Endocrine* 1996;**5**:285–90.
64. Quintanar-Stephano A, Kovacs K, Berczi I. Effects of neurointermediate pituitary lobectomy on humoral and cell-mediated immune responses in the rat. *Neuroinmunomodulation* 2004;**11**:233–40.
65. Campos-Rodríguez R, Quintanar Stephano A, Drago-Serrano ME, Abarca-Rojano E, Berczi I, Ventura-Juárez J, Kormanovski A. The central nervous system modulates the immune response to *Salmonella*. In: Kumar Y, editor. *Salmonella – a diversified superbug*. Croacia: InTech Publication; 2012. p. 375–98.
66. Campos-Rodríguez R, Quintanar-Stephano A, Jarillo-Luna RA, Oliver-Aguillón G, Ventura-Juárez J, Rivera-Aguilar V, Berczi I, Kovacs K. Hypophysectomy and neurointermediate pituitary lobectomy reduce serum IgM. IgG and intestinal IgA responses to *Salmonella enterica* serovar typhimurium infection in rats. *Infect Immun* 2006;**3**:1883–9.
67. Edwards III CK, Lorence RM, Dunham DM, Arkins S, Yunger LM, Greager JA, Walter RJ, Dantzer R, Kelley KW. Hypophysectomy inhibits the synthesis of tumor necrosis factor alpha by rat macrophages: partial restoration by exogenous growth hormone or interferon gamma. *Endocrinology* 1991;**128**:989–96.
68. Edwards III CK, Yunger LM, Lorence RM, Dantzer R, Kelley KW. The pituitary gland is required for protection against lethal effects of *Salmonella typhimurium*. *Proc Natl Acad Sci* 1991;**88**:2274–7.
69. Edwards III CK, Arkins S, Yunger LM, Blum A, Dantzer R, Kelley KW. The macrophage-activating properties of growth hormone. *Cell Mol Neurobiol* 1992;**12**:499–510.
70. Edwards III CK, Ghiasuddin SM, Yunger LM, Lorence RM, Arkins S, Dantzer R, Kelley KW. *In vivo* administration of recombinant growth hormone or gamma interferon activates macrophages: enhanced resistance to experimental *Salmonella typhimurium* infection is correlated with generation of reactive oxygen intermediates. *Infect Immun* 1992;**60**:2514–21.
71. Quintanar-Stephano A, Abarca-Rojano E, Jarillo-Luna RA, Rivera-Aguilar V, Ventura-Juárez J, Berczi I, Kovacs K, Campos-Rodríguez R. Hypophysectomy and neurointermediate pituitary lobectomy decrease humoral immune responses to T-independent and T-dependent antigens. *J Physiol Biochem* 2010;**66**:7–13.

72. Martin R, McFarland HF. Immunological aspects of experimental allergic encephalomyelitis and multiple sclerosis. *Crit Rev Clin Lab Sci* 1995;**32**:121–82.
73. Simmons S, Mason D. Experimental allergic encephalomyelitis in the rat. Lefkovits I, editor. *Immunology methods manual: the comprehensive sourcebook of techniques*, vol. 3. San Diego: Academic Press; 1997. p. 1743–51.
74. Al-Sabbagh A, Weiner HL. Rat experimental autoimmune encephalomyelitis. In: Cohen IR, Miller A, editors. *Autoimmune Disease Models*. San Diego: Academic Press; 1994. p. 15–21.
75. Quintanar-Stephano A, Chavira-Ramírez R, Kovacs K, Berczi I. Neurointermediate pituitary lobectomy decreases the incidence and severity of experimental autoimmune encephalomyelitis in Lewis rats. *J Endocrinol* 2005;**184**:51–8.
76. Quintanar-Stephano A, Organista-Esparza A, Chavira-Ramírez R, Kovacs K, Berczi I. Effects of neurointermediate pituitary lobectomy and desmopressin on acute experimental autoimmune encephalomyelitis in Lewis rats. *Neuroimmunomodulation* 2012;**19**:148–57.
77. Jeneway CA, Travers P, Walport M, Shlomishik MJ. *Immuno Biology*. New York: Garland Publishing; 2001.
78. Quintanar-Stephano A, Organista-Esparza A, Tinajero-Ruelas M, Chavira-Ramírez R, Delgado-González E, García D, Sánchez-Herrera IO, Berczi I, Kovacs K. Blockade of immune cell-vasopressin receptors by V1a antagonist inhibit the peak of the experimental autoimmune encephalomyelitis (EAE) in Lewis rats. Experimental Biology, FASEB Meeting. San Francisco, California, USA 2006; p. 1–5.
79. Sawyer WH, Manning M. The use of antagonists of vasopressin in studies of its physiological functions. *Fed Proc* 1985;**44**:78–80.
80. Kinter LB, Caltabiano S, Huffman WF. Anomalous antidiuretic activity of antidiuretic hormone antagonists. *Biochem Pharmacol* 1993;**45**:1731–7.
81. Jackson EK. Vasopressin and other agents affecting the renal conservation of water. In: Hardman JG, Limbird LE, editors. *Goodman and Gilman's The Pharmacological Basis of Therapeutics*. New York: McGraw-Hill; 2001. p. 789–808.
82. Decaux G, Soupart A, Vassart G. Non-peptide arginine-vasopressin antagonists: the vaptans. *Lancet* 2008;**371**:1624–32.
83. Manning M, Stoev S, Chan WY, Sawyer WH. Receptor-specific antagonists of vasopressin and oxytocin. A current perspective. *Ann NY Acad Sci* 1993;**689**:219–32.
84. Yamamura Y, Ogawa H, Yamashita H, Chihara T, Miyamoto H, Nakamura S, Onogawa T, Yamashita T, Hosokawa T, Mori T. Characterization of a novel aquaretic agent, OPC-31260, as an orally effective, nonpeptide vasopressin V2 receptor antagonist. *Br J Pharmacol* 1992;**4**:787–91.
85. Wada K, Matsukawa U, Fujimori A, Arai Y, Sudoh K, Sasamata M, Miyata K. A novel vasopressin dual V1A/V2 receptor antagonist, conivaptan hydrochloride, improves hyponatremia in rats with syndrome of inappropriate secretion of antidiuretic hormone (SIADH). *Biol Pharm Bull* 2007;**1**:91–5.
86. Veeraveedu PT, Palaniyandi SS, Yamaguchi K, Komai Y, Thandavarayan RA, Sukumaran V, Watanabe K. Arginine vasopressin receptor antagonists (vaptans): pharmacological tools and potential therapeutic agents. *Drug Discov Today* 2010;**15**:826–41.
87. Contreras-Romo MC, Martínez-Archundia M, Deeb O, S'lusarz MJ, Ramírez-Salinas G, Garduño-Juárez R, Quintanar-Stephano A, Ramírez-Galicia G, Correa-Basurto J. Exploring the ligand recognition properties of the human vasopressin V1a receptor using QSAR and molecular modeling studies. *Chem Biol Drug Des* 2014;**83**(2):207–23.
88. Contreras-Romo Mc, Correa-Basurto J, Padilla-Martínez I, Martínez-Archundia M, Martínez-Ramos F, S'lusarz Js, López-Pérez G, Quintanar-Stephano A. Theoretical Studies, Synthesis, And Biological Activity Of 1-[(4-Methylphenyl)Sulfonyl]-5-Oxo-2,3,4,5-Tetrahydro-1H-1-Benzazepine-4-Carbonitrile (C9) As A Non-Peptide

Antagonist Of The Arginine Vasopressin V1a And V2 Receptors. *Med Chem Res* 2014;**23**(3):1581-L1590.
89. Williams RO. Rodent models of arthritis: relevance for human disease. *Clin Exp Immunol* 1998;**114**:330–2.
90. Carol M, Pelegria C, Castellote C, Franch A, Castell M. Immunohistochemical study of lymphoid tissues in adjuvant arthritis (AA) by image analysis; relationship with synovial lesions. *Clin Exp Immunol* 2000;**120**:200–8.
91. Matsuno K, Ueta H, Shu Z, Xue-Dong X, Sawanobori Y, Kitazawa Y, Bin Y, Yamashita M, Shi C. The microstructure of secondary lymphoid organs that support immune cell trafficking. *Arch Histol Cytol* 2010;**73**:1–21.
92. McInnes IB, Schett G. Cytokines in the pathogenesis of rheumatoid arthritis. *Nat Rev Immunol* 2007;**7**:429–42.
93. Chikanza IC, Grossman AS. Hypothalamic-pituitary-mediated immunomodulation: arginine vasopressin is a neuroendocrine immune mediator. *Br J Rheumatol* 1998;**37**: 131–6.
94. Chikanza IC. Prolactin and neuroimmunomodulation: *in vitro* and *in vivo* observations. *Ann NY Acad Sci* 1999;**876**:119–30.
95. Adán N, Guzmán-Morales J, Ledesma-Colunga MG, Perales-Canales SI, Quintanar-Stephano A, López-Barrera F, Méndez I, Moreno-Carranza B, Triebel J, Binart N, Martínez de la Escalera G, Thebault S, Clapp C. Prolactin promotes cartilage survival and attenuates inflammation in inflammatory arthritis. *J Clin Invest* 2013;**123**:3902–13.
96. Macciò A, Madeddu C, Chessa P, Panzone F, Lissoni P, Mantovani G. Oxytocin both increases proliferative response of peripheral blood lymphomonocytes to phytohemagglutinin and reverses immunosuppressive estrogen activity. *In Vivo* 2010;**24**:157–63.
97. Ndiaye K, Poole DH, Pate JL. Expression and regulation of functional oxytocin receptors in bovine T lymphocytes. *Biol Reprod* 2008;**78**:786–93.
98. Oliveira-Pelegrin GR, Saia RS, Cárnio EC, Rocha MJ. Oxytocin affects nitric oxide and cytokine production by sepsis-sensitized macrophages. *Neuroimmunomodulation* 2013;**20**:65–71.
99. Shulman DI, Hu CS, Duckett G, Lavalle-Grey M. Effects of short term growth hormone therapy in rats undergoing 75% small intestinal resection. *J Pediatr Gastroenterol Nutr* 1992;**14**:3–11.
100. Jeschke MG, Herndon DN, Finnerty CC, Bolder U, Thompson JC, Mueller U, Wolf SE, Przkora R. The effect of growth hormone on gut mucosal homeostasis and cellular mediators after severe trauma. *J Surg Res* 2005;**127**:183–9.
101. Zhang X, Li J, Li N. Growth hormone improves graft mucosal structure and recipient protein metabolism in rat small bowel transplantation. *Chin Med J* 2002;**115**:732–5.
102. Cárdenas-Jaramillo LM, Quintanar-Stephano A, Jarillo-Luna RA, Rivera-Aguilar V, Oliver-Aguillón G, Campos-Rodríguez R, Kovacs K, Berczi I. Regulation of intestinal morphology and GALT by pituitary hormones in the rat. *Ann NY Acad Sci* 2012;**1**:72–8.

CHAPTER 6

Integration and Coordination of Bodily Functions by NISS

Istvan Berczi*,, Andrés Quintanar-Stephano****
*Department of Immunology, Faculty of Medicine, The University of Manitoba, Winnipeg, Manitoba, Canada
**Departamento de Fisiología y Farmacología, Centro de Ciencias Básicas, Universidad Autónoma de Aguascalientes, Aguascalientes, Estado Libre y Soberano de Aguascalientes, Mexico City, Mexico

1 INTRODUCTION

1.2 Initial Discoveries

Sporadic suggestions of neuroimmune interactions date back to the late nineteenth century. The students and followers of Pavlov in Russia thoroughly investigated the nature of immunoregulation by neuroimmune mechanisms and eventually concluded that the hypothalamus is a likely center of immunoregulation.[1,2]

Using modern technology, Szentivanyi and coworkers observed that hypothalamic lesions inhibited the development of anaphylactic shock in immunized animals.[3] Tuber cinereum lesions (TBLs) in the hypothalamus inhibited anaphylaxis in preimmunized guinea pigs and (in later experiments) also in rabbits. Anaphylaxis was induced in immunized animals by intravenous (IV) injection of the immunizing antigen. TBLs also inhibited antibody production if the lesions were created prior to immunization. TBLs did not affect antigen–antibody reactions, nor did the release of tissue materials mediating anaphylaxis. Hypothalamic lesions temporarily increased histamine resistance and inhibited the anaphylactic reaction even when the animals were provided with passively transferred antibodies, which elicited lethal shock in control animals. The Schultz–Dale test, which was performed with small pieces of intestine *in vitro,* was also inhibited by TBLs. The Arthus reaction, turpentine-induced inflammation, and the Sanarelli–Shwartzman phenomenon were unaffected. Lesions of other areas of the hypothalamus or of the central nervous system (CNS) were ineffective in modulating immune phenomena. Further, electrical stimulation of the mammillary region of the hypothalamus had an inhibitory effect on the anaphylactic response and increased the resistance of animals to histamine.[4–6] In 1964, Korneva

Insights to Neuroimmune Biology
http://dx.doi.org/10.1016/B978-0-12-801770-8.00006-9

and Khai[7] confirmed that hypothalamic lesions in rabbits, guinea pigs, and rats inhibited the production of complement-fixing antibodies.

Hans Selye described the stress syndrome in 1936.[8] He established that the hypothalamus–pituitary–adrenal (HPA) axis was involved in mediating the stress response, which could be induced by various "nocuous" agents. The adrenal gland was stimulated by stress and the thymus and lymph nodes were involuted by glucocorticoids (GCs). He also observed gastrointestinal lesions.[9] Subsequently, Selye concluded that the stress response was a host defense reaction, which he called the "general adaptation syndrome."[10] These observations clearly implied that the pituitary gland plays an important role in host defense.

While I was in Selye's laboratory, I decided to learn more about the interaction of the pituitary gland and of the brain with immune host defense. We established that adaptive immune (ADIM) function is regulated by pituitary hormones: for example, prolactin (PRL) and growth hormone (GH) stimulated immune function, whereas the HPA axis was inhibitory and antagonized the stimulatory effect of PRL and GH. Human placental lactogen (HPL) was as efficient as PRL or GH in restoring ADIM function in this system. Initially we used young (100 g birth weight) hypophysectomized (Hypox) rats, and the antigen was always injected without adjuvants 7–10 days after operation.

Treatment of intact animals with the dopamine-agonist drug bromocriptine was as effective as Hypox in suppressing ADIM function.[11–13] Others showed that immune-derived cytokines (CTKs) are capable of stimulating the HPA axis. This discovery revealed that ADIM function is regulated by the CNS and the HPA axis, but also PRL and GH, receive feedback regulatory signals from immune-derived CTKs.[14,15]

Ader and coworkers, and Gorczynski and Kennedy showed that immune function is subject to Pavlovian conditioning.[16,17] These observations imply that the host is able to activate immune defense when danger is anticipated. However, the mechanisms involved in immune conditioning are complex and poorly understood.[18]

1.3 Recent Developments in Neuroimmune Biology

1.3.1 The Concept of Neuroimmune Biology

During the past three decades, the field of neuroimmune interactions grew steadily, and has now evolved into a sophisticated and credible area of biology. A number of terms have been proposed to name this multidisciplinary science. Since this science deals with the physiology and pathophysiology of higher organisms, in 2000 we coined the term *neuroimmune biology* (*NIB*),[19] which has been accepted by the scientific community. It is clear that the CNS, the endocrine system (ES), and the immune system (IS) form a regulatory circuit, which integrates, coordinates, and regulates all functions in higher organisms from conception till death.

By now a lot of information has been accumulated about the recognition systems and signaling in this neuroimmune supersystem (NISS).[20] The mediators for signaling are hormones, neurotransmitters, neuropeptides, CTKs, and chemokines, which are shared within the NISS. Indeed, sharing occurs throughout the entire organism, which makes it possible to signal efficiently within the animal or human being.

Additional communication is mediated by innervation and recirculating leukocytes (Figure 6.1). Forward and feedback (or stimulatory and inhibitory) signaling is the rule. Under physiologic conditions (homeostasis), all vital activities and values are maintained at "normal" or physiologic levels. Pathophysiologic conditions (*allostasis*) occur in response to various insults by pathogenic microbes and agents. Pathologic events may also be caused by endogenous abnormalities, defects, and malfunctions. Acute febrile illness leads to the acute phase response (APR), which is initiated by CTKs (e.g., IL-1β, TNF-α, IL-6) released from the *innate immune* (*INIM*) *system*. Under these conditions, hypothalamic corticotropin-releasing hormone (CRH) and vasopressin (VP) stimulate the HPA axis and also induce sympathetic outflow. GCs and catecholamines (CATs) stimulate suppressor/regulatory T lymphocytes (Tsrs) and amplify INIM. Tsr in turn suppresses *ADIM function*. The synthesis of acute-phase proteins is rapidly amplified in the liver, as is the synthesis of natural antibodies in specialized CD4+ B lymphocytes. The CNS, bone marrow, liver, and leukocytes are activated, whereas the function of other organs is reduced and catabolism prevails. *Fever* is a constant symptom of APR. APR is analogous to Selye's general adaptation syndrome. It is an emergency host defense reaction against diverse pathogenic agents, which leads to healing in the overwhelming majority of cases.[21–28]

In the absence of the HPA axis, there is excessive CTK response after immune activation/inflammation, which may have lethal consequences for the host. Even immunization with complete Freund's adjuvant, which contains mycobacterial antigens (acts on TLR-4), would kill Hypox animals (Berczi and Nagy, 1978, unpublished). Further, the sensitivity to bacterial lipopolysaccharide (LPS, also acts on TLR-4) of adrenalectomized (ADRX) mice is elevated by 500–1000 times when compared to normal animals. Tumor necrosis factor (TNF) levels, induced by the same dose of LPS, were

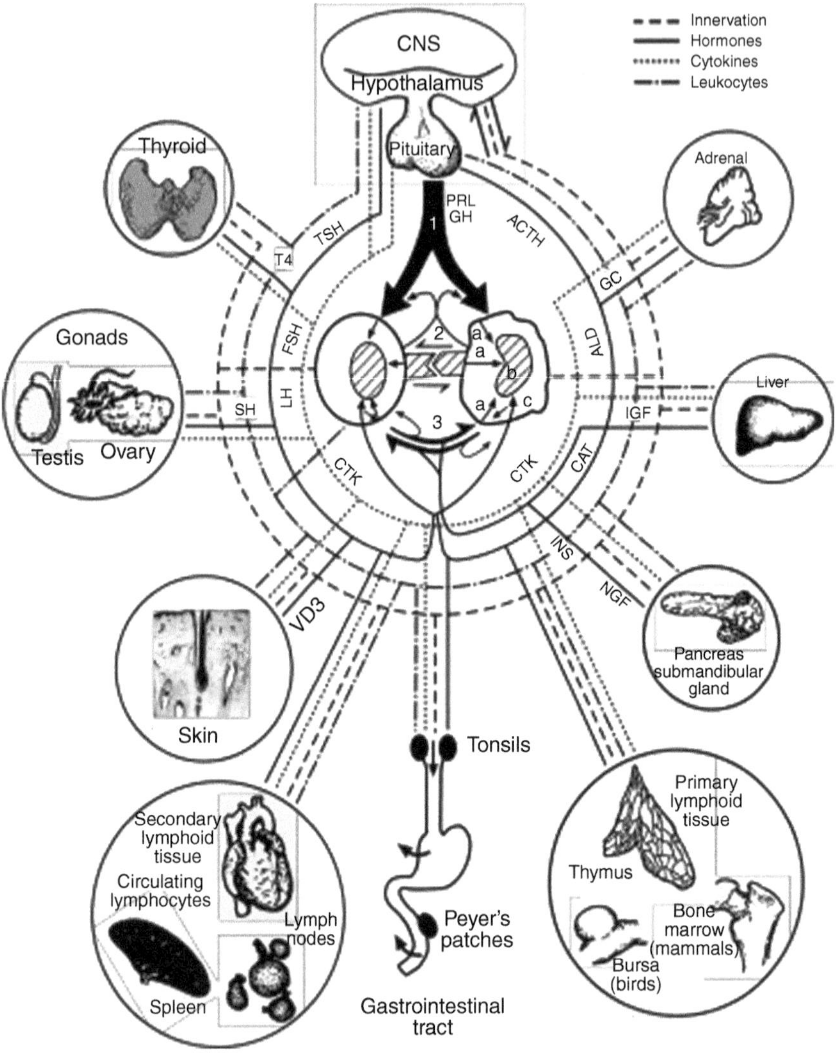

Figure 6.1 ***The Neuroimmune Supersystem.*** ACTH, adrenocorticotropic hormone; ALD, aldosterone; APP, acute phase proteins; CAT, catecholamines; CTK, cytokine; CRH, corticotropin-releasing hormone; GC, glucocorticoids; GH, growth hormone; GMCSF, granulocyte–macrophage colony stimulating factor; IL, interleukin; INIM, innate immunity; INIR, innate immune receptor; INS, insulin; PRL, prolactin; SH, steroid hormone; NGF, nerve growth factor; NPEP, neuropeptide; TNF, tumor necrosis factor; T4, thyroxin; Tsr, suppressor/regulatory T cells; TSH, thyroid stimulating hormone; VD3, vitamin D3; VP, vasopressin.

60 times higher in ADRX mice than in intact controls. Dexamethasone restored the resistance of ADRX mice to LPS.[29] These experiments indicate that stimulation of the INIM system in the absence of the HPA axis results in excessive CTK production, which could kill the host. Further, Korneva and Novikova showed that after immunization, the C-fos gene is expressed in the hypothalamus. C-fos expression indicated that the cells were activated in the hypothalamus after immunization, which was true for several different types of antigens.[30]

These experiments demonstrate that there is a mutual and continuous regulatory interaction between the hypothalamus and the IS during immunization. VP is the dominant hypothalamic regulatory hormone during chronic inflammatory conditions. VP stimulates both PRL and the HPA axis, and thus it is capable of maintaining both the ADIM and the INIM systems in homeostasis and harmony. Indeed, our experiments revealed that VP maintains adaptive immunocompetence. On this basis it has been suggested that recovery from disease is regulated by VP.[28]

1.3.2 Innate–Adaptive Immune Cooperation

It is very well established that monocytes and macrophages, which are related cells, play a fundamental regulatory role in the INIM system and also present and express Fc receptor for the binding of antibodies to their surface, and use such antibodies for better identification of pathogens/noxious agents. Here INIM cells T lymphocytes are present within the ADIM system. The related dendritic cells are regarded as "professional" antigen-presenting cells of the ADIM system. Monocytes recirculate in the blood and macrophages are distributed in tissues throughout the entire host organism. These phagocytic cells recognize infections and other noxious agents through their innate immune receptors (INIRs), engulf the microbe/foreign material, digest (process) it, and present peptides via their major histocompatibility-II (MHC-II) surface antigens to T lymphocytes of the ADIM system. This is the *phagocytic pathway* of antigen presentation. Therefore, monocyte–macrophage activation within the INIM system almost inevitably activates T cells of the ADIM system, so that the entire NISS is mobilized to fight the intruding pathogen/noxious agent. This conclusion is further supported by the fact that *all nucleated cells* are capable of presenting cytosolic antigen via their surface MHC-I to T cells.[20–22]

Once the ADIM system is activated, cytotoxic and helper-effector T cells are produced. As a rule, antibody formation also follows after B lymphocytes present antigen to helper T cells and such T cells stimulate B lymphocytes to form antibodies. Immunoglobulins D (IgD) and M (IgM) are formed as surface immunoglobulins by B cells. After an immune response, IgM is secreted first, and later the same B cell may switch to making IgG, IgA, or IgE, depending on the regulatory environment of the cell. All the different classes of antibodies made by one B cell are specific for a single epitope of the antigen to which the cell was responding.[21,22] IgM, IgG, and IgA are able to fix complement after combining with the antigen. *Complement* is an enzyme system belonging to the INIM system. Moreover, all leukocytes use ADIM antibodies. The complement system has three pathways for activation: by antigen–antibody complexes (ADIM, classical pathway); by microbes (INIM, alternative pathway); and by lectins (INIM).[21,22]

We proposed that the suppressor/regulatory T lymphocyte is a member of the INIM system. The evidence comes from the activation of Tsr by CATs and GCs. These hormones activate INIM cells and suppress ADIM cells.[31] During severe trauma or infectious disease, APR develops whereby ADIM is suppressed and the INIM system takes over command for the highly coordinated and very intensive immunologic battle to save the host organism.[26–28]

The INIM system, which is with us at birth, initiates most of the immune responses against microbes and other noxious agents; it regulates adaptive immunity by antigen presentation and by Tsr cells that suppress ADIM. Therefore, the INIM system is the first to protect the host and it continues to protect the host till the last moment of survival.[27]

Figure 6.1 shows the major systemic neuroimmune regulatory pathways during homeostasis. In the center, two cells are interacting, which is the rule for all tissues and organs, where stromal and parenchymal cells interact. As an example, we use here the phenomenon of antigen presentation to a naive T lymphocyte, which is an ADIM cell, by a macrophage, a cell of the INIM system. Macrophages are phagocytic: they recognize infectious agents and foreign materials via INIRs, engulf the microbe/foreign material, digest it, and present peptides (epitopes) of the antigen on their MHC-II surface molecules to T cells. T lymphocytes proliferate upon exposure to the antigen (phagocytic pathway). Monocytes, dendritic cells, and B lymphocytes are also antigen-presenting cells. Further, all nucleated cells can present cytosolic antigens to T lymphocytes (cytosolic pathway).

The primary immune response is dependent on pituitary PRL and GH. These pituitary hormones regulate cell growth throughout the entire organism according to this principle. Insulin-like growth factor (IGF) 1 acts like a second messenger in most situations. The HPA axis inhibits ADIM and stimulates natural immunity (NATIM). CATs support the immunoregulating activity of the HPA axis.

Type I CTKs produced within the IS will gradually take over the role of PRL/GH, and the secondary response gradually becomes independent of pituitary hormones. Memory T cells, B lymphocytes, and natural killer (NK) cells survive severe illnesses by homeostatic proliferation (HP) and regenerate immune function during healing. The hypothalamic hormone VP stimulates PRL and the HPA axis and thereby regulates the process of healing, and maintains ADIM and NATIM in homeostatic harmony.

Figure 6.1 illustrates that the entire host organism communicates and cooperates with the central regulatory network. Innervation, hormones, CTKs, chemokines, and recirculating leukocytes serve as communication pathways. These pathways of communication are open at all times. No immune response, inflammation, or infection can occur in any tissue/organ in the body without signaling the hypothalamus (paraventricular nucleus, PVN) about such events. Therefore, infection, inflammation, and immune reactions are under continuous surveillance within the NISS. During homeostasis, this system protects us very efficiently, without disturbing any physiologic functions in the body.[106]

1.3.3 T Lymphocytes in the ADIM System

Moreover, all leukocytes express Fc receptors for the binding of antibodies to their surface, and use such antibodies for better identification of pathogens/noxious agents. Here INIM cells use ADIM antibodies.

1.3.4 Innervation, Neurotransmitters, and Neuropeptides

Lymphoid organs, such as the spleen, thymus, and bone marrow, are innervated.[32] This fact indicates that the CNS is in permanent contact with the IS. In addition, neurotransmitters and neuropeptides regulate inflammation and immunity.[33] The mediators involved may stimulate or inhibit immune and inflammatory processes. Substance P, neurokinin A and B (collectively known as *tachykinins*), calcitonin gene-related peptide (CGRP), and vasoactive intestinal peptide (VIP) are proinflammatory mediators capable of eliciting an inflammatory response. These mediators also enhance various immune responses. In contrast, somatostatin and galanin are anti-inflammatory and immunosuppressive mediators.[31,33,34]

α-Adrenergic mechanisms stimulate immunity and inflammation, whereas β-adrenergic mechanisms inhibit inflammation, allergy, and asthma exert an immunosuppressive effect. Cholinergic mechanisms are both anti-inflammatory and immunosuppressive.[35–37]

1.3.5 Hormonal Immunoregulation

Adaptive immunity is under pituitary control. Growth and lactogenic hormones stimulate adaptive immunocompetence and the HPA axis is inhibitory. The secondary antibody response is partially independent of the pituitary. This observation suggests that memory cells are able to operate independent of pituitary hormones.[38] GCs and CATs promote NATIM and inhibit the ADIM system by the stimulation of suppressor/regulatory T cells. Numerous other hormones, CTKs, neurotransmitters, and peptides also modulate immune function.[28]

Innate immunity functions in the absence of the pituitary gland. From our experiments, it appears that VP stimulates the INIM system. The HPA axis exerts an important moderating effect on CTK production. Excessive CTK production can occur after INIM stimulation if the HPA axis is not functional.

1.3.6 Immune Feedback Signals to the Hypothalamus

Healthy people and animals show only trace amounts of CTKs in their serum, which is not sufficient to deliver immune-derived signals to the hypothalamus. Sensory nerves fulfill this function, as they are stimulated locally by CTKs released in immune organs and at other sites where immune/inflammatory reactions occur.

Sensory nerves transmit information about inflammation and immune reactions to the hypothalamic PVN, which is the center regulating ADIM. The vagus was also shown to have similar capabilities. However, recent results indicate that the vagus contains sensory nerve fibers, which serve as immunosensors and transmit information toward the hypothalamus.[33,39]

1.3.7 Cytokine Feedback

CTKs act on the brain; on nerve terminals; on the pituitary, adrenals, gonads, thyroid, epithelium, endothelium, leukocytes, and fibroblasts; and more. In fact, the entire organism uses CTKs to communicate (Figure 6.1). On this basis it is possible to suggest that CTKs are able to activate/suppress the major regulatory circuits in the body, at any level, in case of acute febrile illness. The stress hormones PRL and GH are elevated soon after LPS injection and during the initial phase of stress.[13] Subsequently, it was discovered that during acute stress, both adaptive and INIM functions are significantly boosted within peripheral tissues, which leads to long-lasting immunity. This was termed the *physiologic stress response*.[40] The enhancement of adaptive immunity is fully justified in this case, as both GH and PRL levels are increased; so is the HPA axis, which explains the augmentation of NATIM.

During acute febrile illness, the concentrations of interleukin 1β (IL-1β), TNF-α, IL-6, and GM-CSF, are elevated in the serum, which initiates APR. Later, during the course of the disease, numerous other CTKs will contribute. Additional CTKs that were found to activate the HPA axis were IL-2, -11, -12, and IFN.[13] Gp 130 CTKs, such as ciliary neurotropic factor (CNTF), leukemia inhibitory factor (LIF), oncostatin, IL-11, and cardiolipin, potentiated the activation of the HPA axis by IL-1.[39,41] Leptin plays a role in regulation of the gonadotropic axis.[42]

The CTKs IL-1, IL-2, and IL-6 stimulate the pituitary release of PRL, whereas IFN-γ and endothelin-3 are inhibitory.[43] In rats, administration of exogenous IL-1β stimulated the expression of IL-1β mRNA in the hypothalamus by 99%, but not that of IL-6. IL-1 also significantly elevated the plasma levels of adrenocorticotropic hormone (ACTH), PRL, corticotropin, and cortisol production in the adrenal glands.[44] Rats treated with IFN-γ for 5 days (10 IU/kg body weight) showed a 43% increase in circulating PRL.[45] Two of the most potent CTKs regulating anterior pituitary cell function are *LIF and IL-6,* which belong to the CTK family that uses the common gp130 signal transducer. IL-11 and CNTF exerted a similar stimulatory effect on GH mRNA expression in somatotropic monolayer cell cultures from acromegalic tumors, but these CTKs had no significant influence on GH secretion. CNTF stimulates PRL secretion in lactotropic monolayer cell cultures from patients with prolactinoma. In monolayer cell cultures from normal rat anterior pituitary, IL-11 and CNTF had no significant effect on the release of either GH or PRL, or on GH mRNA. However, when the cells were cultured in aggregates, in which the three-dimensional structure of the cells is reconstituted, both CTKs, in doses at which they had no effect on monolayer cultures, significantly stimulated both PRL and GH secretion.[46]

The presence of L-NAME (1 mM), an inhibitor of NO synthase (NOS), in the incubation medium significantly blunted the inhibition of PRL release produced by TNF-α (50 mg/mL) in female rats. TNF-α increased nitrite release to the incubation medium. The activity of NOS was significantly enhanced when anterior pituitary cells were incubated with TNF-α for 8 h or more. Also, TNF-α-induced iNOS gene expression in anterior pituitary cells.[47] Stromal cell-derived factor-1 (SDF1), via its receptor CXCR4, stimulated the proliferation of the pituitary adenoma cell line GH4C1, and released both PRL and GH through a complex network of intracellular signals.[48]

C3a receptors are expressed in pituitary hormone-secreting and non-hormone-secreting (folliculostellate) cells. Both C3a and C3adesArg (a noninflammatory metabolite) stimulate pituitary cell cultures to release PRL, GH, and adrenocorticotropin. Serum levels of these hormones, together with adrenal corticosterone increase dose dependency with recombinant C3a and C3adesArg administration *in vivo*.[49]

Erythropoetin administration to patients with amyotrophic lateral sclerosis caused a significant reduction of serum PRL levels, maximum 60 min after administration.[50]

GCs and CATs stimulate IL-4, IL-10, and TGF-β, which exert negative feedback on excess CTK production during APR.[51] These CTKs are produced by suppressor/regulatory T lymphocytes.[28] Several mechanisms have been proposed for conveying the CTK signal across the blood–brain barrier (BBB). Once the signal is transmitted across the BBB, the signal travels through neural pathways as follows: area postrema → nucleus tractus solitarius → ventrolateral medulla, → PVN, which initiates HPA axis responses. Descending pathway: PVN → brainstem cell groups, → thoracic spinal cord preganglionic neurons → via sympathetic projections to the end organs such as the thymus and spleen.[52]

Several effects of proinflammatory CTKs exerted in a "healthy" brain are amplified in the CNS by the following mechanisms: (1) peripheral CTKs, such as IL-1, have the capacity to elicit their own synthesis in the brain; and (2) a sustained increase in neuronal activity also induces production of physiologically significant amounts of these mediators. The brain-borne IL-1 and IL-6 fulfill physiologic roles, when their production is not the result of pathologic events in the CNS. These CTKs stimulate the HPA axis and they are involved in physiologic brain mechanisms such as synaptic plasticity, memory formation, and the control of glucose homeostasis.[53] CTKs are now also being investigated for their synaptic and inflammatory action in the CNS. These proteins and their receptors can be synthesized in the brain by glial and neuronal cells and contribute to two main types of action: modulation of neuronal excitability and local inflammatory responses.[54]

1.3.8 Neurogenic Inflammation: Neurokinins, Tachykinins, Mast Cells

There is much evidence for the regulation of inflammation and immunity by neuropeptides. Chemosensitive afferent nerves expressing the capsaicin/TRPV1 receptor play important roles in the initiation and modulation of vascular, inflammatory, and immune reactions. These nerves exert their efferent/local regulatory functions primarily via the release of VIP, CGRP, and SP from their terminals upon antidromic or orthodromic stimulation. The CTKs induced in the neural microenvironment by tissue injury or inflammation promote the release of proinflammatory sensory neuropeptides and lead to augmentation of the inflammatory response. In turn, the release of anti-inflammatory peptides from sensory nerves or from inflammatory cells results in inhibition of the inflammatory vascular reactions.[33]

Mast cells are innervated and thus serve as sensory organs for the CNS and function as effector cells in neurogenic inflammation. The complement split products C3a and C5a (*anaphylatoxins*) cause mast cell discharge and thus signal the brain about complement activation anywhere in the body. Similarly, the inflammatory mediator *bradykinin* signals the brain by the sensory nerve pathway.[31,34]

1.3.9 The Brain as an Immunocompetent Member of the Innate Immune System

Toll-like receptors of innate immunity are expressed by neurons and also by their dendrites.[55] This discovery implies that the nervous system expresses INIRs, is capable of sensing directly the penetration of infectious and noxious agents into tissues, and can respond immediately by inducing neurogenic inflammation and/or by the activation of other immune compartments. Thus, the CNS is an immunocompetent organ, as it is capable of recognizing an antigen directly and responding to it in a polyspecific manner characteristic of innate immunity.[31]

1.4 The Neuroimmune Supersystem

In addition to the CNS, all leukocytes express the innate immune receptor Toll-like receptor (TLR)-1.[55] TLR is expressed in the pituitary gland,[56] adrenal gland,[57] liver,[58] mucosal epithelial cells,[59] endothelial cells,[60] vascular smooth muscle,[61] and also in the cornea.[62] These observations indicate that the entire body participates in INIM reactions. The CNS is capable of directly sensing infectious agents through TLR, and possibly through other INIRs, and can react instantaneously by causing inflammation and mobilizing immune defense mechanisms. Similarly, the pituitary produces POMC in response to LPS (TLR4 is involved), mucosal epithelial TLR participates in inflammation and responds to pathogens, corneal TLR was found to fight infection, and endothelial TLR was observed to play important roles in homeostasis of the heart.[55–62] Therefore, in addition to participating in NATIM, TLRs fulfill important physiologic functions. This is true for

CTKs and also for cellular elements of the IS, whether or not natural or ADIM cells are considered.[54]

At this stage, we may conclude that the entire host participates in APR to noxious stimuli. These stimuli could be immunogenic, but need not be; they may be noxious agents that cause tissue injury, but need not be: it is enough if the host simply senses or anticipates danger. INIM mechanisms have a tremendous capacity to defend the host from physical, chemical, and biological agents that may be harmful to the host. The receptors involved sense antigen presence, inflammation, and injury (e.g., CTKs, nerves), and also receive input from the CNS (e.g., conditioning). The *ADIM system* is largely, but not entirely, concerned with immunogenic agents. Here antigenic receptors dominate. This system is also subject to conditioning. Therefore, the activation of the INIM system, with or without adaptive immunity, is responsible for the profound host resistance to diverse noxious agents. The NISS is capable of mobilizing the entire organism in the interest of host defense in emergency situations (through APR). The same NISS performs important physiologic functions, as is becoming apparent with further research.[53,54]

1.4.1 Hierarchy in Regulation

The regulatory portion of the NISS consists of *hierarchical regulatory circuits*, which are superimposed on each other. For example, regulatory signals from a higher circuit (such as the CNS) are capable of dominating the signals from lower circuits (e.g., pituitary). As we know, pituitary hormones control hormones secreted by the adrenals, gonads, thyroid, liver, and other tissues that secrete IGF. CTKs, which are tissue hormones, represent another regulatory circuit. Our experiments indicate that if a higher regulatory circuit fails, a lower circuit will gain dominance and take precedence over lesser signals that still exist in the system; this way, *INIM function is maintained continuously under any circumstance.*[63] This means that the INIM system continues CTK production during acute illness/injury under any conditions, and these CTKs will signal any regulatory circuits of the host organism that remain active.

Hence, the INIM system never stops protecting the host organism.

1.4.2 Hierarchy of Cellular Receptors

Hierarchy may also be observed among cellular receptors: for example, cell surface receptors operate through long signal-transmission pathways. In contrast, steroid–thyroid hormone and vitamin D receptors are nuclear regulatory proteins. Nuclear receptors are very powerful and are capable of overruling membrane receptors. This is the reason for the tremendous regulatory influence of the HPA axis. Some peptide hormones were proposed to release nuclear regulatory peptides (e.g., PRL) during intracellular processing, but the existence of this phenomenon remains controversial. At the cellular level, adhesion molecules that deliver cell-to-cell signals are dominant for delivering signals that are specific for the position of a single cell in the body. Such positional signaling consists of adhesion and CTK signals that decide the function of a given cell in the tissues.[66] Each cell is bombarded by myriad signals at any time. Activated cells cap the occupied receptors to one pole of the cell where receptor interaction will take place. Positive signals to the cell are transmitted by phosphorylation of receptors by protein kinases, whereas negative signaling involves rapid dephosphorylation of the activated (phosphorylated) sites by phosphatases (Ship). The balance of positive and negative signals decides if the cell will be activated or remain inactive.[67]

1.5 Immunologic Memory

We observed that the secondary antibody response was partially independent of the pituitary. In Hypox rats we obtained a secondary antibody response that was of similar titer to the primary response.[38] This observation shows that memory cells of the ADIM system do not require pituitary hormones for survival and function; they are autonomous. However, their ability to recruit naive lymphocytes to expand the secondary response was impaired. This observation also showed that memory cells survived the severe adaptive immunodeficiency present in Hypox animals. Others also observed that in immunodeficient (lymphopenic) hosts, memory T cells not only survive, but also undergo antigen and MHC-independent HP, the biologic significance of which, is to preserve and regenerate the immunocompetence of the host after severe, debilitating diseases.

1.5.1 Homeostatic Proliferation of Lymphocytes

In lymphopenic/immunodeficient hosts, *naive and memory CD4 and CD8 T cells, naive B cells, and NK cells undergo HP.* This proliferation is MHC and antigen independent, and B cells share many of the inductive and regulatory characteristics established for naive and memory T cells and NK cells.[68–70]

Naive T cells can be induced to undergo HP of variable speed with a few members of the common γ-chain (CD132) family of CTKs. The speed of proliferation depends on the levels of the particular CTK involved.[71]

IL-7 was required for homeostatic expansion of naive CD8+ and CD4+ T cells in lymphopenic hosts and for CD8+ T-cell survival in normal hosts.[72] Unlike naive T cells, HP memory T cells were largely MHC independent and memory CD8+ cells could utilize either IL-7 or IL-15 to undergo HP. In the absence of both IL-7 and IL-15, HP failed to occur. HP of memory CD4+ cells is independent of IL-7 and IL-15 (and also of IL-4).[73] Although HP memory cells responded directly to IL-7 and IL-15, naive T cells required costimulation by dendritic cell-derived CTKs, and selectively respond to IL-4.[74] All human CD8+ T-cell subsets had the ability to respond to IL-15, which suggests a generic influence of this CTK on CD8+ T-cell homeostasis in humans.[75] IL-15 was important for sustained CD8 T-cell proliferation and accumulation in a lymphopenic setting, as revealed by truncated proliferation in IL-15 hosts. At the same time, IL-12 enhanced HP by acting directly on the CD8 T cells, independent of IL-15, suggesting that there are distinct pathways by which CTKs can regulate HP.[76] A novel form of proliferation occurred when naive T cells encountered raised levels of IL-2 and IL-15 *in vivo*. In this situation, CD8 T cells underwent massive expansion and rapid differentiation into effector cells, thus closely resembling the T-cell response to foreign antigens. However, the responses induced by IL-2/IL-15 were not seen in MHC-deficient hosts, implying that the responses were driven by self-ligands. Hence, HP of naive T cells can be either slow or fast, with the quality of the response to self being dictated by the particular CTK (IL-7 versus IL-2/IL-15) concerned.[77] In murine allogeneic bone marrow transplantation (BMT) models, two populations of mature donor T cells could be distinguished: (1) alloreactive T cells with decreased expression of CD127 (IL-7 receptor α chain) and (2) nonalloreactive T cells, which express CD127 and undergo HP. IL-7 administration increased the HP of nonalloreactive T cells, but had no effect on alloreactive T cells or the development of graft-versus-host disease.[78] There is a prominent, nonredundant role for IL-7 in supporting basal HP of CD8T(M). We propose that homeostatic control of antiviral CD4 and CD8 T-cell memory is fundamentally similar and characterized by quantitative, rather than qualitative, differences.[79] IL-7 is a CTK produced predominantly by stromal cells of the thymus and bone marrow and is essential for lymphopoiesis. IL-7 is of particular importance in lymphopenic conditions. Its expression is upregulated in a number of lymphopenic conditions, including marrow ablation prior to BMT, marrow suppression following chemotherapy, and human immunodeficiency virus (HIV) infection. Plasma IL-7 levels inversely correlate with CD4+ T-cell counts in these conditions. Animal models suggest that IL-7 improves immune reconstitution through increasing thymic output and, perhaps more importantly, through antigen-independent homeostatic-driven proliferation in the periphery.[80]

Lymphopenia-induced proliferation depends on low-affinity MHC/self-peptide complexes and on IL-7 T cells proliferating in lymphopenic hosts. These cells do not exhibit a unique gene-expression profile, but instead rely on "traditional" signals for this antigen-independent proliferation, which ultimately results in differentiation to "authentic" memory cells.[81] Memory CD8 T cells retain the ability to respond to dendritic cell-mediated stimulation after adoptive transfer into either TAP (MHC class I-deficient) or wild-type mice. Surprisingly, naive CD8T cells, which fail to undergo HP and erode in number in the absence of MHC class I, also retain the ability to respond to dendritic cell-mediated antigenic stimulation for at least 1 week after transfer into TAP mice.[82] Resting memory CD4 cells are dependent on signals from contact with IL-7 and IL-15, but not MHC class II, for their survival and intermittent HP.[83]

Thymus and oncostatin M (OM)-dependent HP extrathymic pathways of T cells show how the division of labor between primary and secondary lymphoid organs influences the repertoire and homeostasis of T lymphocytes.[84] CD8 T cells express thymic stromal lymphopoietin (TSLP) receptors, and TSLP activates both STAT5 and Akt and induces Bcl-2 in these cells. Correspondingly, TSLP increases CD8 T cell survival *in vitro* as well as in wild-type and T-depleted mice *in vivo*, without altering the HP of these cells. Moreover, TSLP can maintain CD8 T cells even in the absence of IL-7.[85]

Recent work has shown that two members of the common γ chain (γ c) family of CTKs, IL-7 and IL-15, govern homeostasis of memory T cells. It appears that the two types of memory cells do not display identical homeostatic requirements. For antigen-specific memory CD8+ T cells, IL-7 is primarily important for survival, whereas IL-15 is crucial for their background proliferation. For memory CD4+ T cells, IL-7 has an important role, whereas the influence of IL-15 is still unclear.[86] The activation of STAT5 is the primary mechanism underlying both IL-7- and IL-15-dependent HP of naive and memory CD8 T cells and IL-2-dependent development of CD4, CD25 regulatory T cells.[87] There are distinct functions for IL-7 and IL-15 in T lymphocyte development and homeostasis, and stringent regulation of these processes by SOCS1.[88]

Dividing HP memory T cells were present in both lymphoid and non-lymphoid tissues. However, the bone marrow was the preferred site for proliferation and contained a major pool of the most actively dividing memory

CD8 T cells. Adoptive transfer studies indicated that memory cells migrated through the bone marrow and divided there preferentially.[89] Homeostatic expansion permits T cells to re-enter the thymus and deliver antigen in a tolerogenic fashion.[90]

In sublethally irradiated lymphopenic mice, HP T lymphocytes were found to arise in neonatal lymphopenia (CD8+ T cells),[91] which inhibited melanoma and colon carcinoma growth; this protection was effective even against established tumors (CD8 T cells making IFN).[92] HP T cells generated autoimmunity (IL-21-regulated T cell turnover).[93] HP T cells appeared 14 days after burn injury. These cells were CD8,CD44 IL-7R T cells, which induced enhanced allogeneic skin graft rejection in unburned recipient mice. Treatment with GC abolished both the late homeostatic accumulation of CD8 memory-like T cells and enhanced the proinflammatory CD8 T cell response, but not the late enhanced CD8 anti-inflammatory response.[94]

HP T cells transformed chronic rejection to acute rejection of a single MHC class II-mismatched kidney allograft. Such T cells consistently caused reliable rejection even when bona fide memory T cells could not. These functional changes are long-lasting and not restricted to the acute phase of HP.[95] HP cells obtained from T-cell knockout (KO) mice mounted relatively normal acute CD8 T-cell responses to lymphocytic choriomeningitis virus, but with altered T-cell receptor (TCR) repertoires, and they became functional memory cells capable of recall responses.[96]

Current evidence implicates HP in autoimmune diseases and transplant rejection, and suggests that it may represent a barrier to tolerance in protocols that use T-cell depletion.[97]

Clonal competition is a component of homeostasis that may contribute to selection of the peripheral T-cell repertoire.[98] Regulatory T cells (Tregs) play a major role in the control of HP.[99] HP memory CD8 T cells controlled bacterial infection as effectively as "true" memory CD8 T cells, but their protective capacity required the presence of CD4 T cells during HP. The necessity for CD4 help was overcome, however, if the HP memory CD8 T cells lacked expression of TNF-related apoptosis-inducing ligand (TRAIL; also called Apo-2L). Thus, like conventional CD8 memory T cells, the protective function of HP memory CD8 T cells shows dependence on CD4 T-cell help.[100]

Similar factors are required for production of protective memory CD8 T cells via both homeostatic and conventional pathways. Induction of protective HP memory CD8 T cells requires CD4 T-cell "help," which is antigen nonspecific and requires CD40L–CD40 interactions with host cells. The functional competence of HP memory CD8 T cells also requires release of endogenous bacterial components (which follows irradiation-induced lymphopenia), potentially mimicking the role of adjuvants in conventional immune responses. Lymphopenic environments lacking these key factors support similar CD8 T-cell HP and the acquisition of memory phenotype, yet the HP memory cells generated are defective in pathogen elimination.[101] HP memory T cells provide protection without compromising the true-memory population. Differences in HP and true-memory T cells may reveal the basis of competition for limited resources within the memory T-cell compartment.[102]

The significance of this phenomenon is that memory T lymphocytes are able to survive major disasters and bring back immune function once the crisis is over. During APR, the ADIM system is suppressed by Tsr and GCs, but not destroyed. The possibility of recovery is present and ADIM function recovers after acute illness. We proposed that healing is under hypothalamic regulation, with VP as the principal regulator.[28]

1.6 Autonomy and Redundancy

Immune reactions may be induced *in vitro*, which argues for the autonomy of immune function. Indeed, the fully differentiated cells of the INIM system can function under diverse conditions with little regulatory input from the host, even when bona fide memory T cells could not.

Current evidence implicates HP in autoimmune diseases and transplant rejection, and suggests that it may represent a barrier to tolerance in protocols that use T-cell depletion.[97]

Clonal competition is a component of homeostasis that may contribute to selection of the peripheral T-cell repertoire.[98] Regulatory T cells (Tregs) play a major role in the control of HP.[99] HP memory CD8 T cells controlled bacterial infection as effectively as "true" memory CD8 T cells, but their protective capacity required the presence of CD4 T cells during HP. The necessity for CD4 help was overcome, however, if the HP memory CD8 T cells lacked expression of TRAIL. Thus, like conventional CD8 memory T cells, the protective function of HP memory CD8 T cells shows dependence on CD4 T-cell help.[100]

1.7 The Neuroimmune Supersystem Protects Us for a Lifetime

All of us suffer from febrile illness on numerous occasions and recover. This indicates that our INIM system is a very important and efficient defense mechanism. This wonderful defense system is one that we have for life: it is always there for us, never lost, and fights for us until the last moment of our existence.[103] The ADIM system does not decline, either. Rather, it undergoes age-related changes: for example, in older individuals the system is maintained largely by memory cells, so the thymus has little to do; hence it involutes. However, the thymus may be reactivated by severe immune depletion, which indicates that the involuted thymus is still a functional organ that is ready to act when necessary.[28,104] The loss of adaptive immunity is not obligatory with aging; it is well preserved in the case of successful aging. Immunosuppression by Tsr during acute illness is reversible, and it occurs during healing.[28] Moreover, the homeostatic survival of T, B, and NK cells represents a very important and fundamental mechanism of regeneration of immunocompetence during recovery from severe illness. During acute illness, the ADIM system is suppressed due to amplification of the NATIM system by CTKs, GCs, and CATs. This occurs in the majority of individuals who do not age successfully and succumb to various illnesses that activate the NATIM and suppress ADIM. The prolonged activation of NATIM may lead to extreme weight loss (*cachexia*) and eventually death.[26]

The concept of age-related immune deficiency led to doom-and-gloom prospects concerning aging. However, successful aging, without disease, has been observed. Therefore, at this time it is clear that disease-free survival is possible until we reach the natural limit of our lifespan.[105]

REFERENCES

1. Szentivanyi A, Berczi I, Nyanteh H, Goldman A. The immune-neuroendocrine circuitry. History and progress. In: Berczi I, Szentivanyi A, series editors. *Neuroimmune biology*, vol. 3. Amsterdam: Elsevier; 2003. p. 7–14.
2. Korneva EA. On the history of immunophysiology: first steps and main trends. In: Berczi I, editor. *New insights to neuroimmune biology*. Amsterdam: Elsevier; 2010. p. 33–50.
3. Filipp G, Szentivanyi A, Mess B. Anaphylaxis and nervous system. *Acta Med Hung* 1952;**2**:163–73.
4. Szentivanyi A, Filipp G. Anaphylaxis and the nervous system. Part II. *Ann Allergy* 1958;**16**:143–51.
5. Filipp G, Szentivanyi A. Anaphylaxis and the nervous system. Part III. *Allergy* 1958;**16**:306–11.
6. Szentivanyi A, Szekely J. Effect of injury to, and electrical stimulation of, hypothalamic areas on anaphylactic and histamine shock of the guinea pig. A preliminary report. *Ann Allergy* 1956;**14**:259–61.
7. Korneva EA, Khai LM. Effect of destruction of hypothalamic areas on immunogenesis. *Fed Proc* 1964;**23**:T88.
8. Selye H. A syndrome produced by diverse nocuous agents. *Nature (Lond)* 1936;**138**:32.
9. Selye H. Thymus and adrenals in the response of the organism to injuries and intoxication. *Br J Exp Pathol* 1936;**17**:234–48.
10. Selye H. The general adaptation syndrome and the diseases of adaptation. *J Clin Endocrinol* 1946;**6**:117–230.

11. Nagy E, Berczi I. Immunodeficiency in hypophysectomized rats. *Acta Endocrinol* 1978;**89**:530–7.
12. Nagy E, Berczi I, Wren GE, Asa SL, Kovacs K. Immunomodulation by bromocriptine. *Immunopharmacology* 1983;**6**:231–43.
13. Berczi I, editor. Immunoregulation by the pituitary gland. *Pituitary function and immunity. Monograph*. Boca Raton, FL: CRC Press, Inc; 1986 p. 227–240.
14. Besedovsky HO, del Rey A. Processing of cytokine signals at CNS levels: relevance for immune–HPA axis interactions. In: del Rey A, Chrousos G, Besedovsky H, editors. *The hypothalamus–pituitary–adrenal axis*. Berczi I, Szentivanyi A, series editors. *Neuroimmune biology*, vol. 7. Amsterdam: Elsevier; 2008. p. 227–240.
15. Rettori V, Jurcovicova J, McCann SM. Central action of interleukin-1 in altering the release of TSH, growth hormone, and prolactin in the male rat. *J Neurosci Res* 1987;**18**:179–83.
16. Ader R. Letter: behaviorally conditioned immunosuppression. *Psychosom Med* 1974;**36**:183–4.
17. Gorczynski RM, Kennedy M. Associative learning and regulation of immune responses. *Prog Neuropsychopharmacol Biol Psychiatry* 1984;**8**(4–6):593–600.
18. Gorczynski LY, Gorczynski RM. Conditioning of immune reactions. In: Arnason BG, editor. *The brain and host defense*. Berczi I, Szentivanyi A, series editors. *Neuroimmune biology*, vol. 9. Amsterdam: Elsevier; 2010. p. 131–144.
19. Berczi I, Gorczynski R, editors. *New foundation of biology*. In: Berczi I, Szentivanyi A, series editors. *Neuroimmune biology*, vol. 1. Amsterdam: Elsevier; 2001.
20. Berczi I. Integration and regulation of higher organisms by the neuroimmune supersystem. *Int J Integr Biol* 2007;**1**:216–31.
21. Berczi I. Neuroendocrine defence in endotoxin shock. A review. *Acta Microbiol Hung* 1993;**40**:265–302.
22. Berczi I. Neurohormonal host defence in endotoxin shock. *Ann NY Acad Sci* 1998;**840**:787–802.
23. Berczi I. The stress concept and neuroimmunoregulation in modern biology. *Ann NY Acad Sci* 1998;**851**:3–12.
24. Berczi I, Chow DA, Sabbadini ER. Neuroimmunoregulation and natural immunity. *Domest Anim Endocrinol* 1998;**15**:273–81.
25. Berczi I, Bertok L, Chow DA. Natural immunity and neuroimmune host defence. *Ann NY Acad Sci* 2001;**917**:248–57.
26. Berczi I, Szentivanyi A. The acute phase response. In: Berczi I, Szentivanyi A, editors. *The immune–neuroendocrine circuitry. History and progress. Neuroimmmune biology*, vol. 3. Amsterdam: Elsevier; 2003. p. 463–494.
27. Berczi I, Quintanar-Stephano A, Kovacs K. Immunoconversion in the acute phase response. In: Plotnikoff NP, Faith RE, Murgo AJ, Good RA, editors. *Cytokines, stress and immunity*. Boca Raton, FL, USA: CRC Press, Taylor & Francis Group; 2006. p. 215–54.
28. Berczi I, Quintanar-Stephano A, Kovacs K. Neuroimmune regulation in immunocompetence, acute illness and healing. *Ann NY Acad Sci* 2009;**1153**:220–39.
29. Ramachandra RN, Sehon AH, Berczi I. Neuro-hormonal host defence in endotoxin shock. *Brain Behav Immun* 1992;**6**:157–69.
30. Korneva EA, Novikova NS. Brain responses to antigenic challenges. In: Arnason BG, editor. *The brain and host defense*. Berczi I, Szentivanyi A, series editors. *Neuroimmune biology*, vol. 9. Amsterdam: Elsevier; 2010. p. 113–121.
31. Berczi I. Antigenic recognition by the brain: the brain as an immunocompetent organ. In: Berczi I, editor. *New insights to neuroimmune biology*. Amsterdam: Elsevier; 2010. p. 145–59 Elsevier Insights, www.amazon.com.
32. Williams JM, Peterson RG, Shea PA, Schmedtje JF, Bauer DC, Felten DL. Sympathetic innervation of murine thymus and spleen: evidence for a functional link between the nervous and immune systems. *Brain Res Bull* 1981;**6**(1):83–94.

33. Jancso G, Katona M, Horvath V, Santa P, Nagy J. Sensory nerves as modulators of cutaneous inflammatory reactions in health and disease. In: Jancso G, editor. *Neurogenic inflammation in health and disease*. Berczi I, Szentivanyi A, series editors. *Neuroimmune biology*, vol. 8. Amsterdam: Elsevier; 2008. p. 3–38.
34. McKay DM, Bienenstock J. The interaction between mast cells and nerves in the gastrointestinal tract. *Immunol Today* 1994;**15**:533–8.
35. Szentivanyi A. The beta adrenergic theory of the atopic abnormality in bronchial asthma. *J Allergy* 1968;**42**:203–32.
36. Blatteis CM. Fever as a host defense mechanism. In: Arnason BG, editor. *The brain and host defense*. Berczi I, Szentivanyi A, series editors. *Neuroimmune biology*, vol. 9. Amsterdam: Elsevier; 2010. p. 213–235.
37. Pavlov VA, Ulloa L, Metz CN. Infection and sepsis. In: Arnason BG, editor. *The brain and host defense*. Berczi I, Szentivanyi A, series editors. *Neuroimmune biology*, vol. 9, Amsterdam: Elsevier; 2010. p. 309–320.
38. Berczi I, Nagy E, Kovacs K, Horvath E. Regulation of humoral immunity in rats by pituitary hormones. *Acta Endocrinol* 1981;**98**:506–13.
39. Benigni F, Fantuzzi G, Sacco S, Sironi M, Pozzi P, Dinarello CA, et al. Six different cytokines that share GP130 as a receptor subunit, induce serum amyloid A and potentiate the induction of interleukin-6 and the activation of the hypothalamus–pituitary–adrenal axis by interleukin-1. *Blood* 1996;**87**(5):1851–4.
40. Dhabhar FS. Enhancing versus suppressive effects of stress on immune function: implications for immunoprotection and immunopathology. In: del Rey A, Chrousos G, Besedovsky H, editors. *The hypothalamus–pituitary–adrenal axis*. Berczi I, Szentivanyi A, series editors. *Nuroimmune biology*, vol. 7. Amsterdam: Elsevier; 2008. p. 207–224.
41. Chesnokova V, Melmed S. Minireview: neuro-immuno–endocrine modulation of the hypothalamic–pituitary–adrenal (HPA) axis by gp130 signaling molecules. *Endocrinology* 2002;**143**:1571–4.
42. McCann SM, Kimura M, Walczewska A, Karanth S, Rettori V, Yu WH. Hypothalamic control of FSH and LH by FSH-RF, LHRH, cytokines, leptin and nitric oxide. *Neuroimmunomodulation* 1998;**5**:193–202.
43. Chikanza IC. Prolactin and neuroimmunomodulation: *in vitro* and *in vivo* observations. *Ann NY Acad Sci* 1999;**22**:119–30.
44. Skurlova M, Stofkova A, Jurcovicova J. Exogenous IL-1beta induces its own expression, but not that of IL-6 in the hypothalamus and activates HPA axis and prolactin release. *Endocr Regul* 2006;**40**:125–8.
45. Cano P, Cardinali DP, Jimenez V, Alvarez MP, Cutrera RA, Esquifino AI. Effect of interferon-gamma treatment on 24-h variations in plasma ACTH, growth hormone, prolactin, luteinizing hormone and follicle-stimulating hormone of male rats. *Neuroimmunomodulation* 2005;**12**(3):146–51.
46. Perez Castro C, Carbia Nagashima A, Páez Pereda M, Goldberg V, Chervin A, Carrizo G, et al. Effects of the gp130 cytokines ciliary neurotropic factor (CNTF) and interleukin-11 on pituitary cells: CNTF receptors on human pituitary adenomas and stimulation of prolactin and GH secretion in normal rat anterior pituitary aggregate cultures. *J Endocrinol* 2001;**169**:539–47.
47. Theas S, De Laurentiis A, Candolfi M, López SL, Carrasco AE, Zaldivar V, et al. Nitric oxide mediates the inhibitory effect of tumor necrosis factor-alpha on prolactin release. *Neuroendocrinology* 2001;**74**:82–6.
48. Barbieri F, Bajetto A, Porcile C, Pattarozzi A, Schettini G, Florio T. Role of stromal cell-derived factor 1 (SDF1/CXCL12) in regulating anterior pituitary function. *J Mol Endocrinol* 2007;**38**:383–9.
49. Francis K, Lewis BM, Akatsu H, Monk PN, Cain SA, Scanlon MF, et al. Complement C3a receptors in the pituitary gland: a novel pathway by which an innate

immune molecule releases hormones involved in the control of inflammation. *FASEB J* 2003;**17**:2266–8.
50. Markianos M, Kosmidis ML, Sfagos C. Reductions in plasma prolactin during acute erythropoietin administration. *Neuroendocrinol Lett* 2006;**27**:355–8.
51. Elenkov IJ, Chrousos GP. Stress hormones, proinflammatory and antiinflammatory cytokines, and autoimmunity. *Ann NY Acad Sci* 2002;**966**:290–303.
52. Buller KM. Central pathways of immunoregulation. In: Arnason BG, editor. *The brain and host defense.* Berczi I, Szentivanyi A, series editors. *Neuroimmune biology*, vol. 9. Amsterdam: Elsevier; 2010. p. 101–111.
53. Besedovsky HO, del Rey A. The cytokine–HPA axis feed-back circuit. *Z Rheumatol* 2000;**59**(Suppl. 2):II/26–30.
54. Conti B, Tabarean I, Sanchez-Alavez M, Davis C, Brownell S, Behrens M, et al. Cytokine receptors in the brain. In: Phelps C, Korneva EA, Berczi I, Szentivanyi A. *Cytokines and the brain. Neuroimmune biology*, vol. 6. Amsterdam: Elsevier; 2008; p. 21–38.
55. Flo T, Aderen A. Pathogen recognition by toll-like receptors. In: Bertok L, Chow DA, editors. *Natural immunity.* Berczi I, Szentivanyi A, series editors. *Neuroimmne biology*, vol. 5. Amsterdam: Elsevier; 2005. p. 167–182.
56. Iwasaki Y, Taguchi T, Nishiyama M, Asai M, Yoshida M, Kambayashi M, et al. Lipopolysaccharide stimulates proopiomelanocortin gene expression in AtT20 corticotroph cells. *Endocr J* 2008;**55**:285–90.
57. Bornstein SR, Ziegler CG, Krug AW, Kanczkowski W, Rettori V, McCann SM, et al. The role of toll-like receptors in the immune-adrenal crosstalk. *Ann NY Acad Sci* 2006;**1088**:307–18.
58. Katsargyris A, Klonaris C, Alexandrou A, Giakoustidis AE, Vasileiou I, Theocharis S. Toll like receptors in liver ischemia reperfusion injury: a novel target for therapeutic modulation? *Expert Opin Ther Targets* 2009;**13**:427–42.
59. Shang L, Fukata M, Thirunarayanan N, Martin AP, Arnaboldi P, Maussang D, et al. Toll like receptor signaling in small intestinal epithelium promotes B-cell recruitment and IgA production in lamina propria. *Gastroenterology* 2008;**135**:529–38.
60. Curtiss LK, Tobias PS. Emerging role of toll-like receptors in atherosclerosis. *J Lipid Res* 2009;**50**(Suppl.):S340–5.
61. Heo SK, Yun HJ, Noh EK, Park WH, Park SD. LPS induces inflammatory responses in human aortic vascular smooth muscle cells via toll-like receptor 4 expression and nitric oxide production. *Immunol Lett* 2008;**120**:57–64.
62. Pearlman E, Johnson A, Adhikary G, Sun Y, Chinnery HR, Fox T, et al. Toll-like receptors at the ocular surface. *Ocul Surf* 2008;**6**:108–16.
63. Quintanar-Stephano A, Villalobos C, Gonzales E, Organista A, Tinajero M, Chavira R, et al. Role of the neurohypophysis on spleen cytokine MRNA expression and macrophage number in AIA. In: Abstract No. A3, *Proceedings of the seventh INSNIM congress.* Rio de Janeiro; 2008. p. 34.
64. Bertk L. New prospect for the enhancement of natural immunity. In: Bertk L, Chow DA, editors. *Natural immunity.* Berczi I, Szentivanyi A, series editors. *Neuroimmune biology*, vol. 5. Amsterdam: Elsevier; 2005. p. 289–307.
65. Elenkov IJ, Kovács K, Kiss J, Bertók L, Vizi ES. Lipopolysaccharide is able to bypass corticotrophin-releasing factor in affecting plasma ACTH and corticosterone levels: evidence from rats with lesions of the paraventricular nucleus. *J Endocrinol* 1992;**133**:231–6.
66. Berczi I. The role of the growth and lactogenic hormone family in immune function. *Neuroimmunomodulation* 1994;**1**:201–16.
67. Renard V, Cambiaggi A, Vély F, Bléry M, Olcese L, Olivero S, et al. Transduction of cytotoxic signals in natural killer cells: a general model of fine tuning between activatory and inhibitory pathways in lymphocytes. *Immunol Rev* 1997;**155**:205–21.

68. Woodland RT, Schmidt MR. Homeostatic proliferation of B cells. *Semin Immunol* 2005;**17**(3):209–17.
69. Murali-Krishna K, Lau LL, Sambhara S, Lemonnier F, Altman J, Ahmed R. Persistence of memory CD8 T cells in MHC class I-deficient mice. *Science* 1999;**286**:1377–81.
70. Murali-Krishna K, Ahmed R. Cutting edge: naive T cells masquerading as memory cells. *J Immunol* 2000;**16**:1733–7.
71. Ramsey C, Rubinstein MP, Kim DM, Cho JH, Sprent J, Surh CD. The lymphopenic environment of CD132 (common gamma-chain)-deficient hosts elicits rapid homeostatic proliferation of naive T cells via IL-15. *J Immunol* 2008;**180**:5320–6.
72. Schluns KS, Kieper WC, Jameson SC, Lefran çois L. Interleukin-7 mediates the homeostasis of naive and memory CD8 T cells *in vivo*. *Nat Immunol* 2000;**1**:426–32.
73. Tan JT, Ernst B, Kieper WC, LeRoy E, Sprent J, Surh CD. Interleukin (IL)-15 and IL-7 jointly regulate homeostatic proliferation of memory phenotype CD8+ cells but are not required for memory phenotype CD4 cells. *J Exp Med* 2002;**195**:1523–32.
74. Geginat J, Campagnaro S, Sallusto F, Lanzavecchia A. TCR-independent proliferation and differentiation of human CD4 T cell subsets induced by cytokines. *Adv Exp Med Biol* 2002;**512**:107–12.
75. Alves NL, Hooibrink B, Arosa FA, van Lier RA. IL-15 induces antigen-independent expansion and differentiation of human naive CD8 T cells *in vitro*. *Blood* 2003;**102**:2541–6.
76. Sandau MM, Winstead CJ, Jameson SC. IL-15 is required for sustained lymphopeniadriven proliferation and accumulation of CD8 T cells. *J Immunol* 2007;**179**:120–5.
77. Cho JH, Boyman O, Kim HO, Hahm B, Rubinstein MP, Ramsey C, Kim DM, Surh CD, Sprent J. An intense form of homeostatic proliferation of naive CD8 cells driven by IL-2. *J Exp Med* 2007;**204**:1787–801.
78. Alpdogan O, Muriglan SJ, Eng JM, Willis LM, Greenberg AS, Kappel BJ, van den Brink MR. IL-7 enhances peripheral T cell reconstitution after allogeneic hematopoietic stem cell transplantation. *J Clin Invest* 2003;**112**:1095–107.
79. Lenz DC, Kurz SK, Lemmens E, Schoenberger SP, Sprent J, Oldstone MB, et al. IL-7 regulates basal homeostatic proliferation of antiviral CD4 T cell memory. *Proc Natl Acad Sci USA* 2004;**101**:9357–62.
80. Sasson SC, Zaunders JJ, Kelleher AD. The IL-7/IL-7 receptor axis: understanding its central role in T-cell homeostasis and the challenges facing its utilization as a novel therapy. *Curr Drug Targets* 2006;**7**:1571–82.
81. Goldrath AW, Luckey CJ, Park R, Benoist C, Mathis D. The molecular program induced in T cells undergoing homeostatic proliferation. *Proc Natl Acad Sci USA* 2004;**101**:16885–90.
82. Jabbari A, Harty JT. Cutting edge: differential self-peptide/MHC requirement for maintaining CD8 T cell function versus homeostatic proliferation. *J Immunol* 2005;**175**:4829–33.
83. van Leeuwen EM, Sprent J, Surh CD, van Leeuwen EM, Sprent J, Surh CD. Generation and maintenance of memory CD4 T cells. *Curr Opin Immunol* 2009;**21**:167–72.
84. Berczi I, Villalobos Hernández Egina C, Quintanar-Stephano A, Campos R, Kovacs K. The healing power of vasopressin. *Adv Neuroimmune Biol* 2012;**2**:217–24.
85. Blais ME, Louis I, Perreault C. T-cell development: an extrathymic perspective. *Immunol Rev* 2006;**209**:103–14.
86. Rochman Y, Leonard WJ. The role of thymic stromal lymphopoietin in CD8 T cell homeostasis. *J Immunol* 2008;**181**(11):7699–705.
87. Surh CD, Boyman O, Purton JF, Sprent J. Homeostasis of memory T cells. *Immunol Rev* 2006;**211**:154–63.
88. Burchill MA, Goetz CA, Prlic M, O'Neil JJ, Harmon IR, Bensinger SJ, et al. Distinct effects of STAT5 activation on CD4 and CD8 T cell homeostasis: development of CD4, CD25 regulatory T cells versus CD8 memory T cells. *J Immunol* 2003;**171**:5853–64.

89. Ramanathan S, Gagnon J, Leblanc C, Rottapel R, Ilangumaran S. Suppressor of cytokine signaling 1 stringently regulates distinct functions of IL-7 and IL-15 *in vivo* during T lymphocyte development and homeostasis. *J Immunol* 2006;**176**:4029–41.
90. Becker TC, Coley SM, Wherry EJ, Ahmed R. Bone marrow is a preferred site for homeostatic proliferation of memory CD8 T cells. *J Immunol* 2005;**174**:1269–73.
91. Tian C, Bagley J, Iacomini J. Homeostatic expansion permits T cells to re-enter the thymus and deliver antigen in a tolerogenic fashion. *Am J Transplant* 2007;**7**:1934–41.
92. Schüler T, Hämmerling GJ, Arnold B. Cutting edge: IL-7-dependent homeostatic proliferation of CD8 T cells in neonatal mice allows the generation of long-lived natural memory T cells. *J Immunol* 2004;**172**:15–9.
93. Dummer W, Niethammer AG, Baccala R, Lawson BR, Wagner N, Reisfeld RA, et al. T cell homeostatic proliferation elicits effective antitumor autoimmunity. *J Clin Invest* 2002;**110**:185–92.
94. King C, Ilic A, Koelsch K, Sarvetnick N. Homeostatic expansion of T cells during immune insufficiency generates autoimmunity. *Cell* 2004;**117**:265–77.
95. Maile R, Barnes CM, Nielsen AI, Meyer AA, Frelinger JA, Cairns BA. Lymphopenia-induced homeostatic proliferation of CD8 T cells is a mechanism for effective allogeneic skin graft rejection following burn injury. *J Immunol* 2006;**176**:6717–26.
96. Moxham VF, Karegli J, Phillips RE, Brown KL, Tapmeier TT, Hangartner R, et al. Homeostatic proliferation of lymphocytes results in augmented memory-like function and accelerated allograft rejection. *J Immunol* 2008;**180**:3910–8.
97. Lin SJ, Chen AT, Welsh RM. Immune system derived from homeostatic proliferation generates normal CD8 T-cell memory but altered repertoires and diminished heterologous immune responses. *Blood* 2008;**112**:680–9.
98. Hickman SP, Turka LA. Homeostatic T cell proliferation as a barrier to T cell tolerance. *Philos Trans R Soc Lond B Biol Sci* 2005;**360**:1713–21.
99. Troy AE, Shen H. Cutting edge: homeostatic proliferation of peripheral T lymphocytes is regulated by clonal competition. *J Immunol* 2003;**170**:672–6.
100. Shen S, Ding Y, Tadokoro CE, Olivares-Villagómez D, Camps-Ramírez M, Curotto de Lafaille MA, et al. Control of homeostatic proliferation by regulatory T cells. *J Clin Invest* 2005;**115**:3517–26.
101. Hamilton SE, Wolkers MC, Schoenberger SP, Jameson SC. The generation of protective memory-like CD8 T cells during homeostatic proliferation requires CD4 T cells. *Nat Immunol* 2006;**7**:475–81.
102. Hamilton SE, Jameson SC. The nature of the lymphopenic environment dictates protective function of homeostatic-memory CD8 T cells. *Proc Natl Acad Sci USA* 2008;**105**:18484–9.
103. Wolk DA, Klunk W. Update on amyloid imaging: from healthy aging to Alzheimer's disease. *Curr Neurol Neurosci Rep* 2009;**9**(5):345–52.
104. Berczi I, Bertok L, Chow D. Host defence: an interaction of neuroendocrine-, metabolic- and immune mechanisms in the interest of survival. In: Chow D, Bertok L, Berczi I, Szentivanyi A, series editors. *Natural immunity. Neuroimmune biology*, vol. 5. Amsterdam: Elsevier; 2005. p. 3–25.
105. Taub DD, Longo DL. Insights into thymic aging and regeneration. *Immunol Rev* 2005;**205**:72–93.
106. Berczi I, Quintanar-Stephano A, Kovacs K. The brave new world of neuroimmune biology. In: Berczi I, editor. *New insights to neuroimmune biology*. Amsterdam: Elsevier; 2010. p. 4–30 Elsevier Insights, www.amazon.com.

PART 3

Neuropeptides as Immunoregulators

CHAPTER 7

Neuropeptides in Immunoregulation

Istvan Berczi*,, Fabio Rotondo†, Kalman Kovacs†**
*Department of Immunology, Faculty of Medicine, The University of Manitoba, Winnipeg, Manitoba, Canada
**Departamento de Fisiología y Farmacología, Centro de Ciencias Básicas, Universidad Autónoma de Aguascalientes, Aguascalientes, Estado Libre y Soberano de Aguascalientes, Mexico City, Mexico
†Department of Pathology and Laboratory, St. Michael's Hospital, University of Toronto, Toronto, Ontario, Canada

The Discovery of Neurogenic Inflammation

1 INTRODUCTION

Early investigations revealed that sensory nerves influence inflammation but this finding was never followed up. Subsequent investigations into the mechanisms of inflammation largely ignored the possible significance of sensory nerves. In the 1950s Jancsó[1–3] demonstrated that *capsaicin* (CAP), the irritant principle of red pepper (paprika), causes pain and inflammation, which can be prevented by prior repeated local or systemic administration of this agent, this phenomenon was termed "*capsaicin desensitization.*" CAP desensitization or chronic sensory denervation prevented the inflammatory effect of a variety of chemical irritants, including CAP and mustard oil. It was demonstrated that many chemical irritants exert their phlogogenic and pain-producing effects by the stimulation of *CAP-sensitive sensory nerve endings.* These, in turn, release a vasoactive "*neurohumor,*" which induces vasodilation and an increase in vascular permeability, collectively termed the *neurogenic inflammatory response.*[2–4]

2 SENSORY NERVES INDUCE VASCULAR RESPONSES

The discovery of the rather selective neurotoxic action of *CAP and related sensory neurotoxins,*[5–7] permitted the direct morphological identification of CAP-sensitive primary sensory neurons, which mediate the neurogenic inflammatory response. Subsequent studies revealed populations of peptidergic CAP-sensitive sensory neurons,[8–10] and provided evidence for the functional role of some of the peptides contained in these

Insights to Neuroimmune Biology
http://dx.doi.org/10.1016/B978-0-12-801770-8.00007-0

neurons in the mediation of nociceptive and neurogenic inflammatory responses. In addition to earlier observations, experiments on rats treated (neonatally) with CAP led to the finding that CAP-sensitive afferent nerves are involved in the transmission of nociceptive impulses evoked also by heat.[11–13] Recent studies have clearly confirmed the significance of this observation; experiments with cloned CAP receptor confirmed that this receptor might be activated by noxious heat under physiological conditions.[14–16]

Making use of CAP-sensitive primary sensory neurons led to the discovery of a distinct division of the somatosensory system involved with local regulatory functions. Histological studies revealed that CAP-sensitive sensory ganglion cells comprise about 50% of all primary sensory neurons, mostly found in the small-sized cell population and give rise almost exclusively to unmyelinated axons.[17–19] These primary afferent nerves terminate predominantly in the superficial (marginal layer and substantia gelatinosa) layers of spinal and medullary dorsal horn, in some well-defined deeper regions of the spinal cord (lamina V and intermediomedial region), and in visceral sensory nuclei of the solitary tract and area postrema.[20,21] These morphological studies supported the existence of both *somatic and visceral CAP-sensitive sensory nerves*,[20] and strongly suggested[20,22] that CAP-sensitive afferents may have similar functions in somatic and visceral tissues. Immunohistochemical investigations revealed the presence of CAP-sensitive afferent nerves in most somatic and visceral organs and tissues.

During the past few decades, a vast body of experimental evidence has extended and accumulated in support of the original concept by Jancsó,[1–3,23] on the dual function of CAP-sensitive afferent nerves. First, via the *transmission of nociceptive information* toward the central nervous system, these particular afferent nerves fulfill an *afferent function*. Second, through the *secretion of (vasoactive) peptides*, they exert a local regulatory or sensory–efferent function by the modulation of vascular and *inflammatory reactions*[24,25] such as; smooth muscle contractile activity,[26,27] glandular secretion,[28–30] tissue-repair processes,[31] *immune responses*,[32] and the cardiac function.[33–35] Selective chemical lesion of the central branches of (trigeminal) *nociceptive sensory ganglion* neurons, leaving the soma and peripheral branch intact, resulted in a complete *abolition* of the transmission of *nociceptive impulses*, but left the *neurogenic inflammatory response intact*, clearly illustrated the presence of special dual-functional characters of these sensory neurons[36–38] and it was proposed that they should be termed as "*secreto–sensory nociceptive neurons*."[36,37] These findings also suggest that enhanced dorsal root reflexes may not play a decisive role in the development of neurogenic plasma extravasations induced by (chemical) tissue injury.[39,40]

As far as the type(s) of afferent fibers involved in sensory neurogenic cutaneous vascular responses is concerned, it is generally accepted that only CAP-sensitive C-fiber polymodal nociceptors mediate plasma extravasations in rats.[41,42] Neurogenic sensory vasodilation is primarily mediated by CAP-sensitive C-fibers, but the activation of Aδ-afferent fibers may also produce vasodilation.[42] In pigs, cutaneous sensory neurogenic vasodilation, that is, the axon reflex flare is mediated by a specific population of C heat nociceptor afferents and not by polymodal nociceptors.[43] In human skin, the sensory fibers mediating flare response are known to be CAP sensitive[2,3,23] and have been identified as mechano–insensitive C-fiber afferents.[44] Neurogenic sensory vasodilation, at least in the rat dura mater, may also be produced by the stimulation of C-fiber afferents, which are not sensitive to CAP.[45] However, the direct chemical stimulation of C-fiber afferents with CAP at very low (nanomolar) concentrations has also been found to induce a significant increase in dural blood flow in rats.[46]

3 MEDIATORS OF NEUROGENIC INFLAMMATION

3.1 Sensory Neuropeptides

Rats treated neonatally with CAP provided an excellent model for studies on the function of peptidergic afferent neurons. The lack or impairment of sensory functions in these animals may be attributed to the irreversible destruction of CAP-sensitive sensory ganglion neurons[1–3] Experiments in this line indicated that *substance P* (*SP*) is a likely candidate as a mediator of antidromic vasodilation,[47] and neurogenic plasma extravasation evoked by antidromic electrical stimulation of the sensory nerves. Although these reactions were practically completely abolished in rats treated with CAP, this does not necessarily imply that they are mediated by SP, for several other peptides are localized or colocalized in subpopulations of CAP-sensitive sensory neurons. Discovery and localization of the potent vasodilator *calcitonin gene-related peptide* (*CGRP*) in sensory neurons[48] led to a reexamination of this issue. Several lines of evidence indicated that CGRP plays a decisive role in antidromic and neurogenic sensory vasodilation.[49] CGRP elicits marked vasodilation in various organs, and electrical or chemical stimulation of sensory nerves results in the release of the peptide SP. The related *tachykinins neurokinin A* (that is also localized in sensory nerves), is now generally believed to mediate neurogenic plasma extravasation through the activation of *NK1 receptors*.[50–55] It is situated on the endothelial cells of postcapillary venules.[56,57]

Sensory neuropeptides released from sensory nerves may interact with each other or with other mediators of inflammation, resulting in a complex modulation of cutaneous inflammatory reactions. CGRP has been shown to increase plasma extravasation elicited by SP and *histamine*,[58] thereby leaving the effects of serotonin (5-HT) and bradykinin unaffected.[51]

An intriguing aspect relating to the role of sensory neuropeptides in cutaneous inflammation is the indirect modulating role exerted by SP upon the vascular actions of CGRP. Hence, SP released from sensory nerves triggers the release of mast cell mediators, including mast cell proteases, which in turn participate in the degradation of CGRP, leading to a shortening/reduction of its vasodilatory effect.[59]

3.2 Histamine

Studies on the mechanisms of neurogenic cutaneous plasma extravasation revealed vascular responses of similar magnitude in wild-type and mast cell-deficient mice.[60] Similarly, neurogenic plasma extravasation induced by mustard oil, was not affected in mast cell-deficient mice. Pharmacological analysis of CAP-induced mouse ear edema disclosed that it is primarily mediated by SP via NK1 receptors,[61] although the involvement of 5-HT through action on 5-HT2 receptors cannot be excluded.[62] A possible contribution of histamine to the neurogenic inflammatory responses of some tissues cannot be ruled out because there are marked differences between different species and organs with regard to the involvement of mast cells.[9]

3.3 Nitric Oxide

Local control of the cutaneous circulation includes a vasodilator response to local and nonpainful pressure application.[63] This neurally mediated phenomenon was observed in humans,[63] rats,[64] and mice.[65] This mechanism, termed as pressure-induced vasodilation, depends on the activation of CAP-sensitive nerve fibers leading to the release of CGRP, *vasoactive intestinal peptide*, and/or *pituitary adenylate cyclase-activating polypeptide*.[63,64,66]

Neonatal treatment with CAP, inhibition of CGRP, or NOS abolished pressure-induced vasodilation[64] are the results that emphasize the role of nitric oxide in the process but further investigations are necessary to evaluate the cellular mechanisms underlying this phenomenon.[67]

Nitric oxide is also implicated in the release of CGRP from the sensory nerves such as: in the rabbit skin, administration of a *nitric oxide synthase inhibitor* significantly reduced the neurogenic vasodilator response to CAP.[68] Studies utilizing nitric oxide donors indicate that nitric oxide may act through a pathway, which involves prostaglandins.[69]

3.4 Proteinase-Activated Receptors

The proteinase-activated receptors (PARs) are new types of cellular signaling molecules, which are targeted by proteinases such as *thrombin, trypsin, or human mast cell tryptase*. PARs have been localized on *sensory ganglion neurons*.[70,71] In particular, out of the four known subtypes of PAR receptors,[72–74] PAR1 and PAR2 have been shown to be localized on sensory neurons expressing the CAP/TRPV1 receptor,[75,76] which also contain SP,[77] and/or CGRP. PAR-2 is highly expressed in the skin,[78] and its agonists induce many signs of inflammation.[79] PAR agonists have been shown to release sensory neuropeptides in different peripheral tissues, presumably from CAP-sensitive afferent nerve endings.[70] PAR2 agonists have been shown to release these sensory neuropeptides in different peripheral tissues, presumably from CAP-sensitive afferent nerve endings.[70] Further studies have shown that PAR2 may play a role in neurogenic inflammatory responses. The inflammatory effect of PAR2 activation was inhibited by CAP pretreatment and by NK1 receptor antagonists.[71] These findings indicate that the activation of PARs may be involved in the initiation and/or modulation of neurogenically mediated vascular reactions. Also, in an experimental-induced contact dermatitis model of mice swelling responses, plasma extravasation and leukocyte adherence were significantly attenuated in PAR-2 null mutant (PAR-$2^{-}/^{-}$), and PAR2 was capable of the regulation of inflammatory reactions of skin via neurogenic mechanisms.[80] PAR-2 also participates in the pathomechanism of pruritus in inflammatory diseases of the skin, so its modulation may contribute to the treatment of pruritus.[81,82] However, other possible contributions of PARs to the development of skin pathologies (like autoimmune diseases) and the significance of these receptors as putative therapeutic targets await further investigations (Figure 7.1).

Figure 7.1 The visit of Professor Aurelia Jancso to the Institute of Experimental Medicine and Surgery in Montreal. Director Hans Selye said "I was in the Institute while she visited. It was a great pleasure to hear the story of Neurogenic Inflammation from someone who worked on this problem. Aurelia was Nicolas Jancso's wife, having an MD degree. Their son, Gabor Jancso also got an MD degree, and joined with his parents to work on Neurogenic Inflammation. This way Nicholas Jancso managed to build a powerful Research Team at times when research funds were very scarce, or nonexistent."

REFERENCES

1. Jancsó G, Király E. Sensory neurotoxins: chemically induced selective destruction of primary sensory neurons. *Brain Res* 1981;**210**:83–9.
2. Jancsó G, Hökfelt T, Lundberg JM, Király E, Halász N, Nilsson G, Terenius L, Rehfeld J, Steinbusch H, Verhofstad A, Elde R, Said S, Brown M. Immunohistochemical studies on the effect of capsaicin on spinal and medullary peptide and monoamine neurons using antisera to substance P, gastrin/CCK, somatostatin, VIP, encephalin, neurotensin and 5-hydroxytryptamine. *J Neurocytol* 1981;**10**:963–80.
3. Gamse R, Holzer P, Lembeck F. Decrease of substance P in primary afferent neurons and impairment of neurogenic plasma extravasation by capsaicin. *Br J Pharmacol* 1980;**68**:207–13.
4. Jessell TM, Iversen LL, Cuello AC. Capsaicin-induced depletion of substance P from primary sensory neurons. *Brain Res* 1978;**152**:183–8.
5. Jancsó G. Sensory nerves as modulators of inflammatory reactions. In: Chahl LA, Szolcsányi J, Lembeck F, editors. *Antidromic vasodilation and neurogenic inflammation*. Budapest: Akadémiai Kiadó; 1984. p. 207–22.
6. Jancsó G, Jancsó-Gábor A. Effect of capsaicin on morphine analgesia – possible involvement of hypothalamic structures. *Naunyn Schmiedebergs Arch Pharmacol* 1980;**311**:285–8.
7. Jancsó G, Király E, Jancsó-Gábor A. Direct evidence for an axonal site of action of capsaicin. *Naunyn Schmiedebergs Arch Pharmacol* 1980;**313**:91–4.
8. Caterina, Schumacher MA, Tominaga M, Rosen TA, Levine JD, Julius D. The capsaicin receptor: a heat-activated ion channel in the pain pathway. *Nature* 1997;**389**:816–24.
9. Tominaga M, Caterina MJ, Malmberg AB, Rosen TA, Gilbert H, Skinner K, Raumann BE, Basbaum AI, Julius D. The cloned capsaicin receptor integrates multiple pain-producing stimuli. *Neuron* 1998;**21**:531–43.

10. Caterina MJ, Leffler A, Malmberg AB, Martin WJ, Trafton J, Petersen-Zeitz KR, Koltzenburg M, Basbaum AI, Julius D. Impaired nociception and pain sensation in mice lacking the capsaicin receptor. *Science* 2000;**288**:306–13.
11. Barthó L, Benkő R, Patacchini R, Pethő G, Holzer-Petsche U, Holzer P, Lázár Z, Undi S, Illányi L, Antal A, Horváth OP. Effects of capsaicin on visceral smooth muscle: a valuable tool for sensory neurotransmitter identification. *Eur J Pharmacol* 2004;**500**:143–57.
12. Barnes PJ, Belvisi MG, Rogers DF. Modulation of neurogenic inflammation: novel approaches to inflammatory disease. *Trends Pharmacol Sci* 1990;**11**:185–9.
13. Bari F, Jancsó G. Ruthenium red antagonism of capsaicin-induced vascular changes in the rat nasal mucosa. *Eur Arch Otorhinolaryngol* 1994;**251**:287–92.
14. Schultz HD, Roberts AM, Bratcher C, Coleridge HM, Coleridge JC, Davis B. Pulmonary C-fibers reflexly increase secretion by tracheal submucosal glands in dogs. *J Appl Physiol* 1985;**58**:907–10.
15. Baraniuk JN. Sensory, parasympathetic, and sympathetic neural influences in the nasal mucosa. *J Allergy Clin Immunol* 1992;**90**:1045–50.
16. Smith PG, Liu M. Impaired cutaneous wound healing after sensory denervation in developing rats: effects on cell proliferation and apoptosis. *Cell Tissue Res* 2002;**307**:281–91.
17. Nilsson G, Alving K, Ahlstedt S. Effects on immune responses in rats after neuromanipulation with capsaicin. *Int J Immunopharmacol* 1991;**13**:21–6.
18. Ferdinándy P, Csont T, Csonka C, Török M, Dux M, Németh J, Horváth LI, Dux L, Szilvássy Z, Jancsó G. Capsaicin-sensitive local sensory innervation is involved in pacing-induced preconditioning in rat hearts: role of nitric oxide and CGRP? *Naunyn Schmiedebergs Arch Pharmacol* 1997;**356**:356–63.
19. Katona M, Boros K, Sántha P, Ferdinándy P, Dux M, Jancsó G. Selective sensory denervation by capsaicin aggravates adriamycin-induced cardiomyopathy in rats. *Naunyn Schmiedebergs Arch Pharmacol* 2004;**370**:436–43.
20. Jancsó G. Intracisternal capsaic: selective degeneration of chemosensitive primary sensory afferents in the adult rat. *Neurosci Lett* 1981;**27**:41–5.
21. Gamse R, Jancsó G, Király E. Intracisternal capsaicin: a novel approach for studying nociceptive sensory neurons. In: Chahl LA, Szolcsányi J, Lembeck F, editors. *Neurogenic inflammation and antidromic vasodilation*. Budapest: Akadémiai Kiadó; 1984. p. 93–110.
22. Lembeck F. Sir Thomas Lewis's nocifensor system, histamine and substance-P-containing primary afferent nerves. *Trends Neurosci* 1983;**6**:106–8.
23. Louis SM, Jamieson A, Russell NJ, Dockray GJ. The role of substance P and calcitonin gene-related peptide in neurogenic plasma extravasation and vasodilation in the rat. *Neuroscience* 1989;**32**:581–6.
24. Lembeck F, Donnerer J, Tsuchiya M, Nagahisa A. The non-peptide tachykinin antagonist, CP-96,345, is a potent inhibitor of neurogenic inflammation. *Br J Pharmacol* 1992;**105**:527–30.
25. Gamse R, Posch M, Saria A, Jancsó G. Several mediators appear to interact in neurogenic inflammation. *Acta Physiol Hung* 1987;**69**:343–54.
26. Stjarne P, Rinder J, Delay-Goyet P. Effects of NK1 receptor antagonists on vasodilation induced by chemical and electrical activation of sensory C-fibre afferents in different organs. *Acta Physiol Scand* 1992;**152**:153–61.
27. Andrews PV, Helme RD. Tachykinin-induced vasodilation in rat skin measured with a laser-Doppler flowmeter: evidence for receptor-mediated effects. *Regul Pept* 1989;**25**:267–75.
28. Inoue H, Nagata N, Koshihara Y. Effect of the tachykinin receptor antagonists, SR 140333, FK 888, and SR 142801, on capsaicin-induced mouse ear oedema. *Inflamm Res* 1996;**45**:303–7.
29. Inoue H, Asaka T, Nagata N, Koshihara Y. Mechanism of mustard oil-induced skin inflammation in mice. *Eur J Pharmacol* 1997;**333**:231–40.

30. Bowden JJ, Garland AM, Baluk P, Lefevre P, Grady EF, Vigna SR, Bunnett NW, McDonald DM. Direct observation of substance P-induced internalization of neurokinin 1 (NK1) receptors at sites of inflammation. *Proc Natl Acad Sci USA* 1994;**91**:8964–8.
31. Ruocco I, Cuello AC, Shigemoto R, Ribeiro-da-Silva A. Light and electron microscopic study of the distribution of substance P-immunoreactive fibers and neurokinin-1 receptors in the skin of the rat lower lip. *J Comp Neurol* 2001;**432**:466–80.
32. Brain SD, Williams TJ. Inflammatory oedema induced by synergism between calcitonin gene related peptide (CGRP) and mediators of increased vascular permeability. *Br J Pharmacol* 1985;**86**:855–60.
33. Brain SD, Williams TJ. Substance P regulates the vasodilator activity of calcitonin gene related peptide. *Nature* 1988;**335**:73–5.
34. Fromy B, Abraham P, Saumet JL. Non-nociceptive capsaicin-sensitive nerve terminal stimulation allows for an original vasodilatory reflex in the human skin. *Brain Res* 1998;**811**:166–8.
35. Fromy B, Merzeau S, Abraham P, Saumet JL. Mechanisms of the cutaneous vasodilator response to local external pressure application in rats: involvement of CGRP, neurokinins, prostaglandins and NO. *Br J Pharmacol* 2000;**131**:1161–71.
36. Steinhoff M, Vergnolle N, Young SH, Tognetto M, Amadesi S, Ennes HS, Trevisani M, Hollenberg MD, Wallace JL, Caughey GH, Mitchell SE, Williams LM, Geppetti P, Mayer EA, Bunnett NW. Agonists of proteinase-activated receptor 2 induce inflammation by a neurogenic mechanism. *Nat Med* 2000;**6**:151–8.
37. de Garavilla L, Vergnolle N, Young SH, Ennes H, Steinhoff M, Ossovskaya VS, D'Andrea MR, Mayer EA, Wallace JL, Hollenberg MD, Andrade-Gordon P, Bunnett NW. Agonists of proteinase-activated receptor 1 induce plasma extravasation by a neurogenic mechanism. *Br J Pharmacol* 2001;**133**:975–87.
38. Macfarlane SR, Seatter M, Kanke T, Hunter GD, Plevin R. Proteinase-activated receptors. *Pharmacol Rev* 2001;**53**:245–82.
39. Coughlin SR. Thrombin signalling and protease-activated receptors. *Nature* 2000;**407**:258–64.
40. Dery O, Corvera CU, Steinhoff M, Bunnett NW. Proteinase-activated receptors: novel mechanisms of signaling by serine proteases. *Am J Physiol* 1998;**274**:C1429–52.
41. Dai Y, Moriyama T, Higashi T, Togashi K, Kobayashi K, Yamanaka H, Tominaga M, Noguchi K. Proteinase-activated receptor 2-mediated potentiation of transient receptor potential vanilloid subfamily 1 activity reveals a mechanism for proteinase-induced inflammatory pain. *J Neurosci* 2004;**24**:4293–9.
42. Amadesi S, Nie J, Vergnolle N, Cottrell GS, Grady EF, Trevisani M, Manni C, Geppetti P, McRoberts JA, Ennes H, Davis JB, Mayer EA, Bunnett NW. Protease-activated receptor 2 sensitizes the capsaicin receptor transient receptor potential vanilloid receptor 1 to induce hyperalgesia. *J Neurosci* 2004;**24**:4300–12.
43. Steinhoff M, Corvera CU, Thomas MS, Kong W, McAlpine BE, Caughey GH, Ansel JC, Bunnett NW. Proteinase-activated receptor-2 in human skin: tissue distribution and activation of kertinocytes by mast cell tryptase. *Exp Dermatol* 1999;**8**:282–94.
44. Schmelz M, Michael K, Weidner C, Schmidt R, Torebjork HE, Handwerker HO. Which nerve fibers mediate the axon reflex flare in human skin? *Neuroreport* 2000;**11**:645–8.
45. Peitl B, Pethő G, Pórszász R, Németh J, Szolcsányi J. Capsaicin-insensitive sensory-efferent meningeal vasodilation evoked by electrical stimulation of trigeminal nerve fibres in the rat. *Br J Pharmacol* 1999;**127**:457–67.
46. Dux M, Sántha P, Jancsó G. Capsaicin-sensitive neurogenic sensory vasodilation in the dura mater of the rat. *J Physiol* 2003;**552**:859–67.
47. Lembeck F, Holzer P. Substance P as neurogenic mediator of antidromic vasodilation and neurogenic plasma extravasation. *Naunyn Schmiedebergs Arch Pharmacol* 1979;**310**:175–83.

48. Brain SD, Williams TJ, Tippins JR, Morris HR, MacIntyre I. Calcitonin gene-related peptide is a potent vasodilator. *Nature* 1985;**313**:54–6.
49. Louis SM, Jamieson A, Russell NJ, Dockray GJ. The role of substance P and calcitonin gene-related peptide in neurogenic plasma extravasation and vasodilation in the rat. *Neuroscience* 1989;**32**:581–6.
50. Lembeck F, Donnerer J, Tsuchiya M, Nagahisa A. The non-peptide tachykinin antagonist, CP-96,345, is a potent inhibitor of neurogenic inflammation. *Br J Pharmacol* 1992;**105**:527–30.
51. Gamse R, Posch M, Saria A, Jancsó G. Several mediators appear to interact in neurogenic inflammation. *Acta Physiol Hung* 1987;**69**:343–54.
52. Stjarne P, Rinder J, Delay-Goyet P. Effects of NK1 receptor antagonists on vasodilation induced by chemical and electrical activation of sensory C-fibre afferents in different organs. *Acta Physiol Scand* 1992;**152**:153–61.
53. Andrews PV, Helme RD. Tachykinin-induced vasodilation in rat skin measured with a laser-Doppler flowmeter: evidence for receptor-mediated effects. *Regul Pept* 1989;**25**:267–75.
54. Inoue H, Nagata N, Koshihara Y. Effect of the tachykinin receptor antagonists, SR 140333, FK 888, and SR 142801, on capsaicin-induced mouse ear oedema. *Inflamm Res* 1996;**45**:303–7.
55. Inoue H, Asaka T, Nagata N, Koshihara Y. Mechanism of mustard oil-induced skin inflammation in mice. *Eur J Pharmacol* 1997;**333**:231–40.
56. Bowden JJ, Garland AM, Baluk P, Lefevre P, Grady EF, Vigna SR, Bunnett NW, McDonald DM. Direct observation of substance P-induced internalization of neurokinin 1 (NK1) receptors at sites of inflammation. *Proc Natl Acad Sci USA* 1994;**91**:8964–8.
57. Ruocco I, Cuello AC, Shigemoto R, Ribeiro-da-Silva A. Light and electron microscopic study of the distribution of substance P-immunoreactive fibers and neurokinin-1 receptors in the skin of the rat lower lip. *J Comp Neurol* 2001;**432**:466–80.
58. Brain SD, Williams TJ. Inflammatory oedema induced by synergism between calcitonin gene related peptide (CGRP) and mediators of increased vascular permeability. *Br J Pharmacol* 1985;**86**:855–60.
59. Brain SD, Williams TJ. Substance P regulates the vasodilator activity of calcitonin gene related peptide. *Nature* 1988;**335**:73–5.
60. Kowalski ML, Sliwinska-Kowalska M, Kaliner MA. Neurogenic inflammation, vascular permeability, and mast cells. II. Additional evidence indicating that mast cells are not involved in neurogenic inflammation. *J Immunol* 1990;**145**:1214–21.
61. Inoue H, Nagata N, Koshihara Y. Involvement of substance P as a mediator in capsaicin-induced mouse ear oedema. *Inflamm Res* 1995;**44**:470–4.
62. Inoue H, Nagata N, Koshihara Y. Participation of serotonin in capsaicin-induced mouse ear edema. *Jpn J Pharmacol* 1995;**69**:61–8.
63. Fromy B, Abraham P, Saumet JL. Non-nociceptive capsaicin-sensitive nerve terminal stimulation allows for an original vasodilatory reflex in the human skin. *Brain Res* 1998;**811**:166–8.
64. Fromy B, Merzeau S, Abraham P, Saumet JL. Mechanisms of the cutaneous vasodilator response to local external pressure application in rats: involvement of CGRP, neurokinins, prostaglandins and NO. *Br J Pharmacol* 2000;**131**:1161–71.
65. Siguado-Roussel D, Demiot C, Fromy B, Koitka A, Leftheriotis G, Abraham P, Saumet JL. Early endothelial dysfunction severely impairs skin blood flow response to local pressure application in streptozotocin-diabetic mice. *Diabetes* 2004;**53**:1564–9.
66. Fizanne L, Siguado-Roussel D, Saumet JL, Fromy B. Evidence for the involvement of VPAC1 and VPAC2 receptors in pressure-induced vasodilation in rodents. *J Physiol* 2004;**554**:519–28.
67. Garry A, Siguado-Roussel D, Merzeau S, Dumont O, Saumet JL, Fromy B. Cellular mechanisms underlying cutaneous pressure-induced vasodilation: *in vivo* involvement of potassium channels. *Am J Physiol Heart Circ Physiol* 2005;**289**:H174–80.

68. Hughes SR, Brain SD. Nitric oxide-dependent release of vasodilator quantities of calcitonin gene-related peptide from capsaicin-sensitive nerves in rabbit skin. *Br J Pharmacol* 1994;**111**:425–30.
69. Holzer P, Jocic M, Peskar BA. Mediation by prostaglandins of the nitric oxide-induced neurogenic vasodilation in rat skin. *Br J Pharmacol* 1995;**116**:2365–70.
70. Steinhoff M, Vergnolle N, Young SH, Tognetto M, Amadesi S, Ennes HS, Trevisani M, Hollenberg MD, Wallace JL, Caughey GH, Mitchell SE, Williams LM, Geppetti P, Mayer EA, Bunnett NW. Agonists of proteinase-activated receptor 2 induce inflammation by a neurogenic mechanism. *Nat Med* 2000;**6**:151–8.
71. de Garavilla L, Vergnolle N, Young SH, Ennes H, Steinhoff M, Ossovskaya VS, D'Andrea MR, Mayer EA, Wallace JL, Hollenberg MD, Andrade-Gordon P, Bunnett NW. Agonists of proteinase-activated receptor 1 induce plasma extravasation by a neurogenic mechanism. *Br J Pharmacol* 2001;**133**:975–87.
72. Macfarlane SR, Seatter M, Kanke T, Hunter GD, Plevin R. Proteinase-activated receptors. *Pharmacol Rev* 2001;**53**:245–82.
73. Coughlin SR. Thrombin signalling and protease-activated receptors. *Nature* 2000;**407**:258–64.
74. Dery O, Corvera CU, Steinhoff M, Bunnett NW. Proteinase-activated receptors: novel mechanisms of signaling by serine proteases. *Am J Physiol* 1998;**274**:C1429–52.
75. Dai Y, Moriyama T, Higashi T, Togashi K, Kobayashi K, Yamanaka H, Tominaga M, Noguchi K. Proteinase-activated receptor 2-mediated potentiation of transient receptor potential vanilloid subfamily 1 activity reveals a mechanism for proteinase-induced inflammatory pain. *J Neurosci* 2004;**24**:4293–9.
76. Amadesi S, Nie J, Vergnolle N, Cottrell GS, Grady EF, Trevisani M, Manni C, Geppetti P, McRoberts JA, Ennes H, Davis JB, Mayer EA, Bunnett NW. Protease-activated receptor 2 sensitizes the capsaicin receptor transient receptor potential vanilloid receptor 1 to induce hyperalgesia. *J Neurosci* 2004;**24**:4300–12.
77. Steinhoff M, Corvera CU, Thomas MS, Kong W, McAlpine BE, Caughey GH, Ansel JC, Bunnett NW. Proteinase-activated receptor-2 in human skin: tissue distribution and activation of kertinocytes by mast cell tryptase. *Exp Dermatol* 1999;**8**:282–94.
78. Vergnolle N. Proneinase-activated receptor-2-activating peptides induced leukocyte rolling, adhesion, and extravasation *in vivo*. *J Immunol* 1999;**163**:5064–9.
79. Seeliger S, Derian CK, Vergnolle N, Bunnett NW, Naworth R, Schmelz M, von der Weid PY, Buddenkotte J, Sunderkotter C, Metze D, Andrade-Gordon P, Harms E, Vestweber D, Luger TA, Steinhoff M. Proinflammatory role of proteinase-activated receptor-2 in humans and mice during cutaneous inflammation *in vivo*. *FASEB J* 2003;**17**:1871–85.
80. Ständer S, Steinhoff M, Schmelz M, Weisshaar E, Metze D, Luger T. Neurophysiology of puritus: cutaneous elicitation of itch. *Arch Dermatol* 2003;**39**:1463–70.
81. Steinhoff M, Neisius U, Ikoma A, Fartasch M, Heyer G, Skov PS, Luger TA, Schmelz M. Proteinase-activated receptor-2 mediates itch: a novel pathway for pruritus in human skin. *J Neurosci* 2003;**23**:6176–80.
82. Szállási Á, Blumberg PM, Nilsson S, Hökfelt T, Lundberg JM. Visualization by [3H]resiniferatoxin autoradiography of capsaicin-sensitive neurons in the rat, pig and man. *Eur J Pharmacol* 1994;**264**:217–21.
83. Csont T, Csonka C, Kovács P, Jancsó G, Ferdinándy P. Capsaicin-sensitive sensory neurons regulate myocardial nitric oxide and cGMP signaling. *Eur J Pharmacol* 2003;**476**:107–13.

Papers on Neurogenic Inflammation

1 NEUROGENIC INFLAMMATION IN HEALTH AND DISEASE

1.1 Nerves, Mediators, and Receptors Involved in Neurogenic Inflammation

Neurogenic inflammation occurs in many tissues, but has been principally studied in skin. Evidence accumulated over the last two decades supports the hypothesis that neurogenic inflammation is of prime importance in arthritis. Joints are richly innervated by neuropeptides-containing primary afferent and autonomic fibers. Peptides are released into the synovial cavity of arthritic joints and exogenous and endogenous neuropeptides exert proinflammatory actions on articular tissues. Acute and chronic interruption of the articular nerve supply renders joints that are less susceptible to arthritis. Evidence also indicates that neuronal activation and neurogenic inflammation may underlie the symmetrical nature of arthritis in humans and experimental animals. The evidence for neurogenic inflammation in the maintenance of arthritis has been reviewed here, and a hypothesis has been suggested for mechanisms, which spread arthritis.[2]

Blood vessels receive peptidergic perivascular innervations. This perivascular nervous system acts both under the control of the central nervous system and independently, as it responds to local conditions through the release of peptides. Peptides released by perivascular nerves not only act locally by altering blood flow, but also exert a systemic effect by interacting with the components of other body systems, which results in an integrated response. Conversely, dysregulation of peptidergic innervations may be considered as an important contributor to the pathology of a variety of common medical conditions.[3]

The TRPV1 ion channel (formerly known as the "capsaicin receptor") plays a crucial role in the excitation of the large subpopulation of nociceptive primary sensory neurons, which are sensitive to noxious heat and to a large variety of other pain-inducing stimuli. These ion channels are the principal mediators of pain sensations associated with peripheral inflammation, which is by far the most important source of pathological pain in humans. Since the cloning of TRPV1, numerous investigations have resulted in the accumulation of substantial amounts of data concerning the molecular, biophysical, and pharmacological properties of the ion channel. Moreover, the mechanisms through which TRPV1 activity may initiate the perception of pain are also beginning to be elucidated. This review offers a summary of development in the rapidly evolving field of TRPV1 research, and considers how TRPV1 activity may contribute to the development of various acute and chronic pain sensations.[4]

The contribution of sensory nerve fibers to inflammatory responses and immune regulation has been the subject of intense study for many years. Capsaicin has been particularly useful as a tool to probe this connection as it dissects a specific subset of sensory nerve fibers, namely those with unmyelinated fibers that carry polymodal nociceptors. These nerves are sites of release for various proinflammatory neuropeptides that initiate the cascade of neurogenic inflammation, and they convey pain to the central nervous system. The capsaicin receptor, also known as the vanilloid VR1 or TRPV1 receptor, plays a pivotal role in the hyperalgesic response that follows inflammatory challenge in rodent models of pain. TRPV1 is activated by noxious heat, and endogenous proinflammatory molecules, such as arachidonic acid metabolites, nerve growth factors, bradykinins, and protons (tissue acidification). They act in concert to reduce the heat activation threshold of this receptor below body temperature. TRPV1 protein levels are upregulated in experimental models of inflammatory hyperalgesia, and in a number of human disorders (e.g., inflammatory bowel disease). Hence, the development of novel VR1 antagonists could bring about new therapies for the treatment of inflammatory diseases. However, the interaction between sensory nerves and immune cells is even more complex than had been previously thought. For example, functional TRPV1 receptors are expressed on a broad array of cells (e.g., mast cells, lymphocytes, dendritic cells) involved in immune responses, and in turn the mediators released from these cells (e.g., mast cell tryptase via PAR-2 and protein kinase C) can activate TRPV1 in sensory nerves. A better understanding of the complex cross-talk between sensory nerves and immune cells may identify novel indications for pharmacological interventions.[5]

2 MODULATION AND PHYSIOLOGICAL SIGNIFICANCE OF NEUROGENIC INFLAMMATION

The present review demonstrates a promising opportunity for anti-inflammatory drug therapy. Theoretical background of the development of these novel-type compounds is an original observation derived from our laboratory. According to our experimental data, somatostatin is released from the capsaicin-sensitive sensory nerve endings by different stimuli. It reaches distant parts of the body by circulation and exerts systemic anti-inflammatory effects via binding to G-protein-coupled membrane receptors. This receptor family consists of five subtypes (sst_1–sst_5). Presumably, sst_1 and sst_4 are responsible for the mediation of anti-inflammatory actions. Somatostatin influences neurogenic inflammation by pre- and postsynaptic actions. It inhibits the release of proinflammatory neuropeptides from the sensory nerve endings, and also acts on receptors of vascular endothelial, inflammatory, and immune cells. Since somatostatin itself is not a suitable lead molecule for drug development due to its wide spectrum of actions and very short elimination half-life, stable, selective, and well-tolerated somatostatin agonists with anti-inflammatory activity have been synthesized and investigated. TT-232, a heptapeptide anti-inflammatory analogue, is under clinical investigation and several nonpeptide analogs have been preclinically tested. Somatostatin analogue anti-inflammatory agents would mean a novel chance in the pharmacotherapy of chronic inflammatory and immune diseases.[6]

Bradykinin is a potent mediator formed upon tissue damage and inflammation. It can both excite and sensitize nociceptors to heat, mechanical, and chemical stimuli. Two types of bradykinin receptors (B_1 and B_2) have been identified, of which the constitutive B_2 receptors mediate most of the acute effects of bradykinin in uninflamed tissues while the B_1 receptors are induced and become activated during inflammation. These receptor subtypes utilize similar signaling pathways including activation of protein kinase C, elevation of intracellular Ca^{2+} concentration and release of arachidonic acid. Protein kinase C activation is the major mechanism underlying the neuronal excitatory and heat-sensitizing actions of bradykinin, while Ca^{2+} accumulation induces formation of nitric oxide within sensory neurons, which is involved – together with receptor downregulation – in the development of tachyphylaxis of B_2 receptor-mediated effects of bradykinin. However, nitric oxide may also contribute to the excitatory and sensitizing actions of bradykinin. Cyclooxygenase metabolites of arachidonic acid (prostanoids) may also be involved in both excitatory and sensitizing effects of bradykinin. Recently a new signaling mechanism has been revealed for bradykinin, which involves activation of the capsaicin TRPV1 receptor through protein kinase C activation and formation of 12-lipoxygenase products of arachidonic acid. According to a novel hypothesis the neuronal excitatory action of bradykinin is not a separate effect but in fact a heat response as a result of a massive heat sensitization with a threshold drop below the ambient temperature. Prostanoids can also sensitize nociceptors to heat, mechanical and chemical stimuli predominantly via the cyclic adenosine 3′,5′-monophosphate – protein kinase A pathway, which modulates various membrane channels including Ca^{2+}-dependent or voltage-gated K^+ channels, tetrodotoxin-resistant Na^+ channels, as well as ligand or noxious heat-gated ion channels.[7]

The components of neurogenic inflammation include plasma extravasation, vasodilation, smooth muscle contraction, release of mast cell mediators, recruitment and activation of inflammatory cells, and possibly other proinflammatory effects. All these effects can be evoked by neuropeptides released from the peripheral terminals of primary afferent neurons. The neuropeptides involved include the tachykinin substance P (SP), neurokinin A, and calcitonin gene-related peptide. Antagonists blocking their receptors in the target tissue are able to inhibit neurogenic inflammation. Additionally, the prevention of sensitization and excitation of afferent neurons, and modulators of the release of neuropeptides, are known to be effective in the control of neurogenic inflammation. In many inflammatory conditions of the skin, joints, upper respiratory systems, and bronchial tree, as well as intestinal and urinary tract, a neurogenic component may be involved, although in certain instances only partially. The neurogenic components might play a role either in the initiation, perpetuation, or exacerbation of inflammatory processes. Therefore, the overall therapeutic significance of a specific blockade of the neurogenic component in chronic inflammatory conditions is not yet clear.[8]

Neurogenic inflammation of the dura mater encephali is considered to be involved in the generation of primary vascular headaches. Different components of neurogenic inflammation – increased blood flow and vascular permeability, activation of endothelial and mast cells – seem to have a differential significance in the pathophysiology of primary headaches. While the release of vasoactive substances from primary sensory neurons, endothelium or mast cells has clearly been shown to occur during headache attacks, vascular permeability changes seem not to play any role. Clinical and experimental observations indicate that the sensory neuropeptide calcitonin gene-related peptide (CGRP) has a key position in meningeal vasodilation. Beside its direct vasorelaxant effect, CGRP releases the inflammatory mediator histamine from mast cells and interacts with another potent vasodilator, nitric oxide (NO), originating mainly from dural endothelial cells. Parasympathetic nerve fibers of the dura mater are also activated in some types of primary headaches, possibly by a trigeminal–parasympathetic reflex involving brain stem mechanisms. Interactions between different vasoactive regulators of meningeal blood flow may contribute to the pathogenesis of primary headaches.[9]

Neurogenic inflammation occurs in many tissues, but has been principally studied in skin. Evidence accumulated over the last two decades supports the hypothesis that neurogenic inflammation is of prime importance in arthritis. Joints are richly innervated by neuropeptide-containing primary afferent and autonomic fibers, peptides are released into the synovial cavity of arthritic joints and exogenous and endogenous neuropeptides exert proinflammatory actions on articular tissues. Acute and chronic interruption of the articular nerve supply renders joints less susceptible to arthritis. Evidence also indicates that neuronal activation and neurogenic inflammation may underlie the symmetrical nature of arthritis, in both humans and experimental animals. The evidence for neurogenic inflammation in the maintenance of arthritis is reviewed, and a hypothesis is suggested for mechanisms, by which arthritis might start spreading.[2]

Substantial evidence has been described supporting the hypothesis that bradykinin-induced synovial inflammation is dependent on the sympathetic-postganglionic neuron. There is good evidence that a sympathetic-C-fiber afferent neuron integration plays a role in acute and chronic inflammation in the synovium and other tissues. However, in order to firmly establish such integration further studies are needed that employ controlled selective activation of the afferents supplying the synovium, for example, by electrical stimulation of the corresponding dorsal roots. Similar experiments demonstrated sympathetic-C-fiber coupling in skin inflammatory response. In this case sympathetic efferent activity acts on peripheral adrenergic receptors to enhance C-fiber sensitization, thereby augmenting the generation of dorsal root reflexes that produce vasodilation. Furthermore, the evidence suggests that there exists a functionally distinct sympathetic efferent pathway that enables the central nervous system to modulate inflammatory response by acting on components of the immune system. Specifically, at a systemic level activity in the sympathetic neurons modulate immune cell function via innervation of lymphoid organs and by direct contact with lymphocytes and macrophages, and at the local level the sympathetic neurons modulate the sensitivity of nociceptors, directly or via immune cells and their signaling molecules (e.g., cytokines).[10]

NO is an important cardioprotective molecule via its vasodilator, antioxidant, antiplatelet, and antineutrophil actions and it is essential for normal heart function. However, NO is detrimental if it combines with superoxide (O_2^-) to form peroxynitrite ($ONOO^-$), which rapidly decomposes to highly reactive oxidant species. There is a critical balance between cellular concentrations of NO, O_2^-, and superoxide dismutase (SOD), which physiologically favor NO production but in pathological conditions such as ischemia and reperfusion result in $ONOO^-$ formation. However, exposure of the heart to brief episode(s) of ischemia markedly enhances its ability to withstand a subsequent ischemic injury. The activation of this endogenous cardioprotective mechanism known as preconditioning, requires both NO and O_2^- synthesis. However, preconditioning in turn attenuates the overproduction of NO, O_2^- and $ONOO^-$ during subsequent ischemia and reperfusion, thereby protecting the heart. However, still very little is known about the role of NO, O_2^-, and $ONOO^-$ in postconditioning.[11]

Tachykinin-containing projections of capsaicin-sensitive primary afferent neurons widely distributes to visceral organs of the respiratory, gastrointestinal, and genitourinary systems. At this level, tachykinins along with CGRP and other neuropeptides, can be released by adequate stimuli having pathological rather than physiological relevance. Such stimuli may represent chemical irritants to the respiratory tree or alimentary tract, or noxious agents secreted into the urinary system. Tachykinins produced in the visceral organs are part of the responses obtained at somatic level by antidromic activation of capsaicin-sensitive nerve terminals and is collectively known as "neurogenic inflammation," that is, increase in vascular permeability and plasma protein leakage followed by edema. In addition tachykinins produced at visceral level elicit smooth muscle contraction, neuronal stimulation, mucus secretion, and recruitment/activation of immune cells. The physiological significance of these effects is to afford protection toward chemical irritants or noxious stimuli by facilitating their removal from the body. Nonetheless, a large body of preclinical evidence exists, indicating that tachykinins participate in the genesis and/or maintenance of symptoms accompanying various human diseases, such as asthma/bronchial hyperactivity, cystitis of various aetiologies, inflammatory bowel diseases, and irritable bowel syndrome. Tachykinin receptor antagonists, which are proven to prevent/reduce tachykinin-mediated effects in experimental animal models of these pathologies, are expected to afford therapeutically relevant effects in humans.[12]

The vagus nerve has long been considered to be exclusively an efferent communication channel as a physiological regulator. Recently, it was found that the vagus nerve is primarily an afferent nerve and serves a variety of physiological and pathophysiological functions. Abdominal-vagal afferents carry a wealth of information to brainstem, including nutrient homeostasis, feeding and drinking behaviors, noxious challenge of tissues, infection, and inflammation. Vagal afferents participate in a wide range of physiological reactions aimed at maintenance of homeostasis. It is likely that vagal afferents comprise a multitude of neuronal subpopulations that have different sensory modalities and functional implications. One of the challenges faced is to dissect these distinct subpopulations of afferent neurons, to expand the understanding of vagus nerve physiology, and to take use of this knowledge in the design and development of novel therapeutic strategies. Advances in vagal nerve pharmacology offer therapeutic opportunities to address the control of food intake and the treatment of functional gastrointestinal disorders, nausea and emesis, gastroesophageal reflux disease, abdominal pain, hyperalgesia, disturbances of affective–emotional behavior, gastrointestinal infection, and gastrointestinal inflammation.[13]

REFERENCES

1. Jancso G, editor. In: Berczi I, Szentivanyi A, series editors. *Neuroimmune biology*, vol. 8. *Neurogenic inflammation in health and disease.* Elsevier: Amsterdam; 2010. p. 3–36.
2. Donaldson L. Neurogenic mechanisms in arthritis. In: Jancso G, editor. *Neuroimmune biology*, vol. 8. Berczi I, Szentivanyi A, series editors. *Neurogenic inflammation in health and disease.* Elsevier: Amsterdam; 2010. p. 39–48.
3. Scott TM, Scott MM. Peptidergic innervation of blood vessels. In: Jancso G, editor. *Neurogenic inflammation in health and disease.* Berczi I, Szentivanyi A, series editors. Elsevier: Amsterdam; 2010. p. 49–75.
4. Nagy I, Paule CC, White JPM. Molecular mechanisms of TRPV1-medited pain. In: Jancso G, editor. *Neurogenic inflammation in health and disease.* Berczi I, Szentivanyi A, series editors. Elsevier: Amsterdam; 2010 p. 75–99.
5. Cortright DN, Szallasi A. The role of the vanilloid and related receptors in nociceptor function and neuroimmune regulation. In: Jancso G, editor. *Neuroimmune biology*, vol. 8. Berczi I, Szentivanyi A, series editors. *Neurogenic inflammation in health and disease.* Elsevier: Amsterdam; 2010. p. 101–117.
6. Pintér E, Helyes Z, Németh J, Szolcsányi J. Somatostatin as an anti-inflammatory neuropeptide: from physiological basis to drug development. In: Jancso G, editor. *Neuroimmune biology*, vol. 8. Berczi I, Szentivanyi A, series editors. *Neurogenic inflammation in health and disease.* Elsevier: Amsterdam; 2010. p.121–134.
7. Pethő G, Reeh PW. Effects of bradykinin and related mediators on nociceptors. In: Jancso G, editor. *Neurogenic inflammation in health and disease.* Berczi I, Szentivanyi A, series editors. Elsevier: Amsterdam; 2010. p. 135–168.
8. Donnerer J, Holzer-Petsche U. The inhibition of neurogenic inflammation. In: Jancso G, editor. *Neuroimmune biology*, vol. 8. Berczi I, Szentivanyi A, series editors. *Neurogenic inflammation in health and disease.* Elsevier: Amsterdam; 2010. p. 169–189.
9. Dux M, Messlinger K. Neurogenic vascular responses in the dura mater and their relevance for the pathophysiology of headache. In: Jancso G, editor. *Neuroimmune biology*, vol. 8. Berczi I, Szentivanyi, A, series editors. *Neurogenic inflammation in health and disease.* Elsevier: Amsterdam; 2010. p. 193–309.

10. Green PG. Neurogenic regulation of bradykinin induced synovitis. In: Jancso G, editor. *Neuroimmune biology*, vol. 8. Berczi I, Szentivanyi A, series editors. *Neuropetides in health and disease*. Elsevier; Amsterdam: 2010, p. 243–309.
11. Ferdinandy P, Jancsó G. Capsaicin-sensitive sensory nerves in myocardial ischemia-reperfusion injury and ischemic stress adaptation: role of nitric oxide and calcitonin-gene related peptide. In: Jancso G, editor. *Neuroimmune biology*, vol. 8. Berczi I, Szentivanyi A, series editors. *Neurogenic inflammation in health and disease*. Elsevier: Amsterdam; 2010. p. 267–288.
12. Patacchini R, Maggi CA. Tachykinins and neurogenic inflammation at visceral level. In: Jancso G, editor. *Neuroimmune biology*, vol. 8. Berczi I, Szentivanyi A, series editors. *Neurogenic inflammation in health and disease*. Elsevier: Amsterdam; 2010. p. 267–288.
13. Holtzer P. The role of the vagus nerve in afferent signaling and homeostasis during visceral inflammation. In: Jancso G, editor. *Neuroimmune biology*, vol. 8. Berczi I, Szentivanyi A, series editors. *Neurogenic infammation in health and disease*. Elsevier: Amsterdam; 2010. p. 321–338.

Neuropeptides as Immunoregulators

Leptin (LEP) is produced primarily by adipocytes and is structurally related to the GLH cytokine family. LEP signals by a class I cytokine receptor, of which two isoforms are known. LEP regulates energy metabolism, reproductive function and lymphocyte development and function. It stimulates TH-1 mediated immune responses and upregulates phagocytosis and the production of proinflammatory cytokines. During the acute phase response, LEP rises rapidly in response to elevated TNF levels. LEP appears to contribute significantly to survival in sepsis by moderating glucocorticoid and IL-6 production, stimulating IL-1 receptor antagonist, and potentiating the immune response.

Arginine vasopressin (AVP) exerts an immunostimulatory effect through V_1-type receptors. AVP is produced by lymphocytes.

Substance P (SP) is produced in the nervous and immune systems. SP is a major mediator of neurogenic inflammation and is capable of causing mast cell discharge, increased capillary permeability, and smooth muscle construction. In general, SP has a stimulatory effect on immune phenomena. Its effect on immunoglobulin secretion is variable. SP stimulates bone marrow cytokines and the formation of granuloma.

Calcitonin gene related peptide (CGRP) stimulates mast cell discharge and inflammatory response, whereas it has an inhibitory effect on immune reactions. T lymphocytes synthesize CGRP. Somatostatin (SOM) is an antagonist of SP and inhibits inflammatory and immune reactions. Vasoactive intestinal peptide (VIP) is produced within the immune system and the receptors are present on monocytes and lymphocytes. In general, VIP functions as an immunosuppressive neuropeptide and is involved in the immunosuppressive properties of aqueous humor in the eye. Its effect on the immunoglobulin secretion and NK cell activity is variable. Besides VIP downregulates inflammatory response. Pituitary adenylate cyclase activating peptide (PACAP) has a similar immunosuppressive and anti-inflammatory effect.

1 NERVE GROWTH FACTOR

Nerve growth factor (NGF) was discovered in murine submandibular glands as a factor stimulating the growth of sensory and sympathetic ganglia[1]. Low-affinity pan-neurotropic receptor (P75), specific high-affinity receptor tyrosine kinases, gp140 trkA for NGF, gp145 trkB for brain-derived neurotrophic factor (BDNF), and gp145 trkC for neurotropin 3 are known. These receptors belong to the TNF receptor superfamily, which includes the Fas and CD40 receptors on lymphoid tissue. The trkA is expressed in thymus and lymph nodes, trkB in the spleen, and localized primarily to the stroma of these organs. Some expression was also found in splenocytes and thymocytes. B lymphocytes and follicular dendritic cells also express NGF receptors.[2–6]

NGF is mitogenic for human B lymphoblastoid cell lines, which is associated with tyrosine phosphorylation, the activation of mitogen-activated protein kinase (MAPK)[7]. NGF-stimulated cell proliferation in mixed lymphocyte cultures,[8] increased the response of rat spleen cells to T- and B-cell mitogens[9] and the growth of unstimulated human B lymphocytes. The production of IgM and IgG4 by B cells was stimulated by NGF.[10,11] NGF inhibited the induction of IgE by IL-4 in human peripheral B lymphocytes and IL-4 prevented the NGF-induced increase in IgA and IgM production.[12] Murine T and B lymphocytes, and rat peritoneal mast cells are capable of synthesizing biologically active NGF.[13]

NGF treatment increased the number and size of mast cells in neonatal mice. It also stimulated histamine release from mast cells, both *in vivo* and *in vitro*.[14–18] NGF stimulated the development of hematopoietic colonies, as well as the chemotactic and phagocytic activity of neutrophilic leukocytes.[19–21]

The stimulatory effect of NGF on mast cells and neutrophilic leukocytes suggests a proinflammatory role. However, the suppression of inflammation has been observed *in vivo* by systemically applied NGF in several experimental models.[22,23] The reason for this paradox is not well understood. Recent studies revealed that lymphocyte-derived NGF provides protection for the nervous system and to other host tissues during inflammatory reactions. This phenomenon implies the existence of beneficial "autoimmune" reactions.[24]

2 LEPTIN

LEP is produced primarily by fat cells (adipocytes). Structurally, LEP belongs to the GLH/CTK family and signals by a class I cytokine receptor (Ob-R). Two receptors are known, namely, Ob-Ra and Ob-Rb. LEP regulates energy metabolism, reproduction, lymphoid development, and function. Physiologically, LEP is regulated by insulin, cortisol, and sex steroids, mainly testosterone. In rats, centrally administered LEP suppressed the mitogenic response of splenic lymphocytes, mediated through CRF-sympathetic activation. LEP regulates nutritional state and T-cell function. In immunosuppressed starving mice, LEP enhanced TH1-immune responses, inspite of catabolic state of animals. Starving animals that have reduced LEP levels, are sensitive to endotoxin. Fasting mice respond to LPS with a blunted corticosterone and exaggerated TNF production. This could be corrected by LEP treatment[25,26].

During acute phase responses (APR) (e.g., sepsis) LEP rapidly rises in the serum. TNF-α, causes this elevation. LEP inhibits glucocorticoids and IL-6 production. Blood LEP correlates positively with survival of septicaemic patients. LEP stimulates IL-1 receptor antagonist (IL-1ra), which protects the mice from LPS toxicity. In murine glial cells LEP was induced with IL-1β. In animals administered with LEP phagocytosis, proinflammatory cytokines were upregulated. LEP stimulated wound healing and angiogenesis.[27–29]

3 NEUROPEPTIDES

3.1 Arginine Vasopressin

Human peripheral blood monocytes expressed V_1-type AVP receptors. AVP enhanced the autologous mixed lymphocyte culture, stimulated the production of PGE2 by human mononuclear phagocytes, and of β-endorphin by human peripheral blood mononuclear cells. Centrally administered AVP is antipyretic and attenuates fever. AVP was detected in rat thymocytes and splenic lymphocytes.[30–34]

3.2 Substance P

Substance P (SP) receptors are present in thymocytes, B and T lymphocytes, macrophages, mast cells, and astrocytes. SP is a major mediator of neurogenic inflammation and capable of inducing mast cell degranulation, plasma extravasation, and bronchoconstriction. Uterine mast cells of rats released histamine and cytokines in response to SP at dioestrus, but not during proestrus, or on the day 4 of pregnancy. In proestrus preincubation of uterine mast cells with SP, significantly enhanced mediator release by anti-IgE antibodies.[35] SP acts on lymphocytes, macrophages, and neutrophils. Lymphocyte proliferation and lymphokine production are enhanced by SP, whereas the effect on immunoglobulin secretion is variable. SP increased Fcγ and receptors, decreased C3b on eosinophils, released TNF-α from macrophages, and modified macrophage function during stress.[36–38] In polymorphonuclear leukocytes, SP stimulated the respiratory burst, chemotactic, and phagocytic activities.[39,40] SP increased the release of PGE2 and collagenase from rheumatoid synoviocytes,[41] and PGA and thromboxane B2 from astrocytes.[42] SP-induced IL-3 and GM-CSF secretion by bone marrow cells. This was partially mediated by the stimulation of IL-1 and IL-6 in marrow cells.[43] SP activated platelets for cytotoxicity against *Schistosoma mansoni* larvae. The SP receptor was necessary for normal granulomatous response to *S. mansoni*.[44]

3.3 Calcitonin Gene Related Peptide

Mature lymphocytes, macrophages, and bone marrow cells express receptors for calcitonin gene related peptide (CGRP). These receptors are functionally coupled to adenylate cyclase.[45] CGRP inhibited the proliferative response of murine thymus cells to Con A.[46] In the skin, CGRP can trigger mast cell discharge, induce slow onset intense erythema and vasodilation.[47] CGRP has an inhibitory effect on antigen presentation, lymphocyte proliferation, IL-2 production, and mRNA synthesis for TNF-α, -β, and IFN-γ. IFN-γ-induced H_2O_2 production by human monocytes was also suppressed by CGRP.[48,49]

CGRP containing nerve fibers innervated Langerhans cells in the human epidermis and exerted an inhibitory effect on antigen presentation by these cells.[50] Locally applied CGRP inhibited the induction of cutaneous hypersensitivity reactions in mice,[51] but others observed increased cutaneous inflammation to allergens and irritants and an enhanced sensitization process.[52] CGRP inhibited the expression of NF6B and promoted apoptosis. T lymphocytes from rat thymus and mesenteric lymph nodes synthesized CGRP.[53]

3.4 Somatostatin

T and B lymphocytes, and mast cells possess receptors for somatostatin (SOM). SOM inhibits the IgE-dependent mediator release from human basophils and mast cells. SOM is an antagonist of SP and it has beneficial effects in models of autoimmune diseases and of chronic inflammation. Lymphocyte proliferation, endotoxin-induced leukocytosis, IgA secretion, and IFN-γ production are all inhibited by SOM. Its effect on antibody-dependent cytotoxicity is variable. SOM exerts a regulatory influence on macrophages.[54–56]

3.5 Vasoactive Intestinal Peptide and Pituitary Adenylate Cyclase Activating Peptide

Vasoactive intestinal peptide (VIP) receptors are present on monocytes and lymphocytes. Human and murine thymocytes were protected by VIP against the cytolytic effect of prednisolone.[57] VIP regulated T-cell homing to gut associated lymphoid tissue. VIP stimulated macrophage chemotaxis by PKC activation and inhibited lymphocyte chemotaxis through the activation of adenylate cyclase.[58] VIP enhanced phagocytosis by mouse peritoneal macrophages,[59] inhibited lymphocyte mitogenesis, had a variable effect on immunoglobulin secretion, and NK cell activity.[60,61] The addition of VIP to the cultures of human B lymphocytes from the colonic mucosa increased IgA production and reduced significantly IgG levels. *In vitro* VIP-induced IgA production by isotype switching and inhibited the production of IL-2 and IL-4 by murine thymocytes.[62,63] Immunoreactive VIP was detected in rat thymus, spleen, and lymph nodes in both lymphoid and nonlymphoid cells.[64]

VIP and α-MSH contribute to the immunosuppressive properties of aqueous humor in the eye. Antigen presentation in the eye is deviated toward antibody production. Cell-mediated immune responses were suppressed. SP acted in the aqueous humor as an antagonist of VIP. Antigen or TNF-α-induced VIP release into the aqueous humor.[65]

VIP and PACAP inhibited the nuclear translocation of NF6B in stimulated macrophages by inhibiting Jak/Stat phosphorylation. This antagonized the effect of IFN-γ and downregulated the inflammatory response.[66] The production of TGF-β1, IL-4, -6, -12, TNF-α, and nitric oxide were inhibited by both the peptides.[67–70] IL-6 production was enhanced by VIP/PACAP in unstimulated macrophages.[71] VIP and PACAP inhibited antigen-induced apoptosis in CD4+ (but not CD8+) T lymphocytes by downregulating Fas. In humans, topically injected VIP inhibited the development of contact dermatitis to nickel sulfate.[72]

REFERENCES

1. Levi-Montalcini R. The nerve growth factor, its mode of action on sensory and sympathetic nerve cells. *Harvey Lect* 1966;**60**:217–59.
2. Brodie C, Gelfand EW. Functional nerve growth factor receptors on human B lymphocytes: interaction with IL-2. *J Immunol* 1992;**148**:3492–7.
3. Pezzati P, Stanisz AM, Marshall JS. Expression of nerve growth factor receptor immunoreactivity on follicular dendritic cells from human mucosa associated lymphoid tissues. *Immunology* 1992;**76**:485–90.
4. Hannestad J, Levanti MB, Vega JA. Distribution of neurotrophin receptors in human palatine tonsils: an immunohistochemical study. *J Neuroimmunol* 1995;**58**:131–7.

5. Lomen-Hoerth C, Shooter EM. Widespread neutrophin receptor expression in the immune system and other nonneuronal rat tissues. *J Neurochem* 1995;**64**:1780–9.
6. Cosman D. A family of ligands for the TNF receptor superfamily. *Stem Cells* 1994;**12**: 440–55.
7. Franklin RA, Brodie C, Melamed I. Nerve growth factor induces activation of MAP-kinase and p90^{rsk} in human B lymphocytes. *J Immunol* 1995;**154**:4965–72.
8. Manning PT, Russell JH, Simmons B, Johnson EM. Protection from guanethidine-induced neuronal destruction by nerve growth factor: effect of NGF on immune function. *Brain Res* 1985;**340**:61–9.
9. Thorpe LW, Perez-Polo JR. The influence of nerve growth factor on the *in vitro* proliferative response of rat spleen lymphocytes. *J Neurosci Res* 1987;**18**:134–9.
10. Otten U, Ehrhard P, Peck R. Nerve growth factor induces growth and differentiation of human B lymphocytes. *Proc Natl Acad Sci USA* 1989;**86**:10059–63.
11. Kimata H, Yoshida A. Effect of growth hormone and insulin-like growth factor-I on immunoglobulin production by and growth of human B cells. *J Clin Endocrinol Metab* 1994;**78**:635–41.
12. Brodie C, Oshiba A, Renz H, Bradley K, Gefland EW. Nerve growth factor and anti-CD40 provide opposite signals for the production of IgE in interleukin-4-treated lymphocytes. *Eur J Immunol* 1996;**26**:171–8.
13. Santambrogio L, Beneditti M, Chao M, Muzaffar R, Kulig K, Gabellini N, Hochwald G. Nerve growth factor production by lymphocytes. *J Immunol* 1994;**153**:4488–95.
14. Aloe L, Levi-Montalcini R. Mast cells increase in tissues of neonatal rats injected with the nerve growth factor. *Brain Res* 1977;**133**:358–66.
15. Bohm A, Aloe L, Levi-Montalcini R. Nerve growth factor enhances precocious differentiation and numerical increase in mast cells in cultures of rat splenocytes. *Accad Naz Lincei* 1986;**80**:1.
16. Pearce FL, Thompson HL. Some characteristics of histamine secretion from rat peritoneal mast cells stimulated with nerve growth factor. *J Physiol* 1986;**372**:379–93.
17. Marshall JS, Stead RH, McSharry C. The role of mast cell degranulation products in mast cell hyperplasia. I. Mechanism of action of nerve growth factor. *J Immunol* 1990;**144**:1886–92.
18. Bischoff SC, Dahinden CA. Effect of nerve growth factor on the release of inflammatory mediators by mature human basophils. *Blood* 1992;**79**:2662–6.
19. Gee AP, Boyle MDP, Munger KL. Nerve growth factor: stimulation of polymorphonuclear leukocyte chemotaxis *in vitro*. *Proc Natl Acad Sci USA* 1983;**80**:7213–8.
20. Boyle MDP, Lawman MJP, Gee AP, Young M. Nerve growth factor: a chemotactic factor for polymorphonuclear leukocytes *in vivo*. *J Immunol* 1985;**135**:564–8.
21. Cannon JG, Tatro JB, Reichlin S, Dinarello CA. Alpha melanocyte stimulating hormone inhibits immunostimulatory and inflammatory actions of interleukin I. *J Immunol* 1986;**337**:2232–6.
22. Banks BEC, Vernon CA, Warner JA. Nerve growth factor has anti-inflammatory activity in the rat hind paw oedema test. *Neurosci Lett* 1984;**47**:41–5.
23. Amico-Roxas M, Caruso A, Leone MG, Scifo R, Vanella A, Scapagnini U. Nerve growth factor inhibits some acute experimental inflammations. *Arch Int Pharmacodyn Ther* 1989;**299**:269–85.
24. Hohlfeld R, Kerschensteiner M, Stadelmann C, Lassmann H, Wekerle H. The neuroprotective effect of inflammation: implications for the therapy of multiple sclerosis. *J Neuroimmunol* 2000;**107**:161–6.
25. Reinshagen M, Rohm H, Steinkamp M, Lieb K, Geerling I, Von Herbay A, Flamig G, Eysselein VE, Adler G. Protective role of neurotrophins in experimental inflammation of the rat gut. *Gastroenterology* 2000;**119**:368–76.
26. Matarese G. Leptin and the immune system: how nutritional status influences the immune response. *Eur Cytokine Netw* 2000;**11**:7–14.

27. Bornstein SR, Licinio J, Tauchnitz R, Engelmann L, Negrao AB, Gold P, Chrousos GP. Plasma leptin levels are increased in survivors of acute sepsis: associated loss of diurnal rhythm, in cortisol and leptin secretion. *J Clin Endocrinol Metab* 1998;**83**:280–3.
28. Giovambattista A, Chisari AN, Corro L, Gaillard RC, Spinedi E. Metabolic, neuroendocrine and immune functions in basal conditions and during the acute-phase response to endotoxic shock in undernourished rats. *Neuroimmunomodulation* 2000;**7**:92–8.
29. Okamoto S, Irie Y, Ishikawa I, Kimura K, Masayuki S. Central leptin suppresses splenic lymphocyte functions through activation of the corticotropin-releasing hormone-sympathetic nervous system. *Brain Res* 2000;**855**:192–7.
30. Berczi I. *Pituitary Function and Immunity*. Boca Raton, FL: CRC Press; 1986.
31. Malkinson TJ, Bridges TE, Lederis K, Veale WL. Perfusion of the septum of the rabbit with vasopressin antiserum enhances endotoxin fever. *Peptides* 1987;**8**:385–9.
32. Torres BA, Johnson HW. Arginine vasopressin (AVP) replacement of helper cell requirement in IFN-γ production: evidence for a novel AVP receptor on mouse lymphocytes. *J Immunol* 1988;**140**:2179–83.
33. Jessop D, Chowdrey HS, Lightman SL, Larsen PJ. Vasopressin is located within lymphocytes in the rat spleen. *J Neuroimmunol* 1995;**56**:219–23.
34. Jessop D, Murphy D, Larsen PJ. Thymic vasopressin (AVP) transgene expression in rats: a model for the study of thymic AVP hyper-expression in T cell differentiation. *J Neuroimmunol* 1995;**62**:85–90.
35. Cocchiara R, Albeggiani G, Azzolina A, Bongiovanni A, Lampiasi N, DiBlasi F, Geraci D. Effect of substance P on uterine mast cell cytokine release during the reproductive cycle. *J Neuroimmunol* 1995;**60**:107–15.
36. Lee HR, Ho WZ, Douglas SD. Substance P augments tumor necrosis factor release in human monocyte-derived macrophages. *Clin Diagn Lab Immunol* 1994;**1**:223–419.
37. Nair MPN, Schwartz SA. Immunomodulatory effects of corticosteroids on natural killer and antibody-dependent cellular cytotoxic activities of human lymphocytes. *J Immunol* 1984;**132**:2876–82.
38. Chancellor-Freeland C, Zhu GF, Kage R, Beller DI, Leeman SE, Black PH. Substance P and stress-induced changes in macrophages. *Ann NY Acad Sci* 1995;**771**:472–84.
39. Payan DG, McGillis JP, Renold FK, Mitsuhashi M, Goetzl EJ. The immunomodulating properties of neuropeptides. In: Berczi I, Kovacs K, editors. *Hormones and Immunity*. Lancaster, England: MTP Press; 1987.
40. Jancso G, Obal Jr F, Toth-Kasa I, Katona M, Husz S. The modulation of cutaneous inflammatory reactions by peptide-containing sensory nerves. *Int J Tissue React* 1985;**7**:449–57.
41. Lotz M, Carson DA, Vaughan JH. Substance P activation of rheumatoid synoviocytes: neural pathway in pathogenesis of arthritis. *Science* 1987;**235**:893–5.
42. Hartung HP, Heininger K, Schafer B, Toyka KV. Substance P and astrocytes: stimulation of the cyclooxygenase pathway of arachidonic acid metabolism. *Fed Am Soc Exp Bio J* 1988;**2**:48–51.
43. Rameshar P, Ganea D, Gascon P. Induction of IL-3 and granulocyte-macrophage colony-stimulating factor by substance P in bone marrow cells is partially mediated through the release of IL-1 and IL-6. *J Immunol* 1994;**152**:4044–54.
44. Damonneville M, Monte D, Auriault C, Capron A. The neuropeptide substance-P stimulates the effector functions of platelets. *Clin Exp Immunol* 1990;**81**:346–51.
45. Mullins MW, Ciallella J, Rangnekar V, McGillis JP. Characterization of a calcitonin gene-related peptide (CGRP) receptor on mouse bone marrow cells. *Regul Pept* 1993;**49**:65–72.
46. Bulloch K, McEwen BS, Diwa A, Baird S. Relationship between dehydroepiandrosterone and calcitonin gene-related peptide in the mouse thymus. *Am J Physiol* 1995;**268**: E168–73.
47. Foreman JC. Neuropeptides and the pathogenesis of allergy. *Allergy* 1987;**42**:1–11.
48. Nong YH, Titus RG, Ribeiro JMC, Remold HG. Peptides encoded by the calcitonin gene inhibit macrophage function. *J Immunol* 1989;**145**:45–9.

49. Wang J, Klein JR. Thymus-neuroendocrine interactions in extrathymic T cell development. *Science* 1994;**265**:1860–2.
50. Hosoi J, Murphy GF, Egan CL, Lerner EA, Grabbe S, Asahina A, Granstein RD. Regulation of Langerhans cell function by nerves containing calcitonin gene-related peptide. *Nature* 1993;**363**:159–63.
51. Asahina A, Hosoi J, Beissert S, Stratigos A, Granstein RD. Inhibition of the induction of delayed-type and contact hypersensitivity by calcitonin gene-related peptide. *J Immunol* 1995;**154**:3056–61.
52. Gutwald J, Goebeler M, Sorg C. Neuropeptides enhance irritant and allergic contact dermatitis. *J Invest Dermatol* 1991;**96**:695–8.
53. Millet I, Phillips RJ, Sherwin RS, Ghosh S, Voll RE, Flavell RA, Vignery A, Rincon M. Inhibition of NF-kappaB activity and enhancement of apoptosis by the neuropeptide calcitonin gene-related peptide. *J Biol Chem* 2000;**275**:15114–21.
54. Foris G, Gyimesi E, Komaromi I. The mechanism of antibody-dependent cellular cytotoxicity stimulation by somatostatin in rat peritoneal macrophages. *Cell Immunol* 1985;**90**:217–25.
55. Muscettola M, Grasso G. Somatostatin and vasoactive intestinal peptide reduce interferon gamma production by human peripheral blood mononuclear cells. *Immunobiology* 1990;**180**:419–30.
56. Church MK, El-Lati S, Caulfied JP. Neuropeptide-induced secretion from human skin mast cells. *Int Arch Allergy Appl Immunol* 1991;**94**:310–8.
57. Ernstrom U, Gafvelin G, Mutt V. Rescue of thymocytes from cell death by vasoactive intestinal peptide. *Regul Pept* 1995;**57**:99–104.
58. de la Fuente M, Delgado M, del Rio M, Garrido E, Leceta J, Hernanz A, Gomariz RP. Vasoactive intestinal peptide modulation of adherence and mobility in rat peritoneal lymphocytes and macrophages. *Peptides* 1994;**15**:1157–63.
59. Ichinose M, Asai M, Imai K, Sawada M. Enhancement of phagocytosis in mouse macrophages by pituitary adenylate cyclase activating polypeptide (PACAP) and related peptides. *Immunopharmacology* 1995;**30**:217–24.
60. Ottaway CA. *In vitro* alterations of receptors for vasoactive intestinal peptide changes the *in vivo* localization of mouse T cells. *J Exp Med* 1984;**160**:1054–69.
61. O'Dorisio MS. Biochemical characteristics of receptors for vasoactive intestinal polypeptide in nervous endocrine and immune systems. *Fed Proc* 1987;**46**:192–5.
62. Xin Z, Tang H, Ganea D. Vasoactive intestinal peptide inhibits interleukin (IL)-2 and IL-4 production in murine thymocytes activated via the TCR/CD3 complex. *J Neuroimmunol* 1994;**54**:59–68.
63. Boirivant M, Fais S, Annibale B, Agostini D, Delle Fave G, Pallone F. Vasoactive intestinal polypeptide modulates the *in vitro* immunoglobulin: a production by intestinal lamina propria lymphocytes. *Gastroenterology* 1994;**106**:576–82.
64. Leceta J, Martinez MC, Delgado M, Garrido E, Gomariz RP. Lymphoid cell subpopulations containing vasoactive intestinal peptide in the rat. *Peptides* 1994;**15**:791–7.
65. Taylor AW, Streinlein JW, Cousins SW. Immunoreactive vasoactive intestinal peptide contributes to the immunosuppressive activity of normal aqueous humor. *J Immunol* 1994;**153**:1080–6.
66. Ferguson TA, Fletcher S, Herndon J, Griffith TS. Neuropeptides modulate immune deviation induced via the anterior chamber of the eye. *J Immunol* 1995;**155**:1746–56.
67. Delgado M, Ganea D. Inhibition of IFN-gamma-induced janus kinase-1-STAT1 activation in macrophages by vasoactive intestinal peptide and pituitary adenylate cyclase-activating polypeptide. *J Immunol* 2000;**165**:3051–7.
68. Sun W, Tadmori I, Yang L, Delgado M, Ganea D. Vasoactive intestinal peptide (VIP) inhibits TGF-beta1 production in murine macrophages. *J Neuroimmunol* 2000;**107**:88–99.
69. Xin Z, Sriram S. Vasoactive intestinal peptide inhibits IL-12 and nitric oxide production in murine macrophages. *J Neuroimmunol* 1998;**89**:206–12.

70. Kawamura N, Tamura H, Obana S, Wenner M, Ishikawa T, Nakata A, Yamamoto H. Differential effects of neuropeptides on cytokine production by mouse helper T cell subsets. *Neuroimmunomodulation* 1998;**5**:9–15.
71. Martinez C, Delgado M, Pozo D, Leceta J, Calvo JR, Ganea D, Gomariz RP. Vasoactive intestinal peptide and pituitary adenylate cyclase-activating polypeptide modulate endotoxin-induced IL-6 production by murine peritoneal macrophages. *J Leukoc Biol* 1998;**63**:591–601.
72. Lundeberg L, Mutt V, Nordlind K. Inhibitory effect of vasoactive intestinal peptide on the challenge phase of allergic contact dermatitis in humans. *Acta Derm Venereol* 1999;**79**:178–82.

Current Papers on Neuroimmune Regulation by Neuropeptides

1 ANGIOTENSIN II

Angiotensin II (AngII) has been recognized to act as a powerful proinflammatory mediator. Some research has showed that AngII plays important roles in autoimmune diseases including RA, systemic lupus erythematosus, and multiple sclerosis. AngII blockers prove effective in reducing inflammation and autoimmunity in rheumatic diseases and their relative safety, together with their effects for reducing the cardiovascular disease risk, suggest that AngII blockers may at least act as an effective adjunctive therapy for disease control in patients with RA.[1]

The peptide hormone AngII, in tumor bearing mice amplifies self-renewing hematopoietic stem cells (HSCs) and macrophage progenitors. This process occurred in the spleen but not in the bone marrow, and was independent of hemodynamic changes. The effects of AngII required direct hormone ligation on HSCs, depending on S1P1 signaling, and allowed the extramedullary tissue to supply new tumor-associated macrophages throughout cancer progression. Conversely, blocking AngII production prevented cancer-induced HSC and macrophage progenitor amplification and thus restrained the macrophage response at its source. These findings indicate that AngII acts upstream of a potent macrophage amplification program and tumors can remotely exploit the hormone's pathway to stimulate cancer-protecting immunity.[2]

Thus, the adaptive immune response and role of pro- and anti-inflammatory cytokine signaling in hypertension are distinct between the sexes and need be understood to improve therapeutics for hypertension-associated diseases in both men and women.[3]

AngII induces vascular injury in part by activating innate and adaptive immunity. However, the mechanisms are unclear. We investigated the role of interferon (IFN)-γ and interleukin (IL)-23 signaling. We infused AngII into IFN-γ receptor (IFN-γR) knockout (KO) mice and wild-type controls, as well as into mice treated with neutralizing antibodies against IL-23 receptor and IL-17A. AngII-treated IFN-γR KO mice exhibited reduced cardiac hypertrophy, reduced cardiac macrophage, T-cell infiltration, less fibrosis, and less arrhythmogenic electric remodeling independent of blood pressure changes. In contrast, IL-23 receptor antibody treatment did not reduce cardiac hypertrophy, fibrosis, or electric remodeling despite mildly reduced inflammation. IL-17A antibody treatment behaved in a similar manner. In the kidney, IFN-γR deficiency reduced inflammation and tubulointerstitial damage, and improved glomerular filtration rate. Nonetheless, albuminuria was increased compared to AngII–treated wild-type controls. The glomeruli of AngII–treated IFN-γR KO mice exhibited fewer podocytes, less nephrin, and synaptopodin staining, as well as impaired podocyte autophagy.

Thus, IFN-γ blockade, but not IL-23 receptor antibody treatment, protects from Ang II-induced cardiac damage and electric remodeling. In the kidney, IFN-γ signaling acts in a cell type-specific manner. Glomerular filtration rate is preserved in the absence of the IFN-γR, whereas podocytes may require the IFN-γR in the presence of AngII for normal integrity and function.[4]

Selective depletion of T regulatory (Treg) cells using a CD25-specific monoclonal antibody significantly enhances the susceptibility of C57Bl/6 mice to Ang II-induced abdominal aortic aneurysm and promotes aortic rupture (n = 25–44 mice/group). Similar results are observed in AngII-treated *Cd80−/−/Cd86−/−* or *Cd28−/−* mice with impaired Treg cell homeostasis (n = 18–23 mice/group). Treg cell depletion is associated with increased immune cell activation and a blunted IL-10 anti-inflammatory response, suggesting an immunoinflammatory imbalance. Interestingly, *Il-10−/−* mice (n = 20 mice/group) show increased susceptibility to Ang II-induced abdominal aortic aneurysm and aortic rupture, and are insensitive to Treg cell depletion. Finally, reconstitution of *Cd28−/−* Treg-deficient mice with Treg cells (n = 22 mice/group) restores a balance in the immunoinflammatory response, rescues the animals from increased susceptibility to aneurysm, and prevents aortic dissection.

Conclusions: These results identify a critical role for Treg cells and IL-10 in the control of aneurysm formation and its progression to rupture and suggest that therapies targeting Treg responses may be suitable to treat aneurysmal disease.[5]

We investigated sex differences in T-cell modulation of Ang II-induced increases in mean arterial pressure in male (M) and female (F) wild-type and recombination-activating-gene-1-deficient ($Rag1^{-/-}$) mice. Sex differences in peak mean arterial pressure in wild-type were lost in $Rag1^{-/-}$ mice (mmHg: wild-type-F, 136 ± 4.9 vs. wild-type-M, 153 ± 1.7; $P < 0.02$; $Rag1^{-/-}$F, 135 ± 2.1 vs. $Rag1^{-/-}$M, 141 ± 3.8). Peak mean arterial pressure was 31 mm Hg higher after adoptive transfer of male (CD3M→Rag1$^{-/-}$M) versus female (CD3F→Rag1$^{-/-}$M) T cells. CD3M→Rag1$^{-/-}$M mice exhibited higher splenic frequencies of proinflammatory IL-17A (2.4 fold), TNF-α (2.2 fold)-producing T cells, lower plasma levels (13 fold), and renal mRNA expression (2.4 fold) of IL-10, whereas CD3F→Rag1$^{-/-}$M mice displayed a higher activation state in general and T-helper-1-biased renal inflammation. Greater T-cell infiltration into perivascular adipose tissue and kidney associated with increased pressor responses to AngII if the T cell donor was male but not female and these sex differences in T-cell subset expansion and tissue infiltration were maintained for 7–8 weeks within the male host. Thus, the adaptive immune response and role of pro- and anti-inflammatory cytokine signaling in hypertension are distinct between the sexes and need to be understood to improve therapeutics for hypertension-associated disease in both men and women.[6]

Here, we tested the relationship of IL-6-signal transducer and activator of transcription-3 signaling with Th17-induced inflammation in the formation of Ang II-induced dissections in C57BL/6 mice. *Ang II infusion*-induced aortic dissections and CD4+ IL-17A-expressing *Th17 cell accumulation in C57BL/6 mice*. A blunted local Th17 activation, macrophage recruitment, and reduced incidence of aortic dissections were seen in IL-6$^{-/-}$ mice. To determine the pathological roles of Th17 lymphocytes, AngII-infused mice were treated with IL-17A-neutralizing antibody, or infused AngII in genetically deficient IL-17A mice and found decreased aortic chemokine *monocytic chemotactic protein-1(MCP1)* production and macrophage recruitment, leading to a reduction in aortic dissections. This effect was independent of blood pressure in IL-17A-neutralizing antibody experiment. Application of a cell-permeable signal transducer and activator of transcription-3 *(STAT-3) inhibitor* to downregulate the IL-6 pathway decreased aortic dilation and Th17 cell recruitment. Also observed was an increased aortic Th17 infiltration and IL-17 mRNA expression *in patients with thoracic aortic dissections*. Finally, we found that AngII-mediated aortic dissections occurred independent of blood pressure changes. *Conclusions*: Our results indicate that the IL-6-signal transducer and activator of transcription-3 signaling pathway converges on Th17 recruitment and IL-17A signaling upstream of macrophage recruitment, thereby mediating aortic dissections.[7]

Although regulatory T cells (Tregs) have been shown to play a protective role in abdominal aortic aneurysm (AAA) formation, it remains unclear whether expansion of endogenous Foxp3+ Tregs prevents AAA. In the current study, it was determined that the effects of endogenous Foxp3+

Treg expansion or depletion in an experimental model of AAA. We continuously infused 12-week-old apolipoprotein E-deficient mice fed a high-cholesterol diet with AngII ($n = 60$) or normal saline ($n = 12$) by implanting osmotic minipumps and evaluated AAA formation at 16 weeks. The Ang II-infused mice received IL-2/anti-IL-2 monoclonal antibody complex (IL-2 complex; $n = 31$) or PBS ($n = 29$). *Eighty-one percent* of AngII-infused mice developed AAA, with 42% mortality possibly because of aneurysm rupture. IL-2 complex treatment systemically increased the number of Foxp3+ Tregs and significantly decreased the *incidence (52%)* and *mortality (17%)* of AAA. Immunohistochemical analysis showed reduced accumulation of macrophages and increased numbers of Foxp3+ Tregs in aneurysmal tissues, suggesting that the expansion of Tregs may suppress local inflammation in the vessel wall and provide protection against AAA formation. Furthermore, genetic depletion of Foxp3+ Tregs led to a significant increase in the mortality of AAA, suggesting a protective role of Foxp3+ Tregs against AAA. Our findings suggest that *Foxp3+ Tregs* may play a protective role in AAA formation and the promotion of an endogenous regulatory immune response may be a potentially valuable therapeutic approach for preventing AAA.[8]

Immune system activation contributes to the pathogenesis of hypertension and the resulting progression of chronic kidney disease. In this regard, we recently identified a role for proinflammatory Th1 T-lymphocyte responses in hypertensive kidney injury. Th1 cells generate IFN-γ and tumor necrosis factor-α (TNF-α), so it was hypothesized that *IFN-γ and TNF-α propagate renal damage* during hypertension induced by activation of the *renin–Ang system*. Therefore, after confirming that genetic Th1 immunity deficient mice were protected from kidney glomerular injury despite a preserved hypertensive response, we subjected mice lacking in IFN-γ or TNF-α to our model of hypertensive chronic kidney disease. IFN deficiency had no impact on blood pressure elevation or urinary albumin excretion during chronic AngII infusion. Contrastingly, *TNF-deficient (KO) mice* had *blunted hypertensive responses* and reduced end-organ damage in our model. As AngII-infused TNF KO mice had exaggerated endothelial nitric oxide synthase expression in the kidney and enhanced nitric oxide bioavailability, we examined the actions of TNF-α generated from renal parenchymal cells in hypertension by transplanting wild-type or TNF KO kidneys into wild-type recipients before the induction of hypertension. Transplant recipients lacking TNF solely in the kidney had blunted hypertensive responses to AngII and augmented renal endothelial nitric oxide synthase expression, confirming a role for *kidney-derived TNF-α* to promote Ang II-induced blood pressure elevation by limiting renal nitric oxide generation.[9]

We measured the expression of thymic stromal lymphopoietin (TSLP) in primary rat vascular smooth muscle cells (VSMCs) exposed to AngII and evaluated the capacity of TSLP induced by AngII to induce the differentiation of Th17 Cells. We then sought to identify the involved upstream regulatory mechanisms. We found that VSMCs express TSLP in response to AngII in an AT1 receptor/NF-κB manner. TSLP can induce the differentiation of Th17 Cells by triggering DCs. *Conclusions:* Through DC activation, Ang II-induced TSLP enhances Th17-driven immune response in atherosclerosis and hypertension.[10]

2 ANGIOGENIC FACTOR

Hypertension is a health problem affecting over 1 billion people worldwide. How the immune system gets activated under hypertensive stimuli to contribute to blood pressure elevation is a fascinating enigma. Here,a splenic role for *placental growth factor* (PlGF) has been shown, which accounts for the onset of hypertension, through immune system modulation. PlGF repressed the expression of the protein *Timp3* (tissue inhibitor of metalloproteinases 3), through the transcriptional *Sirt1-p53* axis. Timp3 repression allowed co-stimulation of T cells and their deployment toward classical organs involved in hypertension. It shows that the spleen is an essential organ for the development of hypertension through a noradrenergic drive mediated by the *celiac ganglion efferent.* Overall, it has been demonstrated that PlGF mediates the neuroimmune interaction in the spleen, organizing a unique and non-redundant response that allows the onset of *hypertension.*[11]

3 CALCITONIN GENE-RELATED PEPTIDE

Upon its mucosal entry, human immunodeficiency virus type 1 (*HIV-1*) is internalized by *Langerhans cells* (*LCs*) in stratified epithelia and transferred locally to T cells. In such epithelia, LCs are in direct contact with *peripheral neurons* secreting calcitonin gene-related peptide (*CGRP*). Although CGRP has immunomodulatory effects on LC functions, its potential influence on the interactions between LCs and HIV-1 is unknown. We show that *CGRP acts via its receptor expressed by LCs and interferes with multiple steps of LC-mediated HIV-1 transmission.* CGRP increases *langerin* expression, decreases selected *integrins*, and *activates NF-κ-B*, resulting in *decreased HIV-1 intracellular content, limited formation of LC–T cell conjugates,* and elevated secretion of the CCR5-binding chemokine *CCL3/MIP-1α*. These mechanisms cooperate to efficiently inhibit HIV-1 transfer from LCs to T cells, and T-cell infection. *In vivo*, HIV-1 infection decreases CGRP plasma levels in both vaginally SHIV challenged macaques and HIV-1-infected individuals. CGRP plasma levels return to baseline after highly active antiretroviral therapy. Our results reveal a novel path, by which a peripheral neuropeptide acts at the molecular and cellular levels to limit mucosal HIV-1 transmission and suggest that CGRP receptor agonists might be used therapeutically against HIV-1.[12]

For gene therapy in the induction phase, EAE developed in 50% of mice in the CGRP-transfected group compared to 80% in mock-transfected group, and the mean pathological score for EAON was one in the CGRP-transfected group compared with two in the mock-transfected group ($P < 0.05$). For gene therapy in the effector phase, the mean EAE clinical score (1.5 vs. 3.0) and mean EAON pathological score (1.0 vs. 2.0) were both lower in the CGRP-transfected group compared with the mock-transfected group ($P < 0.05$). Delayed hypersensitivity was suppressed significantly in the CGRP-transfected group. IL-10 production by spleen cells in the CGRP-transfected group increased independent of MOG concentration, compared with the mock-transfected group. Interestingly, the proportion of CD4+CD25+Foxp3+ cells increased significantly ($P < 0.05$) in the CGRP-transfected group compared with the mock-transfected group. *Conclusions*: Gene therapy with CGRP-expressing mDC was effective in suppressing the development of EAON and EAE.[13]

Dendritic cells (DCs) play essential roles in both innate and adaptive immune responses. In addition, mutual regulation of the nervous system and immune system is well studied. One of the neuropeptides, that is calcitonin gene-related peptide (*CGRP*), is a potent regulator in immune responses. It particularly has *anti-inflammatory effects in innate immunity*. For instance, a deficiency of the CGRP receptor component RAMP 1 (receptor activity-modifying protein 1) results in higher cytokine production in response to LPS (lipopolysaccharide). On the other hand, how CGRP affects DCs in adaptive immunity is largely unknown. In this study, we show that CGRP suppressed Th1 cell differentiation via inhibition of IL-12 production in DCs using an *in vitro* coculture system and an *in vivo* ovalbumin-induced delayed-type hypersensitivity (DTH) model. CGRP also downregulated the expressions of chemokine receptor CCR2 and its ligands CCL2 and CCL12 in DCs. Intriguingly, the frequency of migrating CCR2+ DCs in draining lymph nodes of RAMP1-deficient mice was higher after DTH immunization. Moreover, these CCR2+ DCs highly expressed IL-12 and CD80, resulting in more effective induction of Th1 differentiation compared with CCR22 DCs. These results indicated that CGRP regulated Th1 type reactions by regulating expression of cytokines, chemokines, and chemokine receptors in DCs.[14]

The airway mucosal epithelium is permanently exposed to airborne particles. A network of immune cells patrols at this interface to the environment. The interplay of immune cells is orchestrated by different mediators. In the current study, we investigated the impact of neuronal signals on key functions of DCs. Using two-photon microscopic timelapse analyses of living lung sections from CD11c-EYFP transgenic mice, the influence of neuropeptides on airway DC motility was studied. Additionally, using a confocal microscopic approach, the phagocytotic capacity of CD11c+ cells after neuropeptide stimulation was determined. Electrical field stimulation (EFS) leads to an unspecific release of neuropeptides from nerves. After EFS and treatment with the neuropeptides *vasoactive intestinal peptide* (VIP) or *calcitonin gene-related peptide* (CGRP), airway DC in living lung slices showed an altered motility. Furthermore, the EFS-mediated effect could partially be blocked by pretreatment with the *receptor antagonist CGRP8–37*. Additionally, the phagocytotic capacity of bone marrow-derived and whole lung CD11c+ cells could be inhibited by neuropeptides CGRP, VIP, and Substance P (SP). We then cross-linked these data with the *in vivo* situation by analyzing DC motility in two different OVA asthma models. In the acute and prolonged OVA asthma models altered neuropeptide amounts and DC motility in the airways could be measured. In summary, our data suggests that neuropeptides modulate key features of motility and phagocytosis of mouse airway DC. Therefore, *altered neuropeptide levels in airways during allergic inflammation* have impact on regulation of airway immune mechanisms and therefore might contribute to the pathophysiology of asthma.[15]

4 BOMBESIN

Lack of enteral stimulation during parenteral nutrition (PN) impairs mucosal immunity. *Bombesin (BBS),* a gastrin-releasing peptide analogue, reverses PN-induced defects in acquired immunity. Paneth cells produce *antimicrobial peptides* (AMPs) of innate immunity for release after cholinergic stimulation. *Objective*: Determine if BBS restores AMPs and bactericidal function during PN. *Methods*: Intravenously cannulated male ICR mice were randomized to Chow, PN, or PN + BBS (15 μg three-times daily, $n = 7$ per group) for 5 days. Ileum was analyzed for AMPs (protein: sPLA2 by fluorescence, lysozyme, and RegIII-γ by western and cryptdin-4 by ELISA; mRNA: all by RT-PCR). Cholinergic stimulated (100 μM bethanechol) ileal specimens assessed *Pseudomonas* bactericidal activity. Ileum (Chow: $n = 7$; PN: $n = 9$; PN + BBS: $n = 8$) was assessed for *Escherichia coli* invasion in *ex vivo* culture. *Results*: PN significantly decreased most AMPs versus Chow while BBS maintained Chow levels (sPLA2: Chow: 107 + 14; PN: 44.6 + 7.2; PN + BBS: 78.7 + 13.4 Fl/min/μL/total protein; lysozyme: Chow: 63.9 + 11.9, PN: 26.8 + 6.2; PN + BBS: 64.9 + 13.8 lysozyme/total protein; RegIII-γ: Chow: 51.5 + 10.0, PN: 20.4 + 4.3, PN + BBS: 31.0 + 8.4 RegIII-γ/total protein; cryptdin-4: Chow: 18.4 + 1.5, PN: 12.7 + 1.6, PN + BBS: 26.1 + 2.4† pg/mg (all $P < 0.05$ vs. PN and †$P < 0.05$ vs. Chow). Functionally, BBS prevented PN loss of bactericidal activity after cholinergic stimulation (Chow: 25.3 + 3.6, PN: 13.0 + 3.2; PN + BBS: 27.0 + 4.7% bacterial killing, $P < 0.05$ vs. PN). BBS reduced bacterial invasion in unstimulated tissue barely missing significance ($P = 0.06$). *Conclusions:*The *enteric nervous system (ENS) controls AMP levels in Paneth cells during PN but mucosal protection by innate immunity requires of ENS and parasympathetic stimulation.*[16]

5 METH AND METYL-BRADYKININ

In terms of infection incidence, the yeast *Candida parapsilosis* is the second after *Candida albicans* as causative agent of candidiases in humans. The major virulence factors of *C. parapsilosis* are the secreted enzyme, aspartic proteases (SAPPs), which help the pathogen to disseminate, acquire nutrients, and *dysregulate the mechanisms of innate immunity* of the host. In the current work, we characterized the action of two major extracellular proteases of *C. parapsilosis*, SAPP1 and SAPP2, on human kininogens, proteinaceous precursors of *vasoactive, and proinflammatory bradykinin-related peptides*, collectively called the *kinins. The kininogens, preferably the form with lower molecular mass, were effectively cleaved by SAPPs*, with the release of two uncommon kinins, *Met-Lys-bradykinin* and *Leu-Met-Lys-bradykinin*. While optimal at acidic pH (4–5), the kinin release yield was only two- to threefold lower at neutral pH. These peptides were able to *interact with cellular kinin receptors of B2 subtype* and to *stimulate the human endothelial cells (HMEC-1)* to increased secretion of proinflammatory ILs, *IL-1 and IL-6*. The analysis of stability of SAPP-generated kinins in plasma suggested that they are biologically equivalent to bradykinin, the best agonist of B2 receptor subtype, and can be quickly converted to des-Arg9-bradykinin, the agonist of inflammation-inducible B1 receptors protecting against *C. parapsilosis*.[17]

6 ENDORPHIN AND DYNORPHIN

Dynorphins constitute a family of opioid peptides manifesting the highest affinity for κ opiate receptors. Immune system cells are known to express a κ receptor similar to that in the central nervous system, and as a consequence dynorphins are involved in the interaction between cells of the nervous and immune systems. Here, data on dynorphin structure are analyzed and generalized, the κ opiate receptor is characterized, and data on the regulation by dynorphins of functioning of the innate and adaptive immunity cells are summarized.[18]

Intraperitoneal injection of β-endorphin in doses of 1, 0.01, and 0.0005 μg/kg under conditions of systemic immunization increased the count of antibody-producing cells in the spleen and the titer of antierythrocyte antibodies in the plasma of experimental animals. Intraperitoneal β-endorphin stimulated proliferative activity of splenocytes in mice, in the presence of both B- and T-cell mitogen, did not change the production of IFN-γ, reduced the level of IL-2, and stimulated the secretion of IL-4, which is the main Th2-polarizing factor.[19]

Body and mind interact extensively with each other to control health. Emerging evidence suggests that chronic neurobehavioral stress can promote various tumor growths and progressions. The biological reaction to stress involves a chemical cascade initiated within the central nervous system and extends to the periphery, encompassing the immune, endocrine, and autonomic systems. Activation of the sympathetic nervous system, such as what happens in the "fight or flight" response, down regulates tumor suppressive genes, inhibits immune functions, and promotes tumor growth. On the other hand, an optimistic attitude or psychological intervention helps cancer patients to survive longer via an increase in b-endorphin neuronal suppression of stress hormone levels and sympathetic outflows and activation of parasympathetic control of tumor suppressor gene and innate immune cells to destroy and clear tumor cells.[20]

7 PROCALCITONIN

Inflammation is considered to be a major effector of arterial damage brought about by salt excess in animal models. In a randomized, single masked, and cross-over study in 32 complicated essential hypertensive patients, we assessed the effect of a short-term and low-salt diet on biomarkers of innate immunity (procalcitonin (PCT), IL-6, C-reactive protein, and tumor necrosis factor-a (TNF-α)), adiponectin (ADPN, an anti-inflammatory cytokine), and leptin. *Methods*: Patients were randomized to either a 10–20 mmol sodium diet and sodium tablets (180 mEq/day) to achieve a 200 mmol intake per day or the same diet and identical placebo tablets, each for 2 weeks. At the end of each of these periods, all patients underwent a 24-h urine collection, a fasting blood sampling, and a 24-h ambulatory blood pressure monitoring. *Results*: Parallely, with expected increase in plasma renin activity and aldosterone ($P < 0.001$), both PCT (R33%) and TNF-α (9%) rose at low-salt intake ($P < 0.007$) while ADPN underwent an opposite change (17%, $P < 0.001$). In a linear regression analysis for repeated measurements, PCT was significantly and inversely related to urinary salt (weighted $r = 0.27$, $P1/40.03$). Changes in inflammation biomarkers did not differ in salt-sensitive ($n1/47$) and salt-resistant ($n1/425$) patients. *Conclusions*: In essential hypertensive patients, a very low salt diet generates a proinflammatory phenotype characterized by an increase in PCT and TNF-α and an opposite effect on an anti-inflammatory cytokine like ADPN.[21]

In essential hypertensive patients, a very low salt diet generates a proinflammatory phenotype characterized by an increase in procalcitonin (PCT) and TNF-α and an opposite effect on an anti-inflammatory cytokine like adiponectin (ADPN).[22]

8 ALPHA-MELANOCYTE-STIMULATING HORMONE

Within the immunosuppressive *ocular microenvironment*, there are constitutively present the immunomodulating neuropeptides alpha-melanocyte-stimulating hormone (α-MSH) and neuropeptide Y (NPY), which promote suppressor functionality in macrophages. In this study, the possibility that α-MSH and NPY modulate phagocytic activity in macrophages was examined. The macrophages treated with α-MSH and NPY were significantly suppressed in their capacity to phagocytize unopsonized *E. coli* and *Staphylococcus aureus* bioparticles, but not antibody-opsonized bioparticles. The neuropeptides significantly suppressed phagolysosome activation, and the FcR-associated generation of reactive oxidative species as well. This suppression corresponds to neuropeptide modulation of macrophage functionality within the ocular microenvironment to *suppress the activation of immunogenic inflammation*.[23]

9 METH-ENKEPHALIN

The aim of this study was to investigate the immunological mechanisms by which synthetic methionine enkephalin (MENK) exerts therapeutic effects on *tumor growth*. The findings *in vivo* or *in vitro* showed that *MENK treatment* either *in vivo* or *in vitro* could *upregulate* the percentages *of CD8+ T cells*, induce markers of activated T cells, increased cytotoxic activity *against mouse S180 tumor cells*, and increase secretion of IFNc. In addition, the adoptively transferred CD8+ T cells, after either *in vitro* or *in vivo* treatment with MENK, resulted in a significantly increased survival of S180 tumor-bearing mice and significant shrinkage in tumor growth. *Opioid receptors are detected on normal CD8+ T cells* and exposure to MENK leads to increased expression of opioid receptors. Interaction between MENK and opioid receptors on CD8+ T cells appears to be essential for the activation of CTL, since the addition of *naltrexone (NTX), an opioid receptor antagonist, significantly inhibits all of the effects of MENK.* The evidence obtained indicates that the MENK-induced T cell signaling is associated with a significant upregulation of Ca^{2+} influx into the cytoplasm and the translocation of NFAT2 into nucleus, and these signaling effects are also inhibited by NTX.[24]

10 INTEGRIN

S. aureus is a major component of the skin microbiota and causes a large number of serious infections. *S. aureus* first interacts with epidermal keratinocytes to breach the epidermal barrier through mechanisms that are not fully understood. By use of primary keratinocytes from mice with epidermis-restricted Ilk gene inactivation and control *integrin-linked kinase (ILK)*-expressing littermates, we investigated the role of ILK in epidermal *S. aureus* invasion. Heat-killed but not live, bacteria were internalized to Rab5- and Rab7-positive phagosomes, and incubation with keratinocyte growth factor increased their uptake 2.5-fold. ILK-deficient mouse keratinocytes internalized bacteria two- to fourfold less efficiently than normal cells. The reduced invasion by live *S. aureus* of ILK-deficient cells was restored in the presence of exogenous, constitutively active Rac1. Thus, Rac1 functions downstream from ILK during invasion. Further, invasion by *S. aureus* of Rac1- deficient cells was 2.5-fold lower than in normal cells. Paradoxically, staphylococcal cutaneous penetration of mouse skin explants with ILK-deficient epidermis was 35-fold higher than that of normal skin, indicating defects in epidermal barrier function in the absence of ILK. Thus, we identified an ILK-Rac1 pathway essential for bacterial invasion of keratinocytes, and established *ILK as a key contributor to prevent invasive staphylococcal cutaneous infection.*[25]

11 GALANINE PEPTIDES

Galanin was first identified 30 years ago as a "classic neuropeptide," with actions primarily as a modulator of neurotransmission in the brain and peripheral nervous system. Other structurally related peptides – *galanin-like peptide and alarin* – with diverse biologic actions in brain and other tissues have since been identified, although, unlike galanin, their cognate receptors are currently unknown. Over the last two decades, in addition to many neuronal actions, a number of nonneuronal actions of galanin and other galanin family peptides have been described. These include actions associated with neural stem cells, nonneuronal cells in the brain such as glia, endocrine functions, effects on metabolism, energy homeostasis, and paracrine effects in bone. Substantial new data also indicates an emerging role for galanin in innate immunity, inflammation, and cancer. Galanin has been shown to regulate its numerous physiologic and pathophysiological processes through interactions with three G protein-coupled receptors, GAL1, GAL2, and GAL3, and signaling via multiple transduction pathways, including inhibition of cAMP/PKA (GAL1, GAL3) and stimulation of phospholipase C (GAL2). In this review, we emphasize the importance of novel galanin receptor-specific agonists and antagonists. Also, other approaches, including new transgenic mouse lines (such as a recently characterized GAL3 KO mouse) represent, in combination with viral-based techniques, and citical tools required to evaluate galanin system physiology better. These in turnwill help identify potential targets of the galanin/galanin-receptor systems in a diverse range of human diseases, including pain, mood disorders, epilepsy, neurodegenerative conditions, diabetes, and cancer.[26]

12 SUBSTANCE P

Traumatic brain injury results in significant morbidity and mortality and is associated with infectious complications, particularly pneumonia. However, whether traumatic brain injury directly impacts the host response to pneumonia is unknown. The objective of this study was to determine the nature of the relationship between traumatic brain injury and the prevalence of pneumonia in trauma patients and investigate the mechanism of this relationship using a murine model of traumatic brain injury with pneumonia. *Design*: Data from the National Trauma Data Bank and a murine model of traumatic brain injury with postinjury pneumonia. *Setting*: Academic medical centers in Cincinnati, OH, and Boston, MA. *Patients/Subjects*: Trauma patients in the National Trauma Data Bank with a hospital length of stay greater than 2 days, age of at least 18 years at admission, and a blunt mechanism of injury. Subjects were female ICR mice 8–10 weeks old. *Interventions*: Administration of a SP receptor antagonist in mice.

Measurements and main results: Pneumonia rates were measured in trauma patients before and after risk adjustment using propensity scoring. In addition, survival and pulmonary inflammation were measured in mice undergoing traumatic brain injury with or without pneumonia. After risk adjustment, we found that *traumatic brain injury patients had significantly lower rates of pneumonia compared to blunt trauma patients without traumatic brain injury.*

A murine model of traumatic brain injury reproduced these clinical findings with mice subjected to traumatic brain injury demonstrating increased bacterial clearance and survival after induction of pneumonia. To determine the mechanisms responsible for this improvement, the *SP receptor was blocked* in mice after traumatic brain injury. This treatment abrogated the traumatic brain injury-associated increases in bacterial clearance and survival.

Conclusions: The data demonstrate that patients with traumatic brain injury have lower rates of pneumonia compared to nonhead-injured trauma patients and suggest that the mechanism of this effect occurs through traumatic brain injury-induced *release of SP*, which improves innate immunity to decrease pneumonia.[27]

Neuropeptides (NPs) and neurotransmitters are a heterogeneous group of soluble factors that make connections within the neuroendocrine and immune systems. NPs, including *SP (SP), vasoactive intestinal peptide (VIP), α-melanocyte-stimulating hormone (α-MSH), and calcitonin gene– related peptide (CGRP),* released by nerves that innervate the skin, can modulate the action of innate and *adaptive skin immunity* as well as the skin cells functions. Their role in several inflammatory skin diseases, such as atopic dermatitis, psoriasis, and vitiligo, and in the isotopic response has been reported. *Further progress in understanding* the various processes that modulate the interactions of the nervous and the skin immune system is essential to develop effective treatment for inflammatory skin conditions with neurogenic components and for understanding signs and symptoms in the isotopic response and, in general, in the control of global and regional immunity.[28]

13 MELANOCORTIN

In this study, we examined antifungal and anti-inflammatory effects of the synthetic melanocortin peptide *(Ac-Cys-Lys-Pro-Val-NH2)2* or *(CKPV)2* against *C. albicans* vaginitis. Our *in vitro* results showed that (CKPV)2 dose-dependently inhibited *C. albicans* colonies formation. In a rat *C. albicans* vaginitis model, (CKPV)2 significantly inhibited vaginal *C. albicans* survival and macrophages sub-epithelial mucosa infiltration. For mechanisms study, we observed that (CKPV)2 inhibited to phagocytose of *C. albicans* by Macrophages. Meanwhile, *(CKPV)2* administration *inhibited macrophage proinflammatory cytokines (TNF-αa, IL-1β, and IL-6) release,* while increasing the arginase activity and *anti-inflammatory cytokine IL-10 production*, suggesting macrophages *M1 to M2* polarization. *Cyclic AMP (cAMP)* production was also *induced* by (CKPV)2 administration in macrophages. These above effects on macrophages by (CKPV)2 were almost reversed by melanocortin receptor-1(MC1R) siRNA knockdown, indicating the requirement of MC1R in the process. Altogether, our results suggest that (CKPV)2 exerted antifungal and anti-inflammatory activities against *C. albicans* vaginitis probably through inducing macrophages M1 to M2 polarization and MC1R activation.[29]

14 NEUROTENSIN 2

Stress affects immunity, but some mechanisms are not known. *Neurotensin (NT)* and corticotropin-releasing hormone *(CRH)* are secreted under stress in various tissues, and have immunomodulatory actions. We had previously shown that *NT augments the ability of CRH to increase mast cell-dependent skin vascular permeability in rodents.* Here we show that *NT triggered human mast cell degranulation and significantly augmented CRH-induced vascular endothelial growth factor (VEGF) release.* Investigation of various signaling molecules indicated that only *NF-κB activation* was involved. These effects were blocked by pretreatment with the NTR antagonist SR48692. *NT-induced expression of CRH receptor-1 (CRHR-1),* as shown by Western blot and FACS analysis. Interestingly, *CRH also induced NTR gene* and protein expression. These results indicate unique interactions among NT, CRH, and mast cells that may contribute to rcovery of autoimmune and inflammatory diseases that worsen with stress.[30]

15 NEUROPEPTIDE Y

Research over the past decades has revealed close interactions between the nervous and immune systems that regulate peripheral inflammation and link psychosocial stress with chronic somatic disease. Besides activation of the sympathetic and the hypothalamus–pituitary–adrenal axis, stress leads to increased *neurotrophin and neuropeptide* production in organs at the self–environment interface. The scope of this short review is to discuss key functions of these stress mediators in the skin, an exemplary stress-targeted and stress-sensitive organ. We will focus on the skin's response to acute and chronic stress in tissue regeneration and pathogenesis of allergic inflammation, psoriasis, and skin cancer to illustrate the impact of local stress-induced neuroimmune interaction on chronic inflammation.[31]

16 NEUROKININ 1

Cell-derived microparticles participate in intercellular communication similar to the classical messenger systems of *small and macro-molecules that bind to specialized membrane receptors*. Microparticles have been implicated in the regulation of a variety of complex physiopathologic processes, such as thrombosis, the control of innate and adaptive immunity, and cancer. The neurokinin 1 receptor (NK1R) is a *Gq-coupled receptor* present on the membrane of a variety of tissues, including *neurons in the central and peripheral nervous system, immune cells, endocrine and exocrine glands, and smooth muscle. The endogenous agonist of NK1R is the undecapeptide SP.* We have previously described intracellular signaling mechanisms that regulate NK1R-mediated rapid cell shape changes in HEK293 cells and U373MG cells. In the present study, we show that the activation of *NK1R in HEK293 cells*, but not in U373MG cells, *leads to formation of sheerstress-induced microparticles* that stain positive with the membrane-selective fluorescent dye FM 2–10. SP-induced microparticle formation is independent of elevated intracellular calcium concentrations and activation of NK1R present on HEK293-derived microparticles triggers detectable calcium increase in SP-induced microparticles. The ROCK inhibitor Y27632 and the dynamin inhibitor dynasore inhibited membrane blebbing and microparticle formation in HEK293 cells, strongly suggesting that *microparticle formation in this cell type is dependent on membrane blebbing.*[32]

SP and *hemokinin-1* are *proinflammatory neuropeptides* with potential to promote type 1 immunity through agonistic binding to *neurokinin-1 receptor* (NK1R). DCs are professional antigen-presenting cells that initiate and regulate the outcome of innate and adaptive immune responses. Immunostimulatory DCs are highly desired for the development of positive immunization techniques. DCs express functional NK1R; however, regardless of their potential DC-stimulatory function, the ability of NK1R agonists to promote immunostimulatory DCs remains unexplored. Here, we demonstrate that *NK1R signaling activates therapeutic DCs capable of biasing type 1 immunity by inhibition of IL-10 synthesis and secretion,* without affecting their low levels of IL-12 production. The potent type 1 effector immune response observed following cutaneous administration of NK1R-signaled DCs required their homing in skin-draining lymph nodes (sDLNs) where they induce inflammation and license endogenous-conventional sDLN-resident and -recruited inflammatory DCs to secrete IL-12. Our data demonstrate that *NK1R signaling promotes immunostimulatory DCs*, and provide relevant insight into the mechanisms used by neuromediators to regulate innate and adaptive immune responses.[33]

17 ALPHA-MELANOCYTE-STIMULATING HORMONE

17.1 Urocortin II

Aims: Phagocytosis plays essential roles during inflammation and immune response. This study aims to explore the underlying mechanism of corticotropin-releasing hormone *(CRH) and urocortin (UCN)-promoted phagocytosis* of rat macrophages. *Main methods:* To induce phagocytosis, rat macrophages were incubated with carboxylated fluorescent microspheres. The phagocytosis activity was evaluated by flow cytometric analysis. Actin reorganization was determined by immunostaining with TRITC-labeled phalloidin and transmission electron microscopy (TEM) analysis. Protein expressions of p-RhoA, p-Rac1, p-extracellular signal-related kinase (ERK)1/2, and GAPDH were examined by Western blotting. Protein kinase C (PKC) and protein kinase A (PKA) activities were examined using PreTag nonradio activity assay. *Key findings:* Administration of CRH or UCN alone significantly enhanced phagocytosis of microspheres by rat macrophages, as well as actin reorganization. Ligation of CRH and UCN with CRH receptor increased the phosphorylation of both RhoA and Rac1. Inhibition of RhoA/Rac1 signal pathway suppressed CRH- or UCN-enhanced phagocytosis and actin reorganization. Blockage of PKA signal by MDL-12330A decreased CRH or UCN-promoted p-RhoA and p-Rac1 expressions. Blockage of PKC signal by cholerythine choride decreased CRH or UCN-promoted p-Rac1 expression and UCN-promoted p-RhoA expression, but increased the CRH-induced p-RhoA expression. ERK1/2 was also activated and served as upstream factor of RhoA/Rac1 signal pathway. *Significance:* The results reveal that CRH and UCN promote phagocytosis of rat macrophages through convergent but dissociable pathways. PKA/PKC–ERK1/2-RhoA/Rac1 signal pathway plays an essential role in CRH- and UCN-enhanced phagocytosis.[34]

CRF1 and CRF2 affect stress response and innate immunity. *Adipose tissue expresses a complete CRF system.* The aim of this study was to examine the role of *CRF neuropeptides* in the immune phenotype of adipocytes assessed by their expression of the *toll-like receptor-4* (TLR4), the production of inflammatory cytokines *IL-6, TNF-α and IL-1β, chemokines IL-8,* monocyte attractant protein-1 (*MCP-1*), and of the adipokines *adiponectin, resistin and leptin.* Our data are as follows: (a) *CRF, UCN2 and UCN3 are expressed in human white adipocytes as well as CRFR1a, CRFR2a, and CRFR2b* but not CRFR2c. *3T3L1 preadipocytes and differentiated adipocytes expressed both CRF1 and CRF2 receptors and UCN3,* while UCN2 was detected only in differentiated adipocytes. *CRF2 was upregulated in mouse mature adipocytes.* (b) *CRF1 agonists suppressed LPS-induced preadipocyte differentiation while* CRF2 receptor agonists had no effect. (c) In mouse preadipocytes, *CRF2 agonists suppressed TLR4 expression* and the *production of IL-6, CXCL1 and adiponectin* while CRF1 agonists had no effect. (d) In *mature mouse adipocytes LPS-induced IL-6 and CXCL1 production and suppressed leptin.* (e) In *human visceral adipocytes LPS-induced IL-6, TNF-α, IL-8, MCP-1, and leptin* production and *suppressed adiponectin and resistin.* (f) *In mouse mature adipocytes CRF1 and CRF2 agonists suppressed basal and LPS-induced production of inflammatory cytokines, TLR4 expression and adiponectin production,* while in *human visceral adipocytes CRF and UCN1 suppressed basal and LPS-induced IL-6, TNF-α, IL-8, and MCP-1 production.*

Conclusions: The effects of the activation of CRF1 and CRF2 may be significant in ameliorating the proinflammatory activity of adipocytes in obesity.[35]

We currently face an alarming resurgence in infectious diseases characterized by antimicrobial resistance and therapeutic failure. The need for developing new therapeutic approaches that include agents with nontraditional modes of action, is urgent. A recent interest focused on approaches based on natural immune defenses, especially on peptides that combine innate antimicrobial activity against diverse pathogens and immunoregulatory functions. In this study, *the antimicrobial activity of the neuropeptide urocortin II (UCNII) against a panel of Gram-positive and Gram-negative bacteria and tropical parasites of the genus Leishmania, has been described for the first time.* Importantly, *this cytotoxicity was selective for pathogens* because UCNII did not affect mammalian cell viability. *Structurally, UCNII has a cationic and amphipathic design that resembles antimicrobial peptides.* Using mutants and UCNII fragments, the structural requirements for the interaction between the peptide and the surface of pathogen, were determined. Following its binding to pathogen, *UCNII caused cell death through different membrane-disrupting mechanisms that involved aggregation and membrane depolarization in bacteria and pore formation in Leishmania.* Noteworthily, UCNII killed the infective form of *Leishmania major* even inside the infected macrophages. Consequently, UCNII prevented mortality caused by polymicrobial sepsis and ameliorated pathological signs of cutaneous leishmaniasis. Besides its presence in body physical and mucosal barriers, it was found that innate immune cells produce UCNII in response to infections. Therefore, UCNII could be considered as an ancient highly conserved host peptide involved in the natural antimicrobial defense and emerge as an attractive alternative to current treatments for microbial disorders with associated drug resistances.[36]

17.2 Vasoactive Intestinal Peptide

The *spontaneous autoimmune peripheral polyneuropathy* (SAPP) model in B7-2 KO nonobese diabetic *mice mimics a progressive and unremitting course of chronic inflammatory demyelinating polyradiculoneuropathy* (CIDP). In this study, bone marrow–derived DCs were transduced to express vasoactive intestinal polypeptide (VIP) using a lentiviral vector (LV–VIP). These transduced DCs (LV–VIP–DCs) were then injected intravenously (i.v.) *into 16-weeks old (before disease onset) and 21-weeks old (after disease onset) SAPP mice* in order to prevent or attenuate the disease. Outcome measures included behavioral tests, clinical and histological scoring, electrophysiology, real-time PCR, flow cytometry analyses, and enzyme-linked immunosorbent assay. LV-VIP-DCs were recruited to the inflamed sciatic nerve and reduced the expression of inflammatory cytokines. *A single injection of LV-VIP-DC delayed the onset of disease, stabilized, and attenuated clinical signs correlating with ameliorated behavioral functions, reduced nerve demyelination, and improved nerve conduction.*

This *proof-of-principle* study is an important step potentially leading to a clinical translational study using DCs expressing VIP in cases of CIDP refractory to standard immunosuppressive therapy.[37]

Cytomegalovirus (CMV) infection, following allogeneic bone marrow transplant (allo-BMT), is controlled by donor-derived cellular immunity. Vasoactive intestinal peptide (VIP) suppresses Th1 immunity. It was hypothesized that blockingVIP-signaling would enhance anti-CMV immunity in murine recipients of allo-BMT. Recipients were transplanted with bone marrow (BM) and T-cells from major histocompatibility complex (MHC)–mismatched VIP KO or wild-type donors, and treated with 7 daily subcutaneous injections of *VI-Phyb (peptidic VIP-antagonist)* or phosphate-buffered saline (PBS). *Genetic and pharmacological blockade of VIP-signaling protected allo-BMT recipients from lethal murine CMV (mCMV) infection, improving survival without increasing graft-versus-host disease*. Mice treated withVIP hy transplanted withVIP-KO allografts had significantly lower viral loads, increased numbers of mCMV-M45-peptide-MHC-tetramer1 CD81 T-cells, with lower PD-1 expression, and enhanced primary and secondary cellular immune responses after mCMV infection than did PBS-treated mice. These results demonstrate that administration of a *VIP antagonist after allo-BMT* is a promising safely therapeutic approach to enhance antiviral cellular immunity.[38]

NPs represent an important category of endogenous contributors to the establishment and maintenance of immune deviation in the immune-privileged organs such as the CNS and in the control of acute inflammation in the peripheral immune organs. *VIP* is a major immunoregulatory neuropeptide widely distributed in the central and peripheral nervous system. In addition to neurons, VIP is synthesized by immune cells, which also express VIP receptors. Here, we review the current information on VIP production and VIP-receptor-mediated effects in the immune system, the role of endogenous and exogenous VIP in inflammatory and autoimmune disorders and the present and future VIP therapeutic approaches.[39]

Sjögren's syndrome (SS) is a chronic autoimmune disease characterized by salivary and lacrimal gland dysfunction. Clinical observations and results from animal models of SS support the role of aberrant epithelial cell apoptosis and immune homeostasis loss in the glands as triggering factors for the autoimmune response. VIP promotes potent anti-inflammatory effects in several inflammatory and autoimmune disease models, including the non-obese diabetic (NOD) mouse model of SS. With the knowledge that VIP modulates monocyte function through vasoactive intestinal peptide receptors (VPAC) and that immune homeostasis maintenance depends strongly upon a rapid and immunosuppressant apoptotic cell clearance by monocytes/macrophages, in this study we explored VPAC expression on monocytes from primary SS (pSS) patients and the ability of VIP to modulate apoptotic cell phagocytic function and cytokine profile. Monocytes isolated from individual pSS patients showed an increased expression of VPAC2 subtype of VIP receptors, absent in monocytes from control subjects, with no changes in VPAC1 expression. VPAC2 receptor expression could be induced further with lipopolysaccharide (LPS) in pSS monocytes and VIP inhibited the effect. Moreover, monocytes from pSS patients showed an impaired phagocytosis of apoptotic epithelial cells, as evidenced by reduced engulfment ability and the failure to promote an immunosuppressant cytokine profile. However, VIP neither modulated monocyte/macrophage phagocytic function nor did it reverse their inflammatory profile. We conclude that monocytes from pSS patients express high levels of VPAC2 and display a deficient clearance of apoptotic cells that *is not modulated by VIP*.[40]

REFERENCES

1. Chang Y, Wei W. Angiotensin II in inflammation, immunity and rheumatoid arthritis. *Clin Exp Immunol* 2014;**179**:137–45.
2. Cortez-Retamozo V, Etzrodt M, Newton A, Ryan R, Pucci F, Sio SW, Kuswanto W, Rauch PJ, Chudnovskiy A, Iwamoto Y, Kohler R, Marinelli B, Gorbatov R, Wojtkiewicz G, Panizzi P, Mino-Kenudson M, Forghani R, Figueiredo J, Chen JW, Xavier R, Swirski F, Nahrendorf M, Weissleder R, Pit MJ. Angiotensin II drives the production of tumor-promoting macrophages. *Cell Immunity* 2013;**38**:396–408.
3. Ji H, Zheng W, Li X, Liu J, Wu X, Zhang MA, Umans JG, Hay M, Speth RC, Dunn SE, Sandberg K. Sex-specific T-cell regulation of angiotensin II–dependent hypertension. *Hypertension* 2014;**64**:573–82.
4. Markó L, Kvakan H, Park J-K, Qadri F, Spallek B, Binger KJ, Bowman EP, Kleinewietfeld M, Fokuhl V, Dechend R, Müller DN. *Interferon-γ* signaling inhibition ameliorates *angiotensin II*-induced cardiac damage. *Hypertension* 2012;**60**:1430–6.
5. Ait-Oufella H, Wang Y, Herbin O, Bourcier S, Potteaux S, Joffre J, Loyer X, Ponnuswamy P, Esposito B, Dalloz M, Laurans L, Tedgui A, Mallat Z. Natural regulatory T cells limit *angiotensin II*-induced aneurysm formation and rupture in mice. *Arterioscler Thromb Vasc Biol* 2013;**33**:2374–9.
6. Ji H, Zheng W, Li X, Liu J, Wu X, Zhang MA, Umans JG, Hay M, Speth RC, Dunn SE, Sandberg K. Sex-specific T-Cell regulation of *angiotensin II*-dependent hypertension. *Hypertension* 2014;**64**:573–82.
7. Ju X, Ijaz T, Sun H, Ray S, Lejeune W, Lee C, Recinos III A, Chuan Guo D, Dianna DM, Tilton RG, Brasier AR. Interleukin-6-signal transducer and activator of transcription-3 signaling mediates aortic dissections induced by *angiotensin II* via the T-helper lymphocyte17-interleukin 17 axis in C57BL/6 mice. *Arterioscler Thromb Vasc Biol* 2013;**33**:1612–21.
8. Yodoi K, Yamashita T, Sasaki N, Kasahara K, Emoto T, Matsumoto T, Kita T, Sasaki Y, Mizoguchi T, Sparwasser T, Hirata K. Foxp3+ regulatory T cells play a protective role in *angiotensin II-induced* aortic aneurysm formation in mice. *Hypertension* 2015;**65**: 889–95.
9. Zhang J, Patel MB, Griffiths R, Mao A, Song Y, Karlovich NS, Sparks MA, Jin H, Wu M, Lin EE, Crowley SD. Tumor necrosis factor-α produced in the kidney contributes to *angiotensin II*-dependent *hypertension*. *Hypertension* 2014;**64**:1275–81.
10. Zhaoa H, Lei ML, Yousu W, Hongcheng S. *Angiotensin II* induces TSLP via an AT1 receptor/NF-KappaB pathway, promoting Th17 differentiation. *Cell Physiol Biochem* 2012;**30**:1383–97.

11. Carnevale D, Pallante F, Fardella V, Fardella S, Iacobucci R, Federici M, Giuseppe Cifelli G, De Lucia M, Lembo G. The *angiogenic factor* PlGF mediates a neuroimmune interaction in the spleen the angiogenic factor PlGF mediates a neuroimmune interaction in the spleen to allow the onset of hypertension. *Immunity* 2014;**41**:737–52.
12. Ganor Y, Drillet-Dangeard A-S, Lopalco L, Tudor D, Tambussi G, Delongchamps NB, Zerbib M, Bomsel M. *Calcitonin gene–related peptide inhibits* Langerhans cell–mediated HIV-1 transmission. *J Exp Med* 2013;**210**:2161–70.
13. Matsuda R, Kezuka T, Yoshihiko C, Usui Y, Matsunaga Y, Okunuki Y, Yamakawa N, Ogawa H, Okumura K, Goto H. Suppression of murine experimental autoimmune optic neuritis by mature dendritic cells transfected with calcitonin *gene-related peptide gene. Invest Ophthalmol Vis Sci* 2012;**53**:5475–85.
14. Mikami N, Sueda K, Ogitani Y, Otani I, Takatsuji M, Wada Y, Watanabe K, Yoshikawa R, Nishioka S, Hashimoto N, Miyagi Y, Fukada S, Yamamoto H, Tsujikawa K. Calcitonin gene-related peptide regulates type IV hypersensitivity through dendritic cell functions. *PLoS ONE* 2014;**9**(1):e86367.
15. Voedisch S, Rochlitzer S, Veres TZ, Spies E, Braun A. Neuropeptides control the dynamic behavior of airway mucosal dendritic cells. *PLoS ONE* 2012;**7**:e45951.
16. Busch RA, Heneghan AF, Pierre JF, Wang X, Kudsk KA. The enteric nervous system neuropeptide, *bombesin*, reverse innate immune impairments during parenteral nutrition. *Ann Surg* 2014;**260**:432–44.
17. Brasa G, Bochenskaa O, Rapala-Kozika M, Guevara-Loraa I, Faussnerb A, Kamyszc,d W, Kozika A. Release of biologically active kinin peptides, *Met-Lys-bradykinin* and Leu-Met-Lys-bradykinin from human kininogens by two major secreted aspartic proteases of *Candida parapsilosis*. *Peptides* 2013;**48**:114–23.
18. Gein SV. *Dynorphins* in regulation of immune system functions. *Biochemistry* 2014;**79**:397–405.
19. Gein SV, Baeva1 TA, Nebogatikov VO, Tendryakova SP. *Beta-endorphin* effects on antibody production, proliferation, and secretion of Th1/Th2 cytokines. *In Vivo Bull Exp Biol Med* 2012;**152**:595–9.
20. Sarkar DK, Zhang C. *Beta-Endorphin* neuron regulates stress response and innate immunity to prevent breast cancer growth and progression. *Vitamins Hormones* 2013;**93**:263–76.
21. Mallamaci F, Leonardis D, Pizzini P, Cutrupi S, Tripepi G, Zoccali C. *Procalcitonin* and the inflammatory response to salt in essential hypertension: a randomized cross-over clinical trial. *J Hypertension* 2013;**31**:1424–30.
22. Tripepia G, Zoccalia C. *Procalcitonin* and the inflammatory response to salt in essential hypertension: a randomized cross-over clinical trial. *J Hypertension* 2013;**31**:1424–30.
23. Phan TA, Taylor AW. The neuropeptides *α-MSH and NPY* modulate phagocytosis and phagolysosome activation in RAW 264.7 cells. *J Neuroimmunol* 2013;**260**:9–16.
24. Li W, Chen W, Ronald B, Herberman RB, Plotnikoff NP, Youkilis G. Immunotherapy of cancer via mediation of cytotoxic T lymphocytes by *methionine enkephalin* (MENK). *Cancer Lett* 2014;**344**:212–22.
25. Sayedyahossein S, Xu SX, Rudkouskaya A, McGavin MJ, McCormick JK, Dagnino L. *Staphylococcus aureus* keratinocyte invasion is mediated by *integrin*-linked kinase and Rac1. *FASEB J* 2015;**29**:711–23.
26. Lang R, Gundlach AL, Holmes FE, Hobson SA, Wynick D, Hökfelt T, Kofler B. Physiology, signaling, and pharmacology of *galanin* peptides and receptors: three decades of emerging diversity. *Pharmacol Rev* 2015;**67**:118–75.
27. Remick DG, Lentsch AB. *Substance P* mediates reduced pneumonia rates after traumatic brain injury. *Crit Care Med* 2014;**42**:2092–100.
28. Lotti T, D'Erme AM, Hercogová J. The role of neuropeptides in the control of egional immunity. *Clin Dermatol* 2014;**2**:633–45.
29. Ji H, Zou Y, Duan J, Jia Z, Li X, Wang Z, Li L, Li Y, Liu G, Tong MQ, Li X, Zhang G, Dai X, Ling He1, Li Z, Cao C, Yong Y. The synthetic *melanocortin (CKPV)2* exerts

anti-fungal and anti-inflammatory effects against *Candida albicans* vaginitis via inducing macrophage M2 polarization. *PLoS ONE* 2013;**8**:e56004.

30. Konstantinos–Dionysios A, Asadi S, Angelidou A, Zhang B, Sismanopoulos N, Yang H, Critchfield A, Theoharides TC. *Neurotensin* and CRH interactions augment human mast cell activation. *PLoS ONE* 2012;**7**:e48934.
31. Peters EMJ, Liezmann C, Klapp BF, Kruse J. The neuroimmune connection interferes with tissue regeneration and chronic inflammatory disease in the skin. *Ann NY Acad Sci* 2012;**1262**:118–26.
32. Chen P, Douglas DS, Meshki J, Tuluc1 F. *Neurokinin 1 receptor* mediates membrane blebbing and sheer stress-induced microparticle formation in HEK293 cells. *PLoS ONE* 2011;**7**:e45322.
33. Janelsins BM, Sumpter TL, Tkacheva OA, Rojas-Canales DM, Erdos G, Mathers AR, Shufesky WJ, Storkus Jr WJ, Falo LD, Morelli AE, Larregina AT. *Neurokinin-1 receptor* agonists bias therapeutic dendritic cells to induce type 1 immunity by licensing host dendritic cells to produce IL-12. *Blood.* 2013;**121**:2923–33.
34. Wang W-C, Zhang X-Z, Liub D-W, Qiu J, Wang X-H, Zhou J-H. Corticotropin-releasing hormone and *urocortin* promote phagocytosis of rat macrophages through convergent but distinct pathways. *Life Sci* 2015;**122**:100–7.
35. Dermitzaki E, Liapakis G, Androulidakin A, Venihaki M, Melissas J, Tsatsanis C, Margioris AN. Corticotrophin-releasing factor (CRF) and the *urocortins* are potent regulators of the inflammatory phenotype of human and mouse white adipocytes and the differentiation of mouse 3T3L1 pre-adipocytes. *PLoS ONE* 2014;**9**:e97060.
36. Campos-Salinas J, Caro M, Cavazzuti A, Forte-Lago I, Beverley SM, O'Valle F, Gonzalez-Rey E. Protective role of the neuropeptide *urocortin II* against experimental sepsis and leishmaniasis by direct killing of pathogens. *J Immunol* 2013;**191**:6040–51.
37. Yalvac ME, William WD, Hussain SRA, Braganza C, Shontz KM, Clark KR, Walker CM, Ubogu EE, Mendell JR, Sahenk Z. VIP-expressing dendritic cells protectmagainst spontaneous autoimmune peripheral polyneuropathy. *Am Soc Gene Cell Ther Mol Ther* 2014;**22**:1353–63.
38. Li JM, Hossain MS, Southerland L, Waller EW. Pharmacological inhibition of VIP signaling enhances antiviral immunity and improves survival in murine cytomegalovirus-infected allogeneic bone marrow transplant recipients. *Blood* 2013;**121**:2347–51.
39. Ganea D, Hooper KM, Kong W. The neuropeptide *vasoactive intestinal peptide*: direct effects on immune cells and involvement in inflammatory and autoimmune diseases. *Acta Physiol* 2015;**213**:442–52.
40. Hauk V, Fraccaroli L, Grasso E, Eimon A, Ramhorst R, Hubscher O, Pérez Leirós C. Monocytes from Sjögren's syndrome patients display increased *vasoactive intestinal peptide receptor 2* expression and impaired apoptotic cell phagocytosis. *Clin Exp Immunol* 2014: 177(3):662–70.

PART 4

Neuroimmune Pathology

CHAPTER 8

Vasopressin, the Acute Phase Response and Healing

István Berczi*,, Andrés Quintanar Stephano****
*Department of Immunology, Faculty of Medicine, The University of Manitoba, Winnipeg, Manitoba, Canada
**Departamento de Fisiología y Farmacología, Centro de Ciencias Básicas, Universidad Autónoma de Aguascalientes, Aguascalientes, Estado Libre y Soberano de Aguascalientes, Mexico City, Mexico

1 INTRODUCTION

The first review paper on the neuroendocrine changes of acute illness was published by Hans Selye. He proposed that "nocuous" stimuli (stressors) induce a stress response. Selye recognized that stressed organisms develop a defense reaction (general adaptation syndrome) against the insults encountered.[1] Today, the response to various noxious insults is known as acute phase response (APR).[1–3] Clinically, APR is characterized by fever, loss of appetite, inactivity, and sleepiness. Changes in sleep are hallmarks of the APR to infectious challenge. The regulation of these responses involves a cytokine cascade within the brain, including IL-1, TNF, IL-6, GM-CSF, and several other substances, such as growth hormone releasing hormone, PRL, nitric oxide (NO), and nuclear factor (NF)-κB. These substances are also involved in the regulation of normal spontaneous sleep.[4,5] Endotoxin, infectious disease, and various forms of injury all elicit a systemic elevation of IL-1β, TNF-α, and IL-6, which are secreted by cells of the innate immune system, primarily by monocyte/macrophages.[2] Much attention has been paid recently to *colony-stimulating factors* (*CSF*), which are cytokines originally described to stimulate bone marrow cells.[6] The bone marrow is recognized today as a fundamental organ in natural immunity as it provides the leukocytes that fulfill this function. The bone marrow is activated during APR, which results in the generation of *leukocytosis*.[7] Recent studies provided evidence that granulocyte macrophage CSF (GM-CSF), macrophage CSF (M-CSF), and granulocyte CSF (G-CSF) are involved in the maintenance of host resistance to infectious disease, parasitic infestation,[8] and cancer.[9,10] At least some of the CSFs show elevated levels in response to infectious disease.[11] GM-CSF has been proposed as a physiological regulator as it did increase host defense without causing or aggravating inflammation, which may lead to complications.[12]

Insights to Neuroimmune Biology
http://dx.doi.org/10.1016/B978-0-12-801770-8.00008-2

In addition, GM-CSF stimulates adaptive immune reactions,[13] and immunological tolerance.[14,15] GM-CSF influences dendritic cells (DC), which present antigen to T lymphocytes, and invariant NK T cells that produce cytokines having important functions in inflammation, host defense, and immunoregulation.[16,17] GM-CSF is produced in the bone marrow, by mucosal epithelial cells in the lung and also in the brain. Receptors are also present in the brain for this cytokine.[18,19] Macrophages play important roles in host defense and also regulate body metabolism.[20] GM-CSF is used currently for immunotherapy and for the production of recombinant and human hepatocytes.[21,22] These findings strongly support the aforementioned hypothesis. One may argue that the most efficient way to fuel intensive systemic effort for survival in APR is by the rapid breakdown of bodily tissues. GH is a powerful anabolic hormone, which supports the T lymphocyte-dependent immune system, and acts as an antagonist of the HPA axis that promotes APR.[21–23] The results of this controlled trial, support the hypothesis that the inhibition of the HPA axis and catabolism by GH treatment in APR hampers the body's defense mechanisms, which may have fatal consequences. The regulation of acute phase protein (APP), *GC*, and *CAT* induces α-MFP synergistically in normal rats.[24] Near-basal concentrations of GC exerted a stimulatory effect on hepatic APR, on the proinflammatory macrophage migration inhibition factor secretion, and on the expression of cytokine/chemokine receptors.[25] Adrenalin induces a high level of IL-6 in rats, which can be antagonized by propranolol. When IL-6 release is blocked, the fast reacting APP, α2-macroglobulin, and cysteine protease inhibitors are strongly depressed. Isoprenalin, a β2-adrenergic receptor agonist, also causes very high levels of IL-6, indicating that β2 receptors are involved.[26] *IL-6* is a major inducer of APP synthesis. Interferon (IFN)-γ, leukemia inhibitory factor, TGF-β, and oncostatin M also induce APP from the liver. IL-6 activates the genes of APP through the DNA-binding protein called *NF-IL-6*. NF-IL-6 is a pleiotropic mediator of many inducible genes involved in the acute, immune, and inflammatory responses, similarly to NF-κB. NF-IL-6 and NF-κB binding sites are present in the inducible genes, such as IL-6, IL-8, and several acute phase genes.[27] In humans, IL-6 exerted a hyperglycemic effect, whereas IL-2 had an opposite effect.[28]

1.1 Immunoconversion in APR

During APR, HPA axis and CAT are activated. This activation results in the suppression of T cell-dependent adaptive immune response, which is supported further by the suppression of the hormones essential for the maintenance of thymus and T lymphocytes (e.g., PRL, GH, IGF-I). CAT and GC, which are released in large quantities during APR, induce apoptosis in the thymus with a striking efficiency. The elevated level of TNF-α and zinc deficiency, which develops during APR, also contributes to thymic involution and to the suppression of the adaptive immune system.[29,30]

1.2 Healing

Febrile illness develops during numerous occasions in a lifetime. Most of these febrile episodes are followed by healing, return to health, and normal immunocompetence. By now, much about APR related to how it develops and what it does has been understood. However, little is known about the recovery phase. Recent observations suggest that *VP* (vasopressin) is the central hypothalamic coordinator during healing, which also leads to normal immune function (*immunoreversion*).[7] VP is primarily a neurohypophysial hormone, produced in magnocellular neurons of the hypothalamic paraventricular and supraoptic nuclei, but parvocellular CRH neurons also coexpress VP. VP acts as a second "releasing factor" for ACTH along with CRH. Aminergic, cholinergic, GABAergic, glutamatergic, and a number of peptidergic inputs have all been implicated in the regulation of CRH/VP neurons. VP is also expressed within the immune system.[29–31]

1.3 Vasopressin and Stress

CRH and VP are released during stress by the elevated levels of epinephrine (EP) and norepinephrine, and seconds later the secretion of ACTH is induced, which elicits the secretion of GC by the adrenal cortex. CRH coordinates endocrinal, autonomic, behavioral, and immune responses to stress, and also acts as a neurotransmitter or neuromodulator in the amygdala, dorsal raphe nucleus, hippocampus, and locus coeruleus. VP, 5-hydroxytryptamine, CAT, substance P, vasoactive intestinal polypeptide, neuropeptide Y, and cholecystokinin are produced in these loci. Cytokines such as IL-1 stimulate CRH and VP gene expression, and are implicated in immune–neuroendocrine regulation. Expression profiles of the CRH and VP genes are not uniform after stress exposure, and the VP gene appears to be more sensitive to GC suppression.[30,31] Cytokine-induced CRF and VP synthesis and/or release is modulated by catecholamines, prostaglandins, and NO.[32]

Compared to Wistar rats, Lewis rats exhibited low VP but identical CRH, and IL-1 priming significantly increased VP without affecting CRH stores. This is consistent with a shift to VP-dominated control of ACTH secretion, as described in Wistar rats under conditions of HPA hyper(re)activity. IL-1 priming of Lewis rats attenuated the ACTH responses to an IL-1 challenge 11 days later. There was no effect on immune reactivity.[33] Immunoreactive CRH was significantly decreased in splenic extracts from arthritic rats. Low levels of VP were found in immune tissues, which significantly increased in spleens from arthritic animals, but thymic contents of VP were not altered.[34] Elevated plasma VP levels were noted in seven patients with status asthmaticus during the acute attack. These values returned to normal with resolution of the acute phase.[35] Thirteen multiple sclerosis (MS) patients were studied at baseline and with provocative tests of HPA axis function (ovine CRH, VP, and ACTH stimulation). Compared to matched controls, patients with MS had significantly higher plasma cortisol levels at baseline but showed normal, rather than blunted plasma ACTH responses to ovine CRH. Patients had blunted ACTH responses to VP stimulation and normal cortisol responses to high- and low-dose ACTH stimulation.[36]

1.4 Vasopressin in Chronic Inflammation

During chronic inflammatory diseases, such as adjuvant-induced arthritis of rats, CRH does not act as the major ACTH-releasing factor. This is also the case for experimental allergic encephalomyelitis, eosinophilia–myalgia syndrome, systemic lupus erythematosus, and leishmaniasis. During chronic inflammation VP takes over as the major regulator of the HPA axis.[37] Chronic intermittent exposure to immobilization, insulin-induced hypoglycemia, or psychological stress stimuli increased the number of CRH cells containing VP and the ratio of VP to CRH within the zona externa of the median eminence.[38–41] In chronically restrained rats, exogenous VP but not CRH was found to increase plasma levels of both ACTH and corticosterone.[39] Chronic inflammation is associated with a much larger stimulation of VP than other stress models. Activation of CRH does not appear to play a role under these conditions. However, only CRH can stimulate proopiomelanocortin (POMC) transcription, not VP. VP has been less potent than CRH in producing ACTH release from rat pituitaries. The effect of VP on CRF-mediated ACTH release is either synergistic or additive.[40,41]

1.5 VP and Cytokines

IL-1β significantly potentiated the acetylcholine-induced VP release. This effect was completely blocked by neutralizing antibodies to IL-1β, atropine, or mecamylamine.[42] In IL-1β-treated rats, the change in body temperature and VP release from the ventral septal area, were negatively correlated.[43] In male rats, the ACTH response to IL-1β was significantly reduced by both anti-CRH and anti-VP antisera, compared to the levels after normal rabbit serum.[44] Similarly in rats, the *IL-6*-induced ACTH response was significantly suppressed by both anti-CRH and anti-VP antibodies. The TNF-α-induced ACTH response was not significantly affected by anti-VP antibody, although anti-CRH antibody suppressed the response.[45] IL-6 potentiated acetylcholine-induced VP release in rats.[46] A single injection of recombinant mouse *IL-2* caused a significant increase in VP and oxytocin (OT) mRNA levels in the hypothalamus of nude mice. This effect was specific to the nude mouse.[47] IL-2 released VP from the hypothalamus and amygdala of rats *in vitro*. The IL-2 and acetylcholine-induced VP release was antagonized by N-methyl-L-arginine, which indicates a role for NO in this VP release.[48] IL-2 caused a dose-dependent stimulation of VP, but not CRH, secretion from both the intact rat hypothalamus *in vitro* and hypothalamic cell cultures.[49] Centrally administered *leukemia inhibitory factor* significantly increased the plasma VP concentration from 5 min to 60 min after the injection.[49]

1.6 Role of Vasopressin in the Acute Phase Response

VP attenuated significantly the febrile response of rabbits to bacterial pyrogen endotoxin,[50] may stimulate the generation of NO, which inhibits CRH and generates carbon monoxide, which modulates the release of VP. These are potential counter regulatory controls of HPA activation.[51]

An inhibitor of central EP synthesis (SKF-64139) enhanced basal medial hypothalamus (BMH) and basal median eminence (BME) VP contents in adult male rats. LPS administration significantly decreased BMH VP in control and peripheral EP inhibitor (SKF-29661)-pretreated rats and diminished BME VP in all groups.[52] In rats, CRH gene transcription was upregulated in the PVN 3–4 h following LPS injection. Transcripts of CRH receptor type A were present in the hypothalamus 6 h after endotoxin treatment. No alterations in cytoplasmic VP mRNA levels were noted in rats treated with LPS. ACTH secretion was stimulated within 30 min. These results suggest that systemic LPS acts first within the median eminence where it stimulates peptidic-nerve terminals.[53] LPS potently stimulated CRH and VP secretion into pituitary portal blood in alert normally behaving ewes and stimulated cortisol and progesterone secretion into peripheral blood. Both CRH and VP rose and fell simultaneously, although the peak of the VP response was approximately 10-fold greater than that of CRH. There was a significant suppression of gonadotropin-releasing hormone and LH pulsatile secretion in these same ewes and fever developed.[54]

In Holstein steers, LPS-induced fever, increased plasma ACTH, and cortisol. Pituitary VP receptor V3 mRNA was decreased at 2, 4, and 12 h, following LPS administration and returned to basal by 24 h. A similar temporal regulation was observed in pituitary CRH receptor R1 mRNA, but not in pituitary POMC mRNA. Downregulation of CRHR1 mRNA was not observed in other brain regions following LPS administration (e.g., cerebellum, hypothalamus)[55] vaccines.[56,57] In turn, cytokines stimulate the hypothalamus, bone marrow, liver, and leukocytes directly or indirectly, thus eliciting an APR.[41,47,58] Profound changes occur in serum hormone levels, the HPA axis is activated, and there is also a *sympathetic outflow*, which raises serum CAT levels. GC exerts a powerful suppressive effect on the adaptive immune system and also controls the levels of inflammatory cytokines. Through the activation of this axis and the sympathetic nervous system, adaptive immune reactions are profoundly suppressed. APPs are induced in the liver, and natural antibody production gets dramatically increased by a specific subset of B lymphocytes. Therefore, the conversion of the immune system from an adaptive mode of reactivity to the amplification of natural immunity takes place.[47,59,60] PRL and GH stimulate the adaptive immune system and usually become elevated within the first hour after endotoxin injection, which is followed by a decline, and the level may rise from low to normal and then to subnormal in serious cases of endotoxin shock. As a rule, luteinizing hormone (LH), follicle-stimulating hormone (FSH), estrogens, androgens, progesterone, and thyroid hormones decline during infection and endotoxin shock. Insulin, glucagon, α-melanocyte-stimulating hormone (MSH), endorphin, leptin, corticotropin-releasing hormone (CRH), and arginine vasopressin are increased during endotoxemia.[41,47,58] It is clear that dynamic and diurnal changes of hormones should be kept in mind when hormonal alterations are taken into consideration during APR. Much remains to be elucidated about the nature and significance of APR. A subset of bone marrow-derived macrophages in the brain, termed as *perivascular cells*, synthesize prostaglandin after systemic cytokine or endotoxin challenges, and play a critical role in the IL-1-induced HPA axis activation. It has also been established that peptidergic sensory nerves present in the vagus and elsewhere provide feedback signals to the brain about the sites of local inflammation.

1.7 Acute Phase Proteins

Protein synthesis is profoundly altered in the liver, which is most characteristic for APR. The synthesis of APPs is dramatically enhanced, whereas the production of some normal serum constituents, such as albumin and transferrin, is significantly decreased. In patients with acute illness, C-reactive protein (CRP) and serum amyloid A (SAA) may increase over 1000-fold within 24–48 h. Fibrinogen, 1-antitrypsin, and certain complementary properdin components (e.g., factor B and C3) show a more moderate increase.[58] *CRP* binds to C-type pneumococcal cell walls in the presence of Ca^{2+}. It is present in multiple species, ranging from mammals to fish and crab. It consists of five identical subunits, which form a ring-shaped molecule named *pentraxin*, which also stands for a protein family. Serum amyloid P also belongs to this family. CRP recognizes *homotopes*, which are mostly present on the surface of bacteria, fungi, parasites, damaged cells, and tissues. After getting combined with specific homotopes, CRP activates its complement through the classical pathway, which leads to the induction of chemotaxis, of enhanced phagocytosis by neutrophil, leukocytes, monocytes, and tumoricidal activity in macrophages. CRP stimulates the synthesis of IL-1 and TNF-α and potentiates the cytotoxicity of T lymphocytes, natural killer (NK) cells, and platelets. CRP localizes *in vivo* at the sites of inflammation. It binds platelet-activating factor (PAF) and blocks its activity. The clinical determination of CRP is diagnostic for the presence of infectious and inflammatory diseases. Denaturation takes place after attachment onto polystyrol plates, and free-CRP subunits attain a third conformation, referred to as *neoCRP*, which is a membrane protein on NK cells and macrophages. It functions as a galactose-specific receptor. It also accumulates at the injured sites of a tissue. Monocytes/macrophages express a specific CRP receptor. Proteolytic fragments of CRP activate macrophages and neutrophils. Human CRP protects mice from an otherwise lethal *Streptococcus pneumoniae* infection. *Lipopolysaccharide-binding protein* (LBP) shows a 100-fold increase (from 0.5 μg/mL to 50 μg/mL) in the serum during an APR. It is capable of opsonizing LPS-bearing particles and may be required for the activation of complement by LPS through the alternate pathway. LBP–LPS complexes are also potent stimulators of cytokines from monocytes and macrophages, after combining with CD14 and toll-like receptor 4 on the surface of these cells. In humans, a highdose of LBP (hd-LBP) suppressed the binding of both R-type and S-type LPS to CD14 and inhibited S-induced nuclear translocation of NF-κB. This inhibitory effect of serum could be mimicked by purified high-density lipoprotein (HDL) in serum-free medium, indicating an LBP-mediated transfer of S-type LPS to plasma lipoproteins, such as HDL. *Haptoglobin* (HG) is an APP that binds hemoglobin, thus preventing iron loss and renal damage. It is an antioxidant, has antibacterial activity, and plays a role in modulating many aspects of APR. HG selectively suppresses the *in vitro* production of TNF-α, IL-10, and IL-12 by LPS-stimulated monocytes but it failed to inhibit the production of IL-6, IL-8, and IL-1 receptor antagonists. HG knockout mice were more sensitive to LPS than were their wild-type counterparts. The APP *α(1)-acid glycoprotein* and *α(1)-antitrypsin* exert anti-apoptotic and anti-inflammatory effects, and contribute to delayed type protection associated with ischemic reconditioning in the kidney and in other sites. *Mannose-binding lectin* (MBL) is an APP. It binds to the repeating sugar arrays on many microbial surfaces through multiple lectin domains. After binding, MBL is able to activate the complement system via an associated serum protease, MASP-2. There is an increased incidence of infections in individuals with mutations of MBL and an association with autoimmune disorders, such as systemic lupus erythematosus and rheumatoid arthritis. The serum concentration of MBL is subject to large individual differences. GH regulates MBL levels. Other APPs are *proteinase inhibitors*, such as α2-macroglobulin, α1-acid glycoprotein, antithrombin III, α1-acute-phase globulin, and α1-proteinase inhibitor, which are abundant in the rat. Kupffer cells stimulate α2-macroglobulin synthesis by hepatocytes *in vitro* in the presence of 10^{-9} M Dex. *Fibrinogen* is also an APP with an important role in blood coagulation and healing. α-Macrofetoprotein (α-MFP) is a strong inhibitor of inflammatory mediators, such as histamine, bradykinin, serotonin, and prostaglandin E2, and inhibits polymorphonuclear chemotaxis.

1.8 Metabolic Effects of APR

Proinflammatory cytokines, especially TNF, induce hyperleptinemia. This is an integral part of APR and is necessary for an integrated communication network to coordinate the energy status of the animal with the ability to fight pathogens. Hyperglycemia and lipolysis are characteristic of APR, and catabolism prevails. In patients with APR, the GH-insulin-like growth factor (IGF)-I axis is suppressed and catabolism prevails. Such patients were treated with GH in a controlled trial. GH treatment increased mortality. The deaths attributed to "septic shock or uncontrolled infection" and it occurred nearly four times more commonly in GH treated patients compared to placebo-receiving patients. Although no data were given regarding immune parameters, the authors suggested that alterations in immune functions may have contributed to these fatalities. The APR is a massive neuroimmune and metabolic response that catabolizes the resources and tissues of a body that is in the interest of host defense and survival. The findings of Takala and coworkers[50] strongly suggest that the suppression of the GH–IGF-I axis in APR is required for intense catabolism to take place. A rapid release of nutrients and energy, is necessary under these conditions in order to maximally support the defense system of the body, which includes the HPA axis, sympathetic nervous system, bone marrow, CD5+ B lymphocytes, leukocytes, and the liver.[47,58]

Recent observations showed that GH inhibits the production of APPs in rats with burn injuries and in human hepatocytes. These findings strongly support the aforementioned hypothesis. One may argue that the most efficient way to fuel the intensive systemic effort for survival in APR is by the rapid breakdown of bodily tissues. GH is a powerful anabolic hormone, which supports the T lymphocyte-dependent immune system, and acts as an antagonist of the HPA axis that promotes APR. Results of this controlled trial support the hypothesis that inhibition of the HPA axis and catabolism by GH treatment in APR hampers the body's defense mechanisms, which may have fatal consequences.

The regulation of APP, GC, and CAT induce α-MFP synergistically in normal rats. Near basal concentrations of GC exerted a stimulatory effect on hepatic APR, on the proinflammatory macrophage migration inhibition factor secretion, and on the expression of cytokine/chemokine receptors. Adrenalin induces a high level of IL-6 in rats, which can be antagonized by propranolol. When IL-6 release is blocked, the fast reacting APP, α2-macroglobulin, and cysteine protease inhibitor are strongly depressed. Isoprenalin, a β2-adrenergic receptor agonist, also causes very high levels of IL-6, indicating that β2 receptors are involved. *IL-6* is a major inducer of APP synthesis. IFN-γ, leukemia inhibitory factor, TGF-β, and oncostatin M also induce APP from the liver. IL-6 activates the genes of APP through the DNA binding protein called *NF-IL-6*. NF-IL-6 is a pleiotropic mediator of many inducible genes involved in acute, immune, and inflammatory responses, similarly to NF-κB. NF-IL-6 and NF-κB binding sites are present in the inducible genes, such as IL-6, IL-8, and several acute phase genes.[61] In humans, IL-6 exerted a hyperglycemic effect, whereas IL-2 had an opposite effect.[62]

1.9 Regulation of Pituitary Hormones by Vasopressin

The ACTH response to exogenous administration of VP was impaired in $V1bR^{-/-}$ mice, while CRH-stimulated ACTH release was not significantly different from that in the $V1bR^{+/+}$ mice. The increase in ACTH after a forced swim stress was significantly suppressed in $V1bR^{-/-}$ mice.[63]

In conscious male rats intracerebroventricular (i.c.v.) infusion of HA-induced PRL secretion, which was inhibited by pretreatment with a specific antiserum to VP. AVP antagonist also inhibited the PRL response to HA and inhibited the PRL response to restraint stress. Pretreatment with a specific OT antagonist had no effect on the HA- or stress-induced PRL release.[64] VP antiserum (VP-Ab) was administered i.v. to lactating rats 15 min before permitting their previously isolated pups to suckle or to continuously suckle. The suckling-induced rise in plasma PRL levels was significantly less in VP-Ab-treated mothers, when compared to rats receiving a similar amount of normal rabbit serum.[61] Anterior pituitary cells derived from juvenile female turkeys were incubated with posterior pituitary extracts or test substances for 3 h. Posterior pituitary extracts contained potent substance(s), which stimulated PRL release in a concentration-dependent manner. Antisera to VP and vasoactive intestinal peptide (1:500) completely abolished the PRL-releasing activities.[62] CRH or VP released ACTH and immunoreactive β-endorphin (β-END) in response to HA and restraint stress. Pretreatment with CRH antiserum abolished the ACTH response to stress and inhibited the β-END response by 60%. Immunoneutralization with VP antiserum had only half the inhibitory effect of that seen with CRH antiserum. CRH (100 pmol i.v.) increased the plasma levels of ACTH and β-END. This effect was abolished by pretreatment with CRH antiserum, whereas pretreatment with VP antiserum prevented the CRH-induced ACTH release and inhibited the β-END response by 50%. VP (24–800 pmol i.v.) stimulated ACTH and β-END in a dose-dependent manner. CRH and VP antisera, each prevented the effect of VP on ACTH secretion, whereas the β-END response to VP was only inhibited by about 60% by the antisera.[64] Endogenous OT contributes to the control of basal GH release probably by stimulating-somatostatin secretion and/or inhibiting GH releasing hormone secretion or by both actions. Endogenous VP and OT play a physiologically significant stimulatory role in the control of basal ACTH release.[65]

The novel high-affinity nonapeptide CRHR1 antagonist R121919 attenuates the stress-induced release of corticosterone, PRL, and OT. The decrease in plasma *testosterone* following exposure to stress is abolished by R121919.[66] In rats with paraventricular nucleus lesions, LPS was able to activate the hyperphysical adrenal system in the absence of hypophysiotrophic neuropeptides of paraventricular origin.[67]

1.10 Vasopressin and Immune Response

Brattleboro diabetes insipidus (DI) rats were derived from Long–Evans (LE) rats; they are homozygous for DI and lack VP. In DI rats, there was a permanent decrease in the number of blood lymphocytes, increase in neutrophil count, reduced activity of macrophages, early involution of the thymus and spleen, and suppression of antibody production.[68] In DI rats, NK cell activity was significantly higher than in LE rats. VP replacement normalized water intake in DI rats but had no significant effect on NK cell activity. DI rats exhibited lower plasma corticosterone levels, which were not elevated by VP replacement.[69] In mice with the disruption of VP receptor 1a (VPR1a) gene, a shift was observed from IgM (high)/IgD (high) to the more mature IgM (low)/IgD (high) B cells. Splenic B cell proliferation was significantly greater after anti-IgM stimulation, and enhanced IgG1 and IgG2b production occurred in response to immune challenge with T-dependent antigen. B-1 cells were increased in $VPR1a^{-/-}$ mice. T cell differentiation and activation were normal in $VPR1a^{-/-}$ mice.[70] The VP-binding nonapeptide has the sequence Thr–Met–Lys–Val–Leu–Thr–Gly–Ser–Pro (binding peptide). VP and its six-amino acid, N-terminal, cyclic ring pressinoic acid (PA) are both capable of replacing the IL-2 requirement for IFN-γ production by mouse splenic lymphocytes. The AVP-binding peptide specifically and reversibly blocked AVP help in IFN-γ production but failed to block the helper signal of PA. Thus the intact VP molecule and not just the N-terminal cyclic ring are important for interaction with the binding peptide.[56,57] VP enhanced the autologous mixed lymphocyte response. Enhanced proliferation appeared to be specific for the arginine residues in position 8 of VP.[58] In rats, the i.c.v. administration of VP suppressed the proliferative response of splenic T cells and NK cytotoxicity in an adrenalin-dependent manner. These effects were completely reversed by i.c.v. preadministration of the VR1 receptor antagonist.[59] VPR1 vasopressin receptor agonist induced a complex intracellular Ca^{2+}-signaling cascade in cortical astrocytes. It dramatically decreased the mRNA level of five cytokines, including IL-1β and TNF-α.[60]

2 CONCLUSIONS

This overview indicates that VP participates in immunoregulation, both by regulating pituitary hormones and by direct effect on immunocytes. It seems certain that VP is required for the maintenance of adaptive immunocompetence. Growth and lactogenic hormones are responsible for the maintenance of thymus function and T-cell-regulated adaptive immune system in a competent state.

It is clear that VP has the capacity to stimulate both HPA axis and PRL in a balanced fashion. This is in contrast with CRH, which stimulates the HPA axis exclusively. In APR, CRH is dominant because of its resistance to GC inhibition, whereas VP is in the background.

However, as it has been established, during chronic inflammatory disease, VP will take over the regulation of the HPA axis. Moreover, VP has the capacity to stimulate PRL synthesis, which is suppressed during APR. PRL is an important immunostimulator, and is capable of restoring adaptive immunocompetence, which sets the stage for recovery and healing. At the same time VP also maintains the HPA function at physiological levels, which maintains innate immunity by acting synergistically with catecholamines. These observations promise a better understanding of illness and recovery and are expected to significantly aid patient management. Additional studies are required to unravel the control of immune function by the hypothalamus via the pituitary gland, in a detailed manner.

REFERENCES

1. Berczi I. Neuro-endocrine response to endotoxin. *Ann NY Acad Sci* 1998;**851**:411–5.
2. Berczi I, et al. Immunoconversion in the acute phase response. In: Plotnikoff NP, editor. *Cytokines, stress and immunity*. Boca Raton, FL: CRC Press, Taylor & Francis Group; 2006. p. 215–54.
3. Paltrinieri S. The feline acute phase reaction. *Vet J* 2008;**177**:26–35.
4. Krueger JM, et al. Sleep in host defense. *Brain Behav Immun* 2003;**17**(Suppl 1):S41.
5. Krueger JM, et al., editors. Cytokines and sleep. Phelps C, Korneva EA, Berczi I, Szentivanyi A, series editors. *Neuroimmune biology*, vol. 6. Cytokines and the brain. Elsevier: Amsterdam; 2008. p. 213–240.
6. Metcalf D. Hematopoietic cytokines. *Blood* 2008;**111**:485–91.
7. Berczi I, Szentivanyi A. The acute phase response. Berczi I, Szentivanyi A, editors. *Neuroimmmune biology*, vol. 3. The immune-neuroendocrine circuitry. History and progress. Elsevier: Amsterdam; 2003. p. 463–494.
8. O'Mahony DS, et al. Differential constitutive and cytokine-modulated expression of human toll like receptors in primary neutrophils, monocytes, and macrophages. *Int J Med Sci* 2008;**5**:1–8.
9. Heit B, et al. HIV and other lentiviral infections cause defects in neutrophil chemotaxis, recruitment, and cell structure: immunorestorative effects of granulocyte-macrophage colony-stimulating factor. *J Immunol* 2006;**177**:6405–14.
10. Ballinger MN. Role of granulocyte macrophage colony-stimulating factor during gram negative lung infection with *Pseudomonas aeruginosa. Am J Respir Cell Mol Biol* 2006;**34**:766–74.
11. Kobayashi SD, et al. Spontaneous neutrophil apoptosis and regulation of cell survival by granulocyte macrophage-colony stimulating factor. *J Leukoc Biol* 2005;**78**:1408–18.
12. Kaur A. Bioimmunotherapy of rodent malaria: co-treatment with recombinant mouse granulocyte-macrophage colony-stimulating factor and an encephalin fragment peptide Tyr-Gly-Gly. *Acta Trop* 2004;**91**:27–41.
13. Everly JJ, Lonial S. Immunomodulatory effects of human recombinant granulocyte macrophage N.P. colony-stimulating factor (rhuGMCSF): evidence of antitumour activity. *Expert Opin Biol Ther* 2005;**5**:293–311.
14. Ezaki K, Tsuzuki M. Cytokine therapy for hematological malignancies. *Gan To Kagaku Ryoho Suppl* 1997;**1**:182–94.
15. Selig C, Nothdurft W. Cytokines and progenitor cells of granulocytopoiesis in peripheral blood of patients with bacterial infections. *Infect Immun* 1955;104–9.

16. Presneill JJ. A randomized phase II trial of granulocyte-macrophage colony-simulating factor therapy in severe sepsis with respiratory dysfunction. *Am J Respir Crit Care Med* 2002;**6**:129–30.
17. Mels AK, et al. Immune-stimulating effects of low-dose perioperative recombinant granulocyte macrophage colony-stimulating factor in patients operated on for primary colorectal carcinoma. *Br J Surg* 2001;**88**:539–44.
18. Ganguly D, et al. Granulocyte-macrophage colony-stimulating factor drives monocytes to CD14low CD83+ DCSIGN-interleukin-10-producing myeloid cells with differential effects on T-cell subsets. *Immunology* 2007;**121**:499–507.
19. Jinushi M, et al. MFG-E8-mediated uptake of apoptotic cells by APCs links the pro- and antiinflammatory activities of GM-CSF. *J Clin Invest* 2007;**117**:1902–13.
20. Han S, et al. Macrophage-colony stimulating factor enhances MHC-restricted presentation of exogenous antigen in dendritic cells. *Cytokine* 2005;**32**:187–93.
21. Bezbradica JS. Granulocyte-macrophage colony-stimulating factor regulates effector differentiation of invariant natural killer T cells during thymic ontogeny. *Immunity* 2006;**25**:487–97.
22. Baleeiro CE, Christensen PJ, Morris SB, et al. GM-CSF and the impaired pulmonary innate immune response following hyperoxic stress. *Am J Physiol Lung Cell Mol Physiol* 2006;**291**:L1246–55.
23. Franzen R, et al. Nervous system injury: focus on the inflammatory cytokine "granulocytemacrophage colony stimulating factor". *Neurosci Lett* 2004;**361**:76–8.
24. Naito M. Macrophage differentiation and function in health and disease. *Pathol Int* 2008;**58**:143–55.
25. Encke J, et al. Genetic vaccination with Flt3-L and GM-CSF as adjuvants: enhancement of cellular and humoral immune responses that results in 2005. Protective immunity in a murine model of hepatitis C virus infection. *World J Gastroenterol* 2006;**12**:7118–25.
26. McCormick S, et al. Manipulation of dendritic cells for host defence against intracellular infections. *Biochem Soc Trans* 2006;**34**(Pt 2):283–6.
27. Berczi I, Nagy E. Neurohormonal control of cytokines during injury. In: Rothwell NJ, Berkenbosch F, editors. *Brain control of responses to trauma*. UK: Cambridge University Press; 1994. p. 32–107.
28. Wexler BC, et al. Effects of bacterial polysaccharide (Piromen) on the pituitary-adrenal axis: adrenal ascorbic acid, cholesterol and histological alterations. *Endocrinology* 1957;**61**:300–8.
29. Wilder RL. Neuroendocrine-immune system interactions and autoimmunity. *Annu Rev Immunol* 1995;**13**:307–38.
30. Torpy DJ, Chrousos GP. The three way interaction between the hypothalamic–pituitary–adrenal and gonadal axes and the immune system, in neuroendocrine immune mechanisms of rheumatic diseases. Chikanza IC, editor. *Bailliere's clinical rheumatology*, vol. 10. London: Bailliere Tindall; 1996. p. 181–98.
31. Berczi I, et al. The immune effects of neuropeptides, in neuroendocrine immune mechanisms of rheumatic diseases. Chikanza IC, editor. *Bailliere's clinical rheumatology*, vol. 10. London: Bailliere Tindall; 1996. p. 227–57.
32. Schiltz JC, Sawchenko PE. Signaling the brain in systemic inflammation: the role of perivascular cells. *Front Biosci* 2003;**8**:S1321–9.
33. Jancso Ed G. Neurogenic inflamation in health and disease. Berczi I, Szentivanyi A, Series, editors. *Neuroimmune biology*, vol. 8. Amsterdam: Elsevier; 2010.
34. Kolb-Bachofen V. A review on the biological properties of C-reactive protein. *Immunobiology* 1991;**183**:133–445.
35. Ballou SP, Kushner I. C-reactive protein and the acute phase response. *Adv Intern Med* 1992;**37**:313–36.
36. Young B, et al. C-reactive protein: a critical review. *Pathology* 1991;**23**:118–24.

37. Mold C, et al. C-reactive protein is protective against *Streptococcus pneumoniae* infection in mice. *J Exp Med* 1981;**154**:1703–8.
38. Raetz CRH, et al. Gram-negative endotoxin: an extraordinary lipid with profound effects on eukaryotic signal transduction. *FASEB J* 1991;**5**:2652–60.
39. Flo TH, Aderem A, Pathogen recognition by Toll-like, receptors. In: Bertok L, Chow DA, editors. *Neuroimmune biology*, vol. 5. Berczi I, Szetivanyi A, series editors. *Natural Immunity.* Elsevier, Amsterdam; 2005. p. 167–182.
40. Hamann L, et al. Acute-phase concentrations of lipopolysaccharide (LPS)-binding protein inhibit innate immune cell activation by different LPS chemotypes via different mechanisms. *Infect Immun* 2005;**73**:193–200.
41. Wassell J. Haptoglobin: function and polymorphism. *Clin Lab* 2000;**46**:547–52.
42. Arredouani MS, et al. Haptoglobin dampens endotoxin-induced inflammatory effects both *in vitro* and *in vivo*. *Immunology* 2005;**114**:263–71.
43. Daemen MA, et al. Functional protection by acute phase proteins alpha(1)-acid glycoprotein and alpha(1)-antitrypsin against ischemia/reperfusion injury by preventing apoptosis and inflammation. *Circulation* 2000;**102**:1420–6.
44. Liu H, et al. Characterization and quantification of mouse mannan-binding lectins (MBL-A and MBL-C) and study of acute phase responses. *Scand J Immunol* 2001;**53**: 489–97.
45. Hansen TK, et al. Intensive insulin therapy exerts antiinflammatory effects in critically ill patients and counteracts the adverse effect of low mannose-binding lectin levels. *J Clin Endocrinol Metab* 2003;**88**:1082–8.
46. Bauer J, et al. Induction of rat alpha2-macroglobulin in vivo and in heaptocyte primary cultures: synergistic action of glucocorticoids and a Kupffer cell derived factor. *FEBS Lett* 1984;**177**:89–94.
47. Finck BN, Johnson RW. Tumor necrosis factor-alpha regulates secretion of the adipocytederived cytokine, leptin. *Microsc Res Tech* 2000;**50**:209–15.
48. Hsieh YC, et al. Metabolic modulators following trauma sepsis: sex hormones. *Crit Care Med* 2007;(9 Suppl):S621–9.
49. Collier B, et al. Glucose control and the inflammatory response. *Nutr Clin Pract* 2008;**23**:3–15.
50. Takala J, et al. Increased mortality associated with growth hormone treatment in critically ill adults. *N Engl J Med* 1999;**341**:785–92.
51. Berczi I, et al. Natural immunity and neuroimmune host defence. *Ann NY Acad Sci* 2001;**917**:248–57.
52. Jeschke MG, et al. Recombinant human growth hormone alters acute phase reactant proteins, cytokine expression, and liver morphology in burned rats. *J Surg Res* 1999;**83**:122–9.
53. Derfalvi B, et al. Interleukin-6-induced production of type II acute phase proteins and expression of jun B gene are downregulated by human recombinant growth hormone *in vitro*. *Cell Biol Int* 2000;**24**:109–14.
54. Berczi I. The influence of pituitary-adrenal axis on the immune system. In: Berczi I, editor. *Pituitary function and immunity*. Boca Raton, FL: CRC Press; 1986. p. 49–132.
55. van Gool J, et al. Glucocorticoids and catecholamines as mediators of acute-phase proteins, especially rat alpha-macrofoetoprotein. *Biochem J* 1984;**220**:125–52.
56. Baker JW, et al. Elevated plasma antidiuretic hormone levels in status asthmaticus. *Mayo Clin Proc* 1976;**51**:31–4.
57. Michelson D, et al. Multiple sclerosis is associated with alterations in hypothalamic-pituitary-adrenal axis function. *J Clin Endocrinol Metab* 1994;**79**:848–53.
58. Harbuz MS, et al. Hypothalamo-pituitary-adrenal axis and chronic immune activation. *Ann NY Acad Sci* 2003;**992**:99–106.

59. DeGoeij D, et al. Repeated stress activation of corticotropin releasing factor neurons enhances vasopressin stores and colocalization with corticotropin-releasing factor in the median eminence of rats. *Neuroendocrinology* 1991;**53**:150–9.
60. DeGoeij D, et al. Chronic intermittent stress enhances vasopressin but not corticotropin releasing factor secretion during hypoglycemia. *Am J Physiol* 1992;**263**:E394–9.
61. Akira S, Kishimoto T. IL-6 and NF-IL6 in acute-phase response and viral infection. *Immunol Rev* 1992;**127**:26–50.
62. Harnish MJ, et al. Differential regulation of human blood glucose level by interleukin-2 and -6. *Exp Clin Endocrinol Diabetes* 2005;**113**:43–8.
63. Yeager MP, et al. Glucocorticoid regulation of the inflammatory response to injury. *Acta Anaesthesiol Scand* 2004;**48**:799–813.
64. Van Gool J, et al. The relation among stress, adrenalin, interleukin 6, and acute phase proteins in the rat. *Clin Immunol Immunopathol* 1990;**57**:200–10.
65. Ekman R, et al. Arginine vasopressin in the cytoplasm and nuclear fraction of lymphocytes from healthy donors and patients with depression or schizophrenia. *Peptides* 2001;**22**:67–72.
66. Itoi K, et al. Regulatory mechanisms of corticotropin-releasing hormone and vasopressin gene expression in the hypothalamus. *J Neuroendocrinol* 2004;**16**:348–55.
67. Carrasco GA, Van deKar LD. Neuroendocrine pharmacology of stress. *Eur J Pharmacol* 2003;**463**:235–72.
68. Turnbull AV, et al. Mechanisms of hypothalamic-pituitary-adrenal axis stimulation by immune signals in the adult rat. *Ann NY Acad Sci* 1998;**840**:434–43.
69. Huitinga I, et al. Priming with interleukin-1beta suppresses experimental allergic encephalomyelitis in the Lewis rat. *J Neuroendocrinol* 2000;**12**:1186–93.
70. Chowdrey HS, et al. Evidence for arginine vasopressin as the primary activator of the HPA axis during adjuvant-induced arthritis. *Br J Pharmacol* 1995;**116**:2417–24.

CHAPTER 9

Chronic Stress and Disease

Nicolas Rohleder
Department of Psychology, Brandeis University, Waltham, MA, USA

1 INTRODUCTION

Chronic stress is cross-sectionally and prospectively associated with disease development and reduced life expectancy. Chronic stress can take many different forms, from extreme or traumatic single life events such as rape, accidents, natural disasters to long-term exposure to stressful conditions, such as caregiving, and finally, to accumulation of daily hassles, in someone's private life as well as in the work environment. As summarized in this chapter, there is now strong evidence linking many different forms of chronic stress with different diseases, with shorter life expectancy. In our quest to understand, and potentially intervene against, the health effects of chronic stress, research has targeted the communication systems linking the central nervous system (CNS), in which stress experience is located, with the periphery of the organism, in which damage accumulates that ultimately leads to disease and death.

Research targeting the activity and reactivity of classical stress systems, such as the hypothalamus–pituitary–adrenal (HPA) axis and the sympathetic and parasympathetic nervous systems (SNS and PNS), was disappointing at first, as earlier findings were often contradicting each other, and counterintuitive, as for example, the finding of lower HPA axis activity in chronically stressed and traumatized individuals. These early difficulties are likely caused by the fact that biologic responses to stressors can be very different among people, and also by the fact that while being a logical first candidate to look at, the classical stress systems are very complex in their regulation and adaptation, and less meaningful as direct indicators of health than originally thought.

While all humans are likely exposed to some of these chronic stressors throughout their lives, it is important to understand that there are larger interindividual differences in how much these stressors affect health and wellbeing. These differences in how much an individual is affected might not necessarily be detectable in self-reports of stress, but in some cases, there

Insights to Neuroimmune Biology
http://dx.doi.org/10.1016/B978-0-12-801770-8.00009-4

might be interindividual differences in biologic responses to stress. Furthermore, as more recent research shows, studying stress systems alone is rarely sufficient for understanding associations between stress and health. Instead, it has become increasingly clear that one must also investigate biologic pathways that are regulated by stress systems, and that have the potential to cause actual damage to tissues.

In this chapter, we will first describe the different forms of chronic stress, followed by a summary of stress system changes in humans exposed to chronic stress. After that, we will describe inflammation as one of the most promising links between chronic stress, stress-system signaling, and disease, before exploring alternative stress-sensitive, disease-relevant systems or mechanisms in the human body.

2 FORMS OF CHRONIC STRESS

The term chronic stress describes an array of conditions to which an organism can be exposed that range from single, but very intense events, to less intense, but repeated experiences. Starting with the most extreme type of events, so-called major life events can be a cause of chronic stress, if the individual continues to experience consequences of such events. Depending on the type of event, and the individual response, the consequences can be more or less chronic. Severe trauma caused by events such as sexual assault, combat exposure, or exposure to natural or man-made disasters, usually have long-term impact on humans, and lead under specific circumstances to psychiatric conditions such as posttraumatic stress disorder (PTSD). While PTSD is a distinct and diagnosable psychiatric disorder, it can and should be regarded as a form of chronic stress, since people suffering from PTSD are typically chronically exposed to the consequences of their traumatic experience and show very characteristic alterations in their stress systems. It is highly likely that subclinical cases, that is, humans who have been exposed to trauma, but do not fulfill all diagnostic criteria, will also experience consequences and memories of their trauma chronically.

Similarly, less nontraumatic, but major, life events such as divorce, loss of a loved one, or loss of employment can be regarded as chronic stressors, if the individual continues to suffer from the consequences, or the memories, of that specific event. Many forms of chronic stress are further caused by truly chronic exposure to stressful life circumstances. A classic and well-studied example is familial caregiving. This type of chronic stress often affects older adults who care for a spouse with Alzheimer's disease, to adults of any age who care for a family member with a chronic disease, such as cancer, or parents of children with disabilities.

In the workplace, chronic stress can result from high demands on someone's time, time pressure, and responsibilities for other people's lives, but also from hostile work environment and inhumane working conditions. Conversely, loss of employment can be equally chronically stressful due to the impact on feelings of subjective self-worth, but also constant pressure to apply for jobs, and escape unemployment. Living in impoverished conditions, as well as exposure to environmental threats can be a truly chronic stressor for families living in dangerous neighborhoods, and similar effects can be caused by racial or other forms of discrimination.

Finally, chronic stress can also be caused by repeated exposure to less-intense stressors, such as repeated daily hassles. The example of daily hassles further highlights that it is not only or necessarily the characteristics, severity, or frequency of events that determine the impact of an individual's health. Instead, prospective studies have shown that differences in affective responses to these daily events determine their long-term health impacts, with stronger affective responses being predictors of mental and physical health.[1,2]

It is likely that this applies to most if not all forms of chronic stress, that is, that the characteristics, intensity, and duration of the stressful event are less important than the individual's psychological and consequential biological response.

3 CHRONIC STRESS AND DISEASE

Without exception, all forms of chronic stress introduced earlier have been found to be associated with disease development or even reduced longevity. Several review articles and meta-analyses have been published in recent decades providing strong evidence in support of the chronic stress–disease relationship. Cohen et al. have summarized the literature in a commentary in the *Journal of the American Medical Association* (JAMA) in 2007. They focused on four disease categories: depression, cardiovascular disease, human immunodeficiency virus (HIV) infection/AIDS, and cancer, and came to the conclusion that this cluster of what could be considered to be the most relevant diseases affecting human life span, are all robustly associated with chronic stress. They also note that links with chronic stress are also emerging with upper respiratory tract infections, asthma, herpes viral infections, autoimmune diseases, and wound healing.[3] Others have specifically addressed one disease category. Rozanski et al. for example focused on psychologic risk factors of cardiovascular disease and identified different chronic stressors, including work stress, low social support, caregiving, marital stress, among others, as predictors of cardiovascular events, and also pathophysiologic mechanisms such as the presence of atherosclerotic plaques.[4]

3.1 PTSD and Disease

Individuals suffering from PTSD experience a higher rate of so-called medically unexplained symptoms,[5] and in addition to frequent comorbid psychiatric conditions, such as major depression, PTSD patients have lower life expectancy than the general population.[6] While substance abuse and depression are the most likely comorbid psychiatric conditions, there is strong evidence showing that cardiovascular diseases, including coronary heart disease (1.5 times higher risk[7]), and the metabolic syndrome (1.8 times higher risk[8]) are the main health consequences of PTSD.

3.2 Caregiving and Disease

Probably the best-documented form of chronic stress both in terms of health consequences and also in terms of mechanisms is familial caregiving. As described earlier, the most prevalent form of familial caregiving is spousal caregiving to a partner suffering from dementia, but different forms of caregiving can also include other relatives such as parents, or children, suffering from typically chronic, life-threatening diseases. Interestingly, while the majority of studies find caregiving related with declining health, there seems to be no simple relationship, that is, specific conditions need to be present to explain the caregiving–disease link. Results from the Nurses Health Study including more than 50,000 female participants revealed that caregiving for a disabled spouse of more than 9 h per week was associated with a 1.8-times higher risk of coronary heart disease.[9] Mausbach et al. reported that caregiving was predictive of earlier CVD development in caregivers who experienced more depression, and reported the caregiving to be more stressful.[10] Von Känel et al. reported a higher heart disease risk score, which includes cholesterol levels, blood pressure, smoking status, diabetes, and other traditional risk factors in caregivers.[11]

Similarly, in a prospective cohort study conducted over 4.5 years, a 63% higher mortality was found in caregivers, but only in those that reported to be significantly stressed by the caregiving experience.[12] Finally, caregiving stress is also related with more depressive and anxiety disorders,[13] and with higher risk for cognitive decline.[14]

3.3 Work Stress and Disease

High job demands, low job control, high job strain, job dissatisfaction, burnout, but also unemployment, are factors through which the work environment can exert a significant amount of stress, mostly chronic in nature, on the individual human being. Several models have been proposed to characterize and describe job-related stress, with the two most prominent being the effort–reward imbalance (ERI) model,[15] and the demand–control/job strain model.[16] Several meta-analyses and reviews show a very strong evidence base supporting a relationship between job strain no matter how it was assessed or conceptualized with cardiovascular disease. For example, job strain was found to be related with hypertension in meta-analyses of nine studies,[17] a meta-analyses of combined cohort studies including more than 175,000 participants revealed a 30% higher risk for cardiovascular disease in participants with high job insecurity,[18] and a systematic review of 51 studies showed that high job demands, lack of support and job strain combined with social isolation are risk factors for cardiovascular disease.[19] Taken together, no matter how job strain is conceptualized, there is very strong support for stress at the workplace predicting cardiovascular morbidity and mortality.

3.4 Daily Hassles and Disease

A very interesting form of chronic stress is that of repeated exposure to less severe, but very frequent stressors of lesser intensity, so-called daily hassles. Such daily hassles have received less attention in chronic stress research, for several reasons, one of them being that severe chronic stressors have been more promising research areas in the past because stronger biologic responses are expected and easier to document, and another one being that each individual daily hassle hardly qualifies as a chronic, meaningful stressor. However, it has been argued that repeated exposure to daily hassles can, under certain conditions, mainly lack of habituation to these stressors, and thereby repeated strong biologic responses, lead to an accumulation of damage. This has been addressed, for example, in the allostatic load model which describes how adaptive short-term stress responses can over time become maladaptive and damage the organism.[20] In extension of this theory, it should be noted that individual perception, processing, and responses to daily hassles are likely important factors to consider in this context. Recent studies have shown that in caregivers and controls, more daily stressors experienced over the most recent 24 h were related with higher plasma concentrations of inflammatory molecules, and that this relationship even partially explained the increased inflammation found in caregivers.[21] In patients with systemic lupus, Pawlak et al. found that daily hassles were associated with symptom flares.[22] Finally and most importantly, two recent studies have shown that stronger affective responses to daily hassles were predictive of mental and physical health several years later.[1,2]

In summary, there is strong evidence showing that different forms of chronic stress, including, not only limited to, PTSD, caregiving, work stress, but also daily hassles, is predictive of a wide range of human diseases. Cardiovascular disease is the best-documented outcome in this context, and seems to be a particularly likely consequence of trauma and work stress. In the next section, we will address potential mechanistic links between stress and these negative health outcomes.

4 MECHANISMS LINKING CHRONIC STRESS AND DISEASE

To explain why chronic stress exposure damages the organism and leads to the development, or faster progression, of life-threatening diseases such as cardiovascular disease or cancer, we need to take a close look at the pathways that communicate stress from the CNS to the periphery of the body. The main candidates, and consequentially, the best-studied pathways here are the two main stress systems: the HPA axis and the SNS with its sub component, the sympathoadrenal medullary (SAM) system. Consequentially, these will be looked at first.

4.1 Stress Systems – Can Changes in Stress-System Activity be Linked Directly to Disease?

A common problem when assessing stress-system activity in humans under chronic stress is that a single blood sample taken at a not well-specified time of day is unlikely to reveal useful information about the status of the HPA axis, and similarly, a single blood draw is equally ill-suited for the detection of SNS activity in chronically stressed people. The main reason for this is that stress-system activity in the absence of an acute stressor follows a circadian rhythm, characterized by cortisol wake-up responses with strong increases in the morning, and a decline of cortisol levels throughout the day.[23] Similarly, plasma catecholamine concentrations indicating SNS activity also follow a circadian pattern, but here, levels tend to be lower in the morning and higher toward the late afternoon.

It is not surprising that early studies trying to understand stress-system activity in chronically stressed humans were very contradictive, finding higher, lower, or unchanged concentrations of cortisol or catecholamines, for example: higher cortisol was reported in caregivers versus controls in a small cross-sectional study by Gallagher-Thompson,[24] Mills et al. reported unchanged cortisol but higher norepinephrine in caregivers compared with matched controls,[25] and Brummett et al. found lower cortisol in caregivers.[26]

A solution to this problem is to either measure urinary stress hormone excretion over 12 or 24 h, or to use repeated blood or saliva samples to be able to characterize circadian rhythms, and analyze potential changes in circadian rhythms related with or potentially caused by chronic stress. In a meta-analyses of chronic stress studies that measured HPA axis activity repeatedly throughout the day using saliva samples, Miller et al. were able to characterize HPA axis changes in chronic stress. They reported that timing of measurement relative to the onset of the chronic stressor is of major importance, with higher HPA axis activity typically found when the onset of stress was more recent, while longer chronic stress experience was related with lower HPA axis activity. Further, they found that in phases with high subjective distress, cortisol levels tend to be higher, while in the posttraumatic phase, for example in PTSD, cortisol tends to be lower. In summary, their analysis supports the conclusion that in long-term chronic stress, HPA axis activity is lower, and characterized by loss of circadian rhythm.[27] This general finding of lower cortisol is also supported by urinary excretion studies, for example in PTSD,[28] while catecholamine excretion is consistently found to be increased in PTSD.[29,30]

Taken together, it is surprisingly difficult to find consistent data on stress-system activity in chronic stress, but if anything, and counterintuitively, a general pattern of overactive stress systems is not consistently found in chronically stressed humans. Instead, changes are more complex, with loss of circadian rhythmicity, and at least for the HPA axis, a mix of unchanged activity with hypoactivity.

4.2 Dependent Systems – Inflammation as a Link Between Stress-System Changes and Disease

As the previous section shows, changes in stress-system activity are more difficult to detect in chronic stress than one would expect, making stress systems alone, less promising candidate pathways for explaining the link between chronic stress and disease. The consequence of this is that we need to take the next step and look at mechanisms that are affected by stress, either through direct signals from the CNS, or via stress systems, and that have relevant links to the pathophysiology of the diseases affected by chronic stress.

One of these mechanisms is inflammation. Inflammation is a subprocess of the immune response and invaluable for the organism's defense against invading pathogens, and equally important in wound healing. However, inflammation is a very powerful tool, and as such, also has the potential to inflict severe damage on the organism, if not properly controlled. In particular, inflammation is a main predictor of cardiovascular morbidity and mortality, and inflammation is a main component of the actual atherosclerotic process leading to most cardiovascular diseases.[31–33] It also plays a role in all of the other diseases found to be related with chronic stress, for example, the metabolic syndrome and type 2 diabetes,[34] cancer,[35] as well as age-related diseases including those of the CNS.[36] Taken together, inflammation and inflammatory processes are linked with the same diseases that are related with chronic stress.

As a further necessary condition, inflammatory processes are activated by stress experiences, likely mediated through stress-system activation. In acute stress studies, we have for example shown that one of the central transcription factors in the inflammatory cascade, nuclear factor (NF)-κB, is activated by acute psychosocial stress through adrenergic signaling,[37] and inhibited by stress-induced cortisol.[38] These findings have been translated to chronic stress as well: Miller et al. have shown that caregivers express a higher number of genes that are activated by NF-kB, and a lower number of genes controlled by the glucocorticoid receptors, indicating that chronic stress alters the organism's regulatory control of inflammatory mechanisms.[39]

In line with these mechanistic studies, chronic stress is frequently found to be associated with increased plasma concentrations of inflammatory molecules such as IL-6 and CRP in cross-sectional studies,[40–43] and even more importantly, in longitudinal studies. For example, Kiecolt-Glaser et al. followed-up a group of Alzheimer's caregivers for 6 years and found that caregivers had a 4-times steeper increase of plasma IL-6 than controls over this time period.[44] This effect does not appear to be restricted to elderly spousal caregivers, as we have found a similar pattern of increasing low-grade inflammation in middle-aged caregivers of brain cancer patients. Interestingly, this group of caregivers was characterized by a lack of change in stress-system activity throughout the entire 44-week observation period, but they showed decreased glucocorticoid sensitivity of their inflammatory response, and decreasing expression of anti-inflammatory genes.[45] This group thereby exemplifies the current conceptualization of how chronic stress translates into disease: instead of seeing pronounced alterations in the actual stress systems, chronic stress induces changes further downstream, that is, in the respective target tissues. In the context of inflammation, it has been shown that chronic stress is accompanied by development of a relative glucocorticoid resistance, in that inflammatory cells become more and more resistant to the anti-inflammatory effects of cortisol.[46,47] This phenomenon of GC resistance can further be tracked down to intracellular processes, that is, changes in pro- and anti-inflammatory gene expression.[39,48–50]

4.3 Dependent Systems – Immune Defense Mechanisms

In the previous section, we have argued that inflammation is an important, or maybe the most important, link between chronic stress and disease, mainly because inflammation is both activated and inactivated by stress systems, is found increased in chronic stress, and plays a well-documented central role in the development of most of the diseases relevant for human morbidity and mortality.

Interestingly, this marked overactivity of one component of the immune system under chronic stress can be accompanied by an underactivity, or lack of efficiency of another component of the immune responses, that is, adaptive/specific immune function. While much of the current research focuses on inflammatory overactivity, earlier studies, in particular again in caregivers, have targeted measures of adaptive immune function, for example, responses to vaccination and simpler measures such as antibody titers.

This earlier caregiver literature showed that spousal caregivers of Alzheimer's disease patients had markedly lower antibody responses to different vaccinations, such as influenza vaccination,[51,52] or pneumococcal vaccination.[53] These lower antibody responses were further found to be associated with lower acute cytokine responses, that is, lower *in vitro* IL-2 or IL-1β responses to virus stimulation,[51] which shows that the mechanism is a lack of immune response to the virus/vaccine exposure. These findings were later replicated also in younger caregivers,[54,55] which means that this effect does not only occur in the already declining immune system of the elderly, but is a general phenomenon also affecting the healthy immune system. Together with findings of slower wound healing in chronic stress,[56] and higher susceptibility to induced rhinovirus infections of chronically stressed humans, it can be concluded that chronic stress does not only promote a disinhibition of the inflammatory response leading to low-grade inflammation and its health consequences, but that chronic stress also suppresses the function of the adaptive immune system, making chronically stressed humans more susceptible to infectious diseases, such as influenza and common cold. In particular in the elderly, the lack of response to vaccination is not only a mechanistic test of their immune function, it also means that vaccinations are really less effective in chronically stressed older adults.

In summary, the immune system changes observed under chronic stress resemble those seen during normal aging, only that chronic stress can elicit these aging-like changes already in young human beings, and that chronic stress can accelerate normal immune system aging in older adults, thereby leading to the dangerous situation of overactive inflammatory systems combined with inefficient adaptive immunity. This condition translates into an increased risk for cardiovascular disease, type 2 diabetes, cancer, and cognitive decline on the one hand, and heightened susceptibility to infections on the other hand.

5 SUMMARY AND CONCLUSIONS

Taken together, the state of the research summarized in this chapter strongly supports the conclusion that chronic psychosocial stress fosters the development of major diseases such as cardiovascular disease and cancer, in humans. Many different types of chronic stress, ranging from mild but frequent daily hassles, overwork stress, familial caregivers, up to the aftermath of psychologic trauma, are associated with disease. A key finding of the past two to three decades is that the main stress systems HPA axis and SNS play a less-important role in the translation of the chronically stressed CNS state into physical disease than logic would dictate. The literature of the relationships of cortisol and catecholamine levels with chronic stress is highly heterogenous and up to the present, study findings are inconsistent with regard to the cortisol or catecholamine levels in chronic stress (higher, lower, or unchanged), and with regard to altered stress hormones predicting disease development. However, inflammation has emerged as a major factor linking chronic stress and disease, because of the now undisputable role of inflammation in the development of major human diseases, and because of the role of stress-systems regulating inflammation. One of the most important discoveries in the field of stress research was that while stress systems might not change consistently under chronic stress, the ability of inflammatory cells to respond to stress-system signals changes in a very consistent way under chronic stress, in that the ability to respond to the suppressive cortisol signal declines under chronic stress (see also Refs [46,47,57]) In addition to this loss of ability to exert inhibitory control over inflammation, chronic stress further negatively affects the ability of the adaptive immune system to fight infections, leading to an increased susceptibility to infections in chronically stressed humans. While we have begun to unravel the intracellular signaling mechanism underlying glucocorticoid resistance, we have yet to understand how the signal of this CNS state of "chronic stress" is able to affect these intracellular mechanisms to show long-term functional alterations. It is unlikely that this can be explained by simple receptor-downregulation, for example, since there is no evidence for increased signal; in other words, in the context of unchanged or even lower cortisol concentrations, ligand-induced receptor downregulation is not a logical consequence. Further, we also need to expand our understanding into SNS-to-inflammation signaling mechanisms and include behavioral factors, such as altered eating habits, lack of activity, etc. to fully understand how chronic stress leads to disease.

REFERENCES

1. Piazza JR, Charles ST, Sliwinski MJ, Mogle J, Almeida DM. Affective reactivity to daily stressors and long-term risk of reporting a chronic physical health condition. *Ann Behav Med* 2013;**45**:110–20.
2. Charles ST, Piazza JR, Mogle J, Sliwinski MJ, Almeida DM. The wear and tear of daily stressors on mental health. *Psychol Sci* 2013;**24**:733–41.
3. Cohen S, Janicki-Deverts D, Miller GE. Psychological stress and disease. *JAMA* 2007;**298**:1685–7.
4. Rozanski A, Blumenthal JA, Davidson KW, Saab PG, Kubzansky L. The epidemiology, pathophysiology, and management of psychosocial risk factors in cardiac practice: the emerging field of behavioral cardiology. *J Am Coll Cardiol* 2005;**45**:637–51.
5. Ford JD, Campbell KA, Storzbach D, Binder LM, Anger WK, Rohlman DS. Posttraumatic stress symptomatology is associated with unexplained illness attributed to Persian Gulf War military service. *Psychosom Med* 2001;**63**:842–9.
6. Nemeroff CB, Bremner JD, Foa EB, Mayberg HS, North CS, Stein MB. Posttraumatic stress disorder: a state-of-the-science review. *J Psychiatr Res* 2006;**40**:1–21.
7. Edmondson D, Kronish IM, Shaffer JA, Falzon L, Burg MM. Posttraumatic stress disorder and risk for coronary heart disease: a meta-analytic review. *Am Heart J* 2013;**166**:806–14.
8. Rosenbaum S, Stubbs B, Ward PB, Steel Z, Lederman O, Vancampfort D. The prevalence and risk of metabolic syndrome and its components among people with posttraumatic stress disorder: a systematic review and meta-analysis. *Metabolism* 2015;**64**:926–33.
9. Lee S, Colditz GA, Berkman LF, Kawachi I. Caregiving and risk of coronary heart disease in U.S. women: a prospective study. *Am J Prev Med* 2003;**24**:113–9.
10. Mausbach BT, Patterson TL, Rabinowitz YG, Grant I, Schulz R. Depression and distress predict time to cardiovascular disease in dementia caregivers. *Health Psychol* 2007;**26**:539–44.
11. von Känel R, Mausbach BT, Patterson TL, Dimsdale JE, Aschbacher K, Mills PJ, et al. Increased Framingham Coronary Heart Disease Risk Score in dementia caregivers relative to non-caregiving controls. *Gerontology* 2008;**54**:131–7.
12. Schulz R, Beach SR. Caregiving as a risk factor for mortality: the Caregiver Health Effects Study. *JAMA* 1999;**282**:2215–9.
13. Russo J, Vitaliano PP, Brewer DD, Katon W, Becker J. Psychiatric disorders in spouse caregivers of care recipients with Alzheimer's disease and matched controls: a diathesis-stress model of psychopathology. *J Abnorm Psychol* 1995;**104**:197–204.

14. Vitaliano PP, Murphy M, Young HM, Echeverria D, Borson S. Does caring for a spouse with dementia promote cognitive decline? A hypothesis and proposed mechanisms. *J Am Geriatr Soc* 2011;**59**:900–8.
15. Siegrist J. Effort-reward imbalance at work and cardiovascular diseases. *Int J Occup Med Environ Health* 2010;**23**:279–85.
16. Karasek R, Baker D, Marxer F, Ahlbom A, Theorell T. Job decision latitude, job demands, and cardiovascular disease: a prospective study of Swedish men. *Am J Public Health* 1981;**71**:694–705.
17. Babu GR, Jotheeswaran AT, Mahapatra T, Mahapatra S, Kumar A, Detels R, et al. Is hypertension associated with job strain? A meta-analysis of observational studies. *Occup Environ Med* 2014;**71**:220–7.
18. Virtanen M, Nyberg ST, Batty GD, Jokela M, Heikkilä K, Fransson EI, et al. Perceived job insecurity as a risk factor for incident coronary heart disease: systematic review and meta-analysis. *BMJ* 2013;**347**:f4746.
19. Eller NH, Netterstrøm B, Gyntelberg F, Kristensen TS, Nielsen F, Steptoe A, et al. Work-related psychosocial factors and the development of ischemic heart disease: a systematic review. *Cardiol Rev* 2009;**17**:83–97.
20. McEwen BS, Stellar E. Stress and the individual. Mechanisms leading to disease. *Arch Intern Med* 1993;**153**:2093–101.
21. Gouin JP, Glaser R, Malarkey WB, Beversdorf D, Kiecolt-Glaser J. Chronic stress, daily stressors, and circulating inflammatory markers. *Health Psychol* 2012;**31**:264–8.
22. Pawlak CR, Witte T, Heiken H, Hundt M, Schubert J, Wiese B, et al. Flares in patients with systemic lupus erythematosus are associated with daily psychological stress. *Psychother Psychosom* 2003;**72**:159–65.
23. Pruessner JC, Wolf OT, Hellhammer DH, Buske-Kirschbaum A, von Auer K, Jobst S, et al. Free cortisol levels after awakening: a reliable biological marker for the assessment of adrenocortical activity. *Life Sci* 1997;**61**:2539–49.
24. Gallagher-Thompson D, Shurgot GR, Rider K, Gray HL, McKibbin CL, Kraemer HC, et al. Ethnicity, stress, and cortisol function in Hispanic and non-Hispanic white women: a preliminary study of family dementia caregivers and noncaregivers. *Am J Geriatr Psychiatry* 2006;**14**:334–42.
25. Mills PJ, Ziegler MG, Patterson T, Dimsdale JE, Hauger R, Irwin M, et al. Plasma catecholamine and lymphocyte beta 2-adrenergic receptor alterations in elderly Alzheimer caregivers under stress. *Psychosom Med* 1997;**59**:251–6.
26. Brummett BH, Boyle SH, Siegler IC, Kuhn CM, Surwit RS, Garrett ME, et al. HPA axis function in male caregivers: effect of the monoamine oxidase-A gene promoter (MAOA-uVNTR). *Biol Psychol* 2008;**79**:250–5.
27. Miller GE, Chen E, Zhou ES. If it goes up, must it come down? Chronic stress and the hypothalamic-pituitary-adrenocortical axis in humans. *Psychol Bull* 2007;**133**:25–45.
28. Yehuda R, Southwick S, Giller EL, Ma X, Mason JW. Urinary catecholamine excretion and severity of PTSD symptoms in Vietnam combat veterans. *J Nerv Ment Dis* 1992;**180**:321–5.
29. Kosten TR, Mason JW, Giller EL, Ostroff RB, Harkness L. Sustained urinary norepinephrine and epinephrine elevation in post-traumatic stress disorder. *Psychoneuroendocrinology* 1987;**12**:13–20.
30. Pitman RK, Orr SP. Twenty-four hour urinary cortisol and catecholamine excretion in combat-related posttraumatic stress disorder. *Biol Psychiatry* 1990;**27**:245–7.
31. Danesh J. Smoldering arteries? Low-grade inflammation and coronary heart disease. *JAMA* 1999;**282**:2169–71.
32. Ridker PM, Cushman M, Stampfer MJ, Tracy RP, Hennekens CH. Plasma concentration of C-reactive protein and risk of developing peripheral vascular disease. *Circulation* 1998;**97**:425–8.

33. Hansson GK, Libby P. The immune response in atherosclerosis: a double-edged sword. *Nat Rev Immunol* 2006;**6**:508–19.
34. Hotamisligil GS. Inflammation and metabolic disorders. *Nature* 2006;**444**:860–7.
35. Coussens LM, Werb Z. Inflammation and cancer. *Nature* 2002;**420**:860–7.
36. Ershler WB, Sun WH, Binkley N. The role of interleukin-6 in certain age-related diseases. *Drugs Aging* 1994;**5**:358–65.
37. Bierhaus A, Wolf J, Andrassy M, Rohleder N, Humpert PM, Petrov D, et al. A mechanism converting psychosocial stress into mononuclear cell activation. *Proc Natl Acad Sci USA* 2003;**100**:1920–5.
38. Wolf JM, Rohleder N, Bierhaus A, Nawroth PP, Kirschbaum C. Determinants of the NF-kappaB response to acute psychosocial stress in humans. *Brain Behav Immun* 2009;**23**:742–9.
39. Miller GE, Chen E, Sze J, Marin T, Arevalo JM, Doll R, et al. A functional genomic fingerprint of chronic stress in humans: blunted glucocorticoid and increased NF-kappaB signaling. *Biol Psychiatry* 2008;**64**:266–72.
40. Lutgendorf SK, Garand L, Buckwalter KC, Reimer TT, Hong SY, Lubaroff DM. Life stress, mood disturbance, and elevated interleukin-6 in healthy older women. *J Gerontol A Biol Sci Med Sci* 1999;**54**:M434–M439.
41. von Kanel R, Dimsdale JE, Patterson TL, Grant I. Association of negative life event stress with coagulation activity in elderly Alzheimer caregivers. *Psychosom Med* 2003;**65**:145–50.
42. von Kanel R, Mills PJ, Mausbach BT, Dimsdale JE, Patterson TL, Ziegler MG, et al. Effect of Alzheimer caregiving on circulating levels of C-reactive protein and other biomarkers relevant to cardiovascular disease risk: a longitudinal study. *Gerontology* 2012;**58**:354–65.
43. Lovell B, Moss M, Wetherell M. The psychosocial, endocrine and immune consequences of caring for a child with autism or ADHD. *Psychoneuroendocrinology* 2012;**37**:534–42.
44. Kiecolt-Glaser JK, Preacher KJ, MacCallum RC, Atkinson C, Malarkey WB, Glaser R. Chronic stress and age-related increases in the proinflammatory cytokine IL-6. *Proc Natl Acad Sci USA* 2003;**100**:9090–5.
45. Rohleder N, Marin TJ, Ma R, Miller GE. Biologic cost of caring for a cancer patient: dysregulation of pro- and anti-inflammatory signaling pathways. *J Clin Oncol* 2009;**27**:2909–15.
46. Cohen S, Janicki-Deverts D, Doyle WJ, Miller GE, Frank E, Rabin BS, et al. Chronic stress, glucocorticoid receptor resistance, inflammation, and disease risk. *Proc Natl Acad Sci USA* 2012;**109**:5995–9.
47. Miller GE, Cohen S, Ritchey AK. Chronic psychological stress and the regulation of pro-inflammatory cytokines: a glucocorticoid-resistance model. *Health Psychol* 2002;**21**:531–41.
48. Miller GE, Chen E, Fok AK, Walker H, Lim A, Nicholls EF, et al. Low early-life social class leaves a biological residue manifested by decreased glucocorticoid and increased proinflammatory signaling. *Proc Natl Acad Sci USA* 2009;**106**:14716–21.
49. Miller GE, Rohleder N, Cole SW. Chronic interpersonal stress predicts activation of pro- and anti-inflammatory signaling pathways 6 months later. *Psychosom Med* 2009;**71**:57–62.
50. Bauer ME, Vedhara K, Perks P, Wilcock GK, Lightman SL, Shanks N. Chronic stress in caregivers of dementia patients is associated with reduced lymphocyte sensitivity to glucocorticoids. *J Neuroimmunol* 2000;**103**:84–92.
51. Kiecolt-Glaser JK, Glaser R, Gravenstein S, Malarkey WB, Sheridan J. Chronic stress alters the immune response to influenza virus vaccine in older adults. *Proc Natl Acad Sci USA* 1996;**93**:3043–7.
52. Vedhara K, Cox NK, Wilcock GK, Perks P, Hunt M, Anderson S, et al. Chronic stress in elderly carers of dementia patients and antibody response to influenza vaccination. *Lancet* 1999;**353**:627–31.
53. Glaser R, Sheridan J, Malarkey WB, MacCallum RC, Kiecolt-Glaser JK. Chronic stress modulates the immune response to a pneumococcal pneumonia vaccine. *Psychosom Med* 2000;**62**:804–7.

54. Gallagher S, Phillips AC, Drayson MT, Carroll D. Caregiving for children with developmental disabilities is associated with a poor antibody response to influenza vaccination. *Psychosom Med* 2009;**71**:341–4.
55. Vedhara K, McDermott MP, Evans TG, Treanor JJ, Plummer S, Tallon D, et al. Chronic stress in nonelderly caregivers: psychological, endocrine and immune implications. *J Psychosom Res* 2002;**53**:1153–61.
56. Kiecolt-Glaser JK, Marucha PT, Malarkey WB, Mercado AM, Glaser R. Slowing of wound healing by psychological stress. *Lancet* 1995;**346**:1194–6.
57. Rohleder N. Acute and chronic stress induced changes in sensitivity of peripheral inflammatory pathways to the signals of multiple stress systems – 2011 Curt Richter Award Winner. Psychoneuroendocrinology 2012;**37**:307–16.

PART 5

Epigenetic Immunoregulation

CHAPTER 10

Biologically Controlled Mutations are Right for Evolution

Istvan Berczi
Department of Immunology, Faculty of Medicine, The University of Manitoba, Winnipeg, Manitoba, Canada; Departamento de Fisiología y Farmacología, Centro de Ciencias Básicas, Universidad Autónoma de Aguascalientes, Aguascalientes, Estado Libre y Soberano de Aguascalientes, Mexico City, Mexico

1 INTRODUCTION

It was proposed by Darwin that mutations are spontaneous, governed by chance, and when mutation is making change, natural selection is taking place. Thus animals and plants may evolve.[1] But evolution is a pretty serious and important business; and it is way too important to do it properly. So we say that evolution cannot rely on the haphazard chances of mutation to do the job.

A better way to insure evolution against this problem is that mutation is a biologically controlled reaction, it is inducible as frequently as it is required, and it is certain that sufficient amount of mitogenic impulses occur as it is needed by the organism. A system like this has already developed within the adaptive-immune system (ADIM), it is the *generation of diversity*. This is an enzymatic process. In the bone marrow, the immunoglobulin genes are enzymatically digested, and mutant immunoglobulin chains are created by random actions of recombinase. B lymphocytes are generated with mutated antigen receptors. When these cells face the processed and presented antigen, they find an antigenic determinant, which stimulates them and ultimately these cells survive. All those cells that are not stimulated die by apoptosis.

The generation of diversity is a biologically controlled mutagenic system and it works, as it does not require any hassles unlike cancer or abnormal cells that are generated. This is an ideal system of *controlled mutation*. Thanks to this controlled mutagenic setup, there is diversity within our immune system. Here, we are dealing with the generation of diversity, and at this stage, everything is considered to be self in the bone marrow. So there is no immune response to mutated cells, so the selection process will deal with them.

Insights to Neuroimmune Biology
http://dx.doi.org/10.1016/B978-0-12-801770-8.00010-0

Earlier it was stated that if one knows immunology very well, and does not know about pregnancy, then one would say that it is not possible to achieve pregnancy. But it is happening and the question is that how does it happen? In pregnancies, 50% of the antigens are foreign substances! Apparently, the innate immune system plays an important role in the defense of the fetus. The INIM system recognizes the *maternal antigens in the baby as "self"* as this is enough to practice self defense. So there is protection, despite the paternal antigens also being there. However, pregnancy is an incredibly complex reaction and it cannot be discussed here in detail. So we can argue that there are ways to protect antigenic changes or mutated cells in the body.[2]

2 STRESS AND EPIGENETIC PROGRAMMING

2.1 Epigenetic Programming

Stressful experiences *in utero* or during early life may increase the risk of neurological and psychiatric disorders, arguably via altered *epigenetic regulations. Epigenetic* mechanisms, such as microRNA (*miRNA*) *expression, DNA methylation, and histone modifications* are prone to the changes in response to stressful experiences and hostile environmental factors. Altered epigenetic regulation may potentially influence fetal endocrine *programming* and brain development across several generations. However only recently, more attention has been paid to possible *transgenerational effects of stress.*[3] Epigenetic mechanisms may contribute to the development of hypoxic-/ischemic-sensitive phenotype. *DNA methylation and active demethylation, histone modifications, and miRNA play a role in the regulation of neuronal and vascular developmental plasticity*, which could be important for fetal stress-induced epigenetic programming of hypoxic-/ischemic-sensitive phenotype in the developing brain.[4]

Epigenetic mechanisms, particularly those affecting the expression of genes implicated in stress responses, mediate the association between early childhood adversity and later risk of depression. The resulting studies provide evidence that is consistent with the idea that social adversity, particularly that involving parent–offspring interactions, alters the *epigenetic* state and expression of a wide range of genes, the products of which regulate *hypothalamic–pituitary–adrenal* (HPA) function.[5]

Hippocampal DNA methylation patterns and expression profiles, of female prenatally stressed 5-Htt ± mice, suggest that distinct molecular mechanisms, some of which are promoter methylation-dependent ones, contribute to the behavioral effects of the 5-Htt genotype, prenatal stress (PS) exposure, and their interactions.[6]

2.2 Maternal Stress

Animal models support this hypothesis. For example in the rat, natural variations in maternal care influence HPA axis stress reactivity in the offspring via long-term changes in tissue-specific gene expressions. Studies *in vivo* and *in vitro* show that maternal licking and grooming increases glucocorticoid receptor (GR) expressions in the offspring via increased hippocampal serotonergic tone accompanied by increased *histone acetylase transferase activity, histone acetylation, and DNA demethylation mediated by the transcription factor nerve growth factor-inducible protein-A.* These effects are reversed by early postnatal cross-fostering and pharmacological manipulations, including trichostatin A (TSA) and l-methionine administration during adulthood.[7]

Stressful events in early life, including acute prenatal stress, *maternal deprivation, separation,* and variation in maternal care may lead to neuroendocrine perturbations, thereby affecting reproductive performance, cognitive functions, and stress responses, as well as the risk of infectious cardio-metabolic and psychiatric diseases in later life. *Results*: findings from recent studies based on both genome-wide and candidate gene approaches highlighted the importance of mechanisms that are involved in *epigenetic* regulation of gene expression, such as DNA methylation, histone modifications, and non-coding RNAs, in the long-term effects of exposure to *stress* in early life.[8] Recent studies have identified that the effects of maternal stress are not confined to the first generation and they can extend over multiple generations. These effects are also evident in humans. Although our understanding of the potential mechanisms, by which *transgenerational programming* of the stress response occurs remain largely undetermined, recent studies have begun to identify potential mechanisms of transfer. These effects are also evident in humans.[9] *Maternal stress experience* is associated with neurodevelopmental disorders, including schizophrenia and autism. Recent studies have examined mechanisms, by which changes in the maternal milieu may be transmitted to the developing embryo and may get potentially translated into *programming of the epigenome. Animal models of PS* have identified important sexual- and temporal-specific effects on offspring stress responses. As dysregulation of stress pathways is a common feature in *most neuropsychiatric diseases*, molecular, and epigenetic analyses at the maternal–embryo interface, especially in the *placenta*, may provide a unique insight into the identification of the much needed predictive biomarkers. In addition, as most neurodevelopmental disorders have a *sexual bias,* examination of sexual differences in the inheritance of phenotypic outcomes may pinpoint gene targets and specific *windows of vulnerability* in neurodevelopment, which have been disrupted.[10]

Adult health is dependent partially on maternal nutrition and growth during early life, which may independently affect insulin sensitivity, body composition, and overall energy homeostasis. Since the publication of the "thrifty phenotype hypothesis,"[11] animal experiments have focused on establishing the mechanisms involved, which include changes in fetal *cortisol, insulin, and leptin* secretion or sensitivity. Intrauterine growth retardation can be induced by either *prolonged modest changes in maternal diet* or by *severe changes in uterine blood supply near to term.* These contrasting challenges result in different amounts of cellular stress in the offspring. In addition, shifts in the transcriptional activity of DNA may produce sustained metabolic adaptations. Within tissues and organs that *control metabolic homeostasis (e.g., hypothalamus, adipose tissue, stomach, skeletal muscle, and heart),* a range of phenotypes can be induced by sustained changes in maternal diet by the modulation of genes that control *DNA methylation and histone acetylation, which suggests epigenetic programming.* We now need to understand how changes in *maternal diet affect DNA* and how they are conserved on exposure to *oxidative stress.* A main challenge will be to establish how the dietary environment interacts with a programmed phenotype to trigger the development of *metabolic diseases.* This may aid in the establishment of *nutrigenomic strategies* to prevent the metabolic syndrome.[12]

We now need to understand how changes in *maternal diet affect DNA* and how they are conserved on exposure to *oxidative stress.* A main challenge will be to establish how the dietary environment interacts with the programmed phenotype to trigger the development of *metabolic diseases.* This may aid in the establishment of *nutrigenomic strategies* to prenatal nutrition.

We propose that early nutritional input together with stress hormones and sensory stimuli from the mother during the perinatal period act synergistically to program the adult brain, possibly via epigenetic mechanisms. We hypothesize that stress during gestation or lactation, affects the intake of macro- and micronutrients, including dietary methyl donors, and/or impairs the dam's metabolism, thereby altering nutrient composition and intake by the offspring.[13]

Maternal smoking during pregnancy (MSDP) programs offspring HPA (dys)regulation. Epigenetic regulation of placental GR may serve as a novel underlying mechanism. Results may have implications for delineating pathways to adverse outcomes from MSDP.[14]

2.3 Prenatal Stress

PS is associated with an increased vulnerability to neurodevelopmental disorders, including autism and schizophrenia. To determine the critical time window when fetal antecedents may induce a disease predisposition, we examined behavioral responses in offsprings exposed to stress during early, mid, and late gestation. We found that male offspring exposed to stress early in gestation displayed maladaptive behavioral stress response, anhedonia, and an increased sensitivity to selective serotonin reuptake inhibitor treatment. Long-term alterations in central corticotropin-releasing factor (CRF) and GR expression, as well as increased HPA axis response, were present in these mice and supposedly contributed to elevated stress sensitivity. Changes in CRF and GR gene methylation correlated with altered gene expression, providing important evidence of epigenetic programming during early PS. In addition, the core mechanism underlying male vulnerability was found that might involve sex-specific placenta response, where stress early in pregnancy significantly increased expression of peroxisome proliferator-activated receptor alpha (PPAR-α), insulin-like growth factor binding protein 1 (IGFBP-1), hypoxia-inducible factor 3 alpha (HIF3-α), and glucose transporter 4 (GLUT4) in male placentas but not females. Examination of placental epigenetic machinery revealed basal sex differences, providing further evidence that sex-specific programming begins very early in pregnancy, and may contribute to the timing and vulnerability of a developing fetus to maternal perturbations. Overall, these results indicated that stress experience early in pregnancy might contribute to male neurodevelopmental disorders through impacts on placental function and fetal development.[15]

2.4 Fetal Stress

Stressful experiences *in utero* or during early life, may increase the risk of neurological and psychiatric disorders, arguably via altered *epigenetic regulation*. Epigenetic mechanisms, such as *miRNA expression, DNA methylation, and histone modifications* are prone to changes in response to stressful experiences and hostile environmental factors. Altered epigenetic regulation may potentially influence fetal endocrine programming and brain development across several generations. However recently, more attention has been paid to possible *transgenerational-effects of stress*.[16] Intrauterine stress induces increased risk of adult disease through fetal programming mechanisms. *Oxidative stress* can be generated by several conditions, such as, prenatal hypoxia, maternal under- and overnutrition, and excessive glucocorticoid (GC) exposure. The role of oxidant molecules as signaling factors in fetal programming via epigenetic mechanisms has been discussed. By linking oxidative stress with dysregulation of specific target genes, we may be able to develop therapeutic strategies that protect against organ dysfunction in the programmed offspring.[17]

Fetal stresses, such as *hypoxia, malnutrition,* and fetal exposure to nicotine, alcohol, cocaine, and GCs may directly or indirectly act at cellular and molecular levels to alter the brain development and result in programming of heightened brain vulnerability to hypoxic–ischemic encephalopathy and the development of neurological diseases in the postnatal life.

GCs may play a crucial role in epigenetic programming of neurological disorders of fetal origins. GCs have central effects on the programming of hypoxic–ischemic–sensitive phenotype in the neonatal brain.[18] *Starting in utero,* stress may influence the programming of brain neurotransmitter systems, sympathetic, and parasympathetic nervous system functioning, and the HPA axis, which in turn may alter neural regulation of immune function. *Epigenetic dysregulation* of gene expression may be a fundamental mechanism for programming of early neuroimmune processes.[19]

In humans and animal models, PS, excess exogenous GCs or inhibition of 11β-hydroxysteroid dehydrogenase type 2 (HSD2; the placental barrier to maternal GCs) reduces birth weight and causes hyperglycemia, hypertension, increased HPA axis reactivity, and increased anxiety-related behavior. Molecular mechanisms that underlie the "developmental programming" effects of excess GCs/PSes, include epigenetic changes in target gene promoters. In case of the intracellular GR, this alters tissue-specific GR expression levels, which has persistent and profound effects on GC signaling in certain tissues (e.g., brain, liver, and adipose). Crucially, changes in gene expression persist long after the initial challenge, predisposing the individual to disease in later life. Intriguingly, effects of a challenged pregnancy appear to be transmitted possibly to one or two subsequent generations, suggesting that these epigenetic effects persist.[20]

2.5 Early Life Stress

Early life stress (ELS) includes prenatal, postnatal, and adolescence stress. ELS can affect the development of the HPA axis, and cause cellular and molecular changes in the developing hippocampus that can result in neurobehavioral changes later in life. Epidemiological data implicate stress to be a cause of seizures in both children and adults. Emerging evidence further indicates that both prenatal and postnatal stress can prime the developing brain for seizures and an increase in epileptogenesis.[21] Overall, we provide evidence of a new role of polycomb complex group (Pc-G) proteins in priming ELS-responsive DNA methylation at the arginine vasopressin (AVP) enhancer, prior to epigenetic programming consistent with the idea that Pc-G proteins are part of a flexible silencing system during neuronal development.[22] ELS comprising daily separation of pups from their dams on postnatal days 1–10 was observed to upregulate the expression of the pituitary proopiomelanocortin (Pomc) gene. Pomc serves as a prohormone for ACTH, a key mediator of the adrenocortical response to stress. Detailed analysis revealed that the increase in Pomc mRNA levels results from a reduction in DNA methylation at a critical regulatory region of the Pomc gene; interestingly, this change occurs with some delay after ELS and persists for up to 1 year. Using a Pomc-expressing pituitary cell line (AtT20), we confirmed a role for DNA methylation in restraining Pomc expression under resting conditions. Specifically, we show that CpG site-specific methylation of the Pomc promoter represses Pomc mRNA transcription. Further, it was shown that high-affinity binding of methyl-CpG binding protein-2 to the distal promoter of Pomc, suggesting that methyl-CpG binding protein-2 acts in association with the chromatin modifiers histone deacetylase 2 and DNA methyltransferase 1 to repress Pomc gene expression.[23]

Stress during early life can impact the developing brain and increase vulnerability to mood disorders later in life. Here, we argue that epigenetic mechanisms can mediate the gene-environment dialog in early life and give rise to persistent epigenetic programming of adult physiology eventually resulting in disease. ELS in mice leads to epigenetic marking of the AVP gene underpinning sustained expression and increased HPA axis activity. This epigenetic memory is laid down in the parvocellular neurons of the paraventricular nucleus and involves Ca^{2+}/calmodulin kinase-mediated phosphorylation of the methyl-CpG binding domain protein MeCP2 leading to dissociation from its DNA-binding site and derepression of the *AVP gene.* The reduced occupancy of MeCP2 during this early stage of life facilitates the development of hypomethylation at the *AVP enhancer*, which sustains derepression throughout later life and thereby serves to hardwire early life experiences. The sequential order of these events may represent a critical time window for the preventive therapy of severe trauma.[24]

Epigenetic mechanisms are involved in programming gene expressions throughout development. In addition, they are the key contributors to the processes by which early life experiences fine-tune the expression levels of key neuronal genes, governing learning and memory throughout life. Here, the long-lasting bidirectional effects of early life experiences on learning and memory have been described. How enriched postnatal experience enduringly augments spatial learning, and how chronic ELS results in persistent and progressive deficits in the structure and function of hippocampal neurons, have been discussed. The existing and emerging roles of epigenetic mechanisms in these fundamental neuroplasticity phenomena are illustrated.[25]

Data indicate that not only perinatal nutritional insults have short-term consequences on the growth velocity of the fetus/neonate, but also they sensitize the development of metabolic adult diseases. The pathophysiological mechanisms involved in the so-called "developmental origins of health and adult diseases" are still largely unknown and depend on the types of alteration (nutritional, psychological, endocrine disruptors, etc.), its intensity and duration, species, sex, and the time during which it is applied. Perinatal stress via disturbances of both HPA axis and sympathoadrenal system (SAS), as well as brain–adipose axis and pancrea alterations could play a crucial role. Interestingly, it has been demonstrated that perinatal insults may be *transmitted transgenerationally*, suggesting that these long-term consequences may be inherited via epigenetic mechanisms. Finally, since the placenta demonstrates sensitivity to perinatal nutritional manipulations, the identification of *placental markers* may thus represent a new important avenue to identify more susceptible babies prone to developing metabolic diseases.[26]

Stressful events in early life, including acute perinatal stress, maternal deprivation, separation, and variation in maternal care, may lead to neuroendocrine perturbations, thereby affecting reproductive performance, cognitive functions, and stress responses, as well as the risk for infectious, cardiometabolic, and psychiatric diseases in later life. *Results*: findings from recent studies based on genome-wide and candidate gene approaches highlighted the importance of mechanisms that are involved in epigenetic regulation of gene expression, such as DNA methylation, histone modifications, and noncoding RNAs, in the long-term effects of exposure to stress in early life.[27]

2.6 Genetics

Novel molecular connections have been reported between EGR1 and Pc-G function. Polycomb-associated histone H3 lysine27 trim ethylation (H3K27me3) blocks chromatin access of EGR1. In addition, EGR1 ablation results in abnormal Ezh2 and Bmi1 expression. Consistent with this functional interaction, we identify a number of coregulated target genes in a chondrogenic gene network. Here, an important role for EGR1 in early chondrogenic epigenetic programming to accommodate *early gene–environment interactions* in chondrogenesis, has been described.[28]

Analyses of the epigenomic miRNA environment detected significant reductions in miR-322, miR-574, and miR-873 in the F2-S male brain levels that were again more similar to those of control females. Increased expression of a common gene target for these three miRNAs and β-glycan, was confirmed in these males. These developmental effects were associated with the transmission of a stress-sensitive phenotype and shortened anogenital distances in adult F2-S males. It has been confirmed that the miRNA environment is responsive to organizational testosterone, neonatal males administered the aromatase inhibitor formestane exhibited dramatic changes in brain miRNA patterns, suggesting that miRNAs may serve a previously unappreciated role in organizing the sexually dimorphic brain.[29]

Methylation status of the GR gene was analyzed in mothers and their children, at 10–19 years after birth. These data have been combined with a retrospective evaluation of maternal exposure to intimate partner violence (IPV). Methylation of the mother's GR gene was not affected by IPV. For the first time, it has been shown that *methylation status of the GR gene of adolescent children is influenced by their mother's experience of IPV during pregnancy*. As these sustained epigenetic modifications are established *in utero*, this is considered to be a plausible mechanism, by which PS may program adult psychosocial functions.[30]

DNA methyltransferase 1 (DNMT1) is an important component of the epigenetic machinery, and is responsible for copying DNA methylation patterns during cell division. Coordination of DNA methylation and DNA replication is critical for maintaining epigenetic programming. Knockdown of DNMT1 leads to the inhibition of DNA replication, but the mechanism has been unclear. Here, the depletion of DNMT1 has been shown with either antisense or small interfering RNA (siRNA) that is specific to DNMT1 and activates a cascade of genotoxic stress checkpoint proteins, resulting in phosphorylation of checkpoint kinases 1 and 2 (Chk1 and 2), gammaH2AX focus formation, and cell division control protein 25a (CDC25a) degradation, in an ataxia telangiectasia-mutated and Rad3-related (ATR)-dependent manner. The siRNA knockdown of ATR blocks the response to DNMT1 depletion; DNA synthesis continues in the absence of DNMT1, resulting in global hypomethylation. Similarly, the response to DNMT1 knockdown is significantly attenuated in human mutant ATR fibroblast cells from a Seckel syndrome patient. This response is sensitive to DNMT1 depletion, independent of the catalytic domain of DNMT1, as indicated by abolition of the response with ectopic expression of either DNMT1 or DNMT1 with the catalytic domain deleted. There is no response to short-term treatment with 5-aza-deoxycytidine (5-aza-CdR), which causes demethylation by trapping DNMT1 in 5-aza-CdR-containing DNA but does not cause disappearance of DNMT1 from the nucleus. These data are consistent with the hypothesis that removal of DNMT1 from replication forks is the trigger for this response.[31]

Hippocampal DNA methylation patterns and expression profiles of female prenatally stressed 5-Htt ± mice suggest that distinct molecular mechanisms, some of which are promoter methylation-dependent, contribute to the behavioral effects of the 5-Htt genotype, PS exposure and their interaction.[32]

2.7 Reproductive Technologies

Assisted reproductive technologies (ART), altered gene expression, and more recently, imprinting disorders relating to ART, have become a focused area of research. *In vitro* stress becomes critically important in mitigating the effects of stresses caused by the *in vitro* environment. Any invasion into the natural, highly orchestrated, and dynamic process of sexual reproduction, could perturb the normal progression of epigenetic programming. Therefore the source of gametes, subsequent culture conditions of gametes, and embryos are critically important and require careful attention.[33]

2.8 Environment

The environment can have a long-lasting influence on an individual's physiology and behavior. Although some environmental conditions can be beneficial and result in adaptive responses, others can lead to pathological behaviors. Many studies have demonstrated that changes induced by the environment are expressed not only by the individuals directly exposed, but also by the offsprings, sometimes across multiple generations. Epigenetic alterations have been proposed as underlying mechanisms for such transmissible effects.[34]

2.9 Pathogenesis

Epidemiological studies of IBS point-to-risk factors, such as familial clustering, sexual abuse, and other forms of childhood trauma, low birth weight, and gastrointestinal infection. Epigenetics focuses on the complex and dynamic interactions between DNA sequence, DNA modifications, and environmental factors, all of which combine to produce the phenotype. Early stress studies in animal models and humans, who have experienced childhood trauma or abuse suggest that these events can lead to long-lasting epigenetic changes in the GR gene brought about by hypermethylation of a key regulatory component. Animal studies also indicate that the microbiota has a pivotal role in programming the core stress system, the HPA axis, and the *immune system* through epigenetic mechanisms. In this perspective, an epigenetic model of IBS is presented that incorporates many current findings regarding IBS, including proinflammatory markers, neuroendocrine alterations, and links with both psychosocial stress, and stress related to infection.[35]

Here, a causal link has been shown between stress-induced motor impairment and associated genetic and epigenetic responses, in the relevant central motor areas in a rat model. Exposure to two weeks of mild restraint stresses altered the expression of 39 genes and 9 *miRNAs* in the cerebellum. In line with persistent behavioral impairments, some changes in gene and miRNA expression were resistant to recovery from stress. Interestingly, stress upregulated the expression of Adipoq and prolactin receptor mRNAs in the cerebellum. Stress also altered the expression of Prlr, miR-186, and miR-709 in hippocampus and prefrontal cortex. In addition, our findings demonstrate that miR-186 targets the gene Eps15. Furthermore, we found an age-dependent increase in EphrinB3 and GabaA4 receptors. Even mild stress results in substantial genomic and epigenomic changes involving miRNA expression and associated gene targets in the motor system. These findings suggest a central role of miRNA-regulated gene expressions in the stress response and in associated neurological functions.[36]

2.10 Risk Factors

Exposure to hostile conditions results in a series of coordinated responses aimed at enhancing the probability of survival. Activation of the HPA axis plays a pivotal role in the stress response. Although short-term activation of the HPA axis allows adaptive responses to the challenge, in the long run this can be devastating for the organism. In particular, life events occurring during the perinatal period have strong long-term effects on the behavioral and neuroendocrine response to stressors. In male and female rats exposed to prenatal restraint stress (PRS), these effects include a long-lasting hyperactivation of the HPA response associated with an altered circadian rhythm of corticosterone secretion. Furthermore, male animals exhibit sleep disturbances. In males, these HPA dysfunctions have been reported in infant, young, adult, and aged animals, thus suggesting a permanent effect of early stress. Interestingly, after exposure to an intense inescapable footshock, female PRS rats durably exhibit a blunted corticosterone secretion response to stress. In male PRS rats exposed to an alcohol challenge, the HPA axis is similarly hyporesponsive. Rats exposed to PRS also show behavioral disturbances. Male and female PRS rats show high anxiety levels and depression-like behavior during adulthood, although some studies suggest that female PRS rats present low anxiety levels. With ageing, male and female PRS rats exhibit memory impairments in hippocampus-dependent tasks, while female PRS rats improve their memory performance during adulthood. Gender effect on the behavior seems to be related to a reduction in hippocampal plasticity in male PRS rats, and an increase in female PRS rats. Despite the permanent imprinting induced by early stress, dysfunctions observed after PRS can be reversed by environmental or pharmacological strategies such as environmental enrichment or antidepressive and neurotrophic treatments. Mechanisms underlying the effects of PRS on the offsprings remain largely unknown. However, previous studies have demonstrated that maternal GCs during pregnancy play an important role in the HPA disturbances reported in male offsprings. Finally, gestational stress has long-lasting effects on the HPA axis and on behavior in the dams. Thus, alterations in maternal behavior could also make a strong contribution to the long-term effects of PRS through epigenetic mechanisms.[37]

Hypertension is a major risk factor for cardiovascular and cerebrovascular diseases. Lifelong environmental factors (e.g., salt intake, obesity, alcohol) and genetic factors clearly contribute to the development of hypertension. But it has also been established that stress *in utero* may program the later development of the disease. This phenomenon, known as fetal programming can be modeled in a range of experimental animal models. In maternal low protein diet rat models of programming, administration of angiotensin converting enzyme inhibitors, or angiotensin receptor antagonists in early life can prevent the development of hypertension, thus implicating the renin–angiotensin system in this process. In this model, the expression of AT(1b) angiotensin receptor gene in the adrenal gland is upregulated by the first week of life resulting in increased receptor protein expression, which is consistent with the increased adrenal angiotensin responsiveness observed by the others. Furthermore, we show that the proximal promoter of AT(1b) gene in the adrenal is significantly undermethylated and *in vitro* AT(1b) gene expression is highly dependent on the promoter methylation. These data suggest a link between fetal insults to epigenetic modification of genes and the resultant alteration of gene expression in adult life, leading ultimately to the development of hypertension. It seems highly probable that similar influences may be involved in the development of human hypertension.[38]

Numerous clinical studies associate an adverse prenatal environment with the development of cardiometabolic disorders and neuroendocrine dysfunction, as well as an increased risk of psychiatric diseases in later life. Experimentally, prenatal exposure to stress or excess GCs in a variety of animal models can disturb the offspring's physiology, resulting in a reduction in birth weight and subsequently increasing the likelihood of disorders of cardiovascular function, glucose homeostasis, HPA axis activity, and anxiety-related behaviors in adulthood. During fetal development, placental 11beta-hydroxysteroid dehydrogenase type 2 (11β-HSD2) provides a barrier to maternal GCs. Reduced placental 11β-HSD2 in human pregnancy correlates with lower birth weight and higher blood pressure in later life. Similarly, in animal models, inhibition or knockout of placental 11β-HSD2 lowers offspring birth weight, in part by reducing glucose delivery to the developing fetus in late gestation. Molecular mechanisms thought to underlie the programming effects of ELS and GCs include epigenetic changes in target chromatin, notably affecting tissue-specific expression of the intracellular GR. As such, excess GCs in early life can permanently alter tissue GC signaling, effects of which may have short-term adaptive benefits but increase the risk of later disease.[39]

2.11 Therapy

Epigenetic modifications are reversible and can be manipulated. This provides a unique chance to develop multiple novel *epigenetic*-based therapeutic strategies against many chronic diseases in early developmental periods.[40]

2.12 Stem Cells

Stem cells are omnipotent cells, which have the capacity to regenerate, as well as rebuild all organs and tissues from injury and trauma. This cell is already handling the genome. This cell is a candidate to handle the issue of mutations and inheritance. As the stem cells are in contact with the genome of the entire organism, and possess tissue- and organ-specific stem cells (multipotent cell), it has a tremendous potential to handle mutant cells. So one has to see if epigenetic factors affect stem cells.

2.13 Stem Cells and Epigenetic Regulation

Findings demonstrate that pluripotent human embryonic stem cells (hESC) have a markedly different overall nuclear architecture, remodeling of which is linked to early epigenomic programming and involves formation of unique PML-defined structures.[41]

DNA methylation is an important epigenetic mechanism, affecting normal development, and playing a key role in reprogramming epigenomes during stem cell derivation. Here we report on DNA methylation patterns in native monkey embryonic stem cells (ESCs), fibroblasts, and ESCs generated through somatic cell nuclear transfer (SCNT), identifying and comparing epigenome programming and reprogramming. Hundreds of regions have been characterized that are hyper- or hypomethylated in fibroblasts, compared to native ESCs, and show that these are conserved in human cells and tissues. Remarkably, a vast majority of these regions are reprogrammed in SCNT ESCs, leading to an almost perfect correlation between the epigenomic profiles of the native and reprogrammed lines. At least 58% of these changes are correlated in *cis* to transcription changes, Polycomb Repressive Complex-2 occupancy, or binding by the CTCF insulator. We also show that, while epigenomic reprogramming is extensive and globally accurate, the efficiency of adding and stripping DNA methylation during reprogramming is regionally variable. In several cases, this variability results in regions that remain methylated in a fibroblast-like pattern even after reprogramming.[42]

ESCs, derived from preimplantation of embryo from embryonic germ (EG) cells, which are derived from embryonic precursors of gametes, called primordial germ cells (PGCs), can differentiate into any cell type in the body. Moreover, ES cells have the capacity to differentiate into PGCs *in vitro*. In the present study, the differentiation capacity of *six EG cell lines to form PGCs in vitro*, in comparison to ES cells has been shown. Cell lines were differentiated via embryoid body (EB) formation using the coexpression of mouse vasa homolog (Mvh) and Oct-4 to identify newly formed PGCs *in vitro*. An increase of PGC numbers was found in almost all analyzed cell lines in 5-day-old EBs, thus suggesting that EG and ES cells have similar efficiency to generate PGCs. Addition of retinoic acid confirmed that the cultures had attained a PGC-like identity and continued to proliferate. Furthermore we have shown that the expression pattern of Prmt5 and H3K27me3 in newly formed PGCs is similar to that observed in embryonic day E11.5 PGCs *in vivo*. By coculturing EBs with Chinese hamster ovary (CHO) cells, some of the PGCs entered into meiosis, as judged by Scp3 expression.[43]

Human adipose tissue-derived stem cells (ASCs) can give rise to a variety of cell types *in vitro* and *in vivo*. However, they have a propensity to get differentiated into primarily mesodermal lineages. Even so, their capacity to differentiate into nonadipogenic mesodermal pathways seems to be restricted. Emerging DNA methylation profiles at adipogenic and nonadipogenic gene promoters in freshly isolated, cultured, or differentiated ASCs, aims to provide an epigenetic explanation for this restrictive differentiation potential. A review of these studies indicates that human ASCs are epigenetically marked by mosaic hypomethylation of adipogenic promoters, whereas nonadipogenic lineage-specific promoters are hypermethylated. Surprisingly, *in vitro* differentiation toward various pathways maintains the overall methylation profiles of undifferentiated cells, raising the hypothesis that ASCs are at least epigenetically preprogrammed for adipogenesis.[44]

Fetal growth restriction and over-growth are associated with global shifts toward *DNA hypermethylation*, targeting *cis*-regulatory elements in proximity with genes involved in glucose homeostasis and *stem cell functions*. A sexually dimorphic response, intrauterine growth restriction is associated with substantially greater epigenetic dysregulation in males, whereas large for gestational age growth predominantly affects females. Extreme fetal growth is associated with increased susceptibility to a range of adult diseases through an unknown mechanism of cellular memory.[45]

A genetic optimal control model has been proposed for adult stem cell regeneration that includes the three fundamental processes, along with cell division and adaptation based on differential fitness of phenotypes. In this model, stem cells with a distribution of *epigenetic states* are required to maximize expected performance after each cell division. Heterogeneous proliferation that depends on the epigenetic states of stem cells can improve the maintenance of stem cell distributions to create balanced populations. When mutation is allowed, apoptosis evolves to maximize the performance during homeostasis after multiple cell divisions. The overall results highlight the importance of cross-talk between genetic and epigenetic regulation and performance objectives during homeostasis in shaping a desirable heterogeneous distribution of stem cells in epigenetic states.[46]

PGCs are the precursors of sperm and eggs, which generate a new organism that is capable of creating endless new generations through germ cells. PGCs are specified during early mammalian postimplantation development, and are uniquely programmed for the transmission of genetic and epigenetic information to subsequent generations. In this Primer, we summarize the establishment of the fundamental principles of PGC specification during early development, and discuss how it is now possible to make mouse PGCs from *pluripotent ESCs*, and indeed somatic cells if they are first rendered pluripotent in culture.[47]

Classically activated macrophages (M1) are known to play a role in the development of chronic inflammation associated with impaired wound healing in type 2 diabetes (T2D). A repressive *histone methylation mark, H3K-27me3*, is decreased at the promoter of the IL12 gene in BM progenitors, and this epigenetic signature is passed down to wound macrophages in a murine model of glucose intolerance (diet-induced obese (DIO)). Hence, *histone demethylase inhibitor*-based therapy may represent a novel treatment option for diabetic wounds.[48]

Epigenetic mechanisms are the key components of dynamic transcriptional programming that occurs along the process of differentiation from normal stem cells to more specialized cells. In the development of cancer and according to the cancer stem cell model, aberrant epigenetic changes may ensure the property of cancer cells to switch *cancer stem cell markers* on and off in order to generate a heterogeneous population of cells. The tumor will then be composed of tumorigenic (cancer stem cells) and nontumorigenic (the side population that constitutes the bulk of the tumor) cells. Characterizing epigenetic landscapes may thus help discriminate aberrant marks (good candidates for tumor detection) from cancer stem cell specific profiles.[49]

Cell fate decisions are largely programmed by interactions between multiple layers of regulation of gene expression. Among these, epigenetic states have been extensively examined, mostly in the context of ESC differentiation. However, recent studies have focused on understanding chromatin-based mechanisms of differentiation of adult progenitor cells into specific lineages but not others. The results point to the view that promoter DNA methylation patterns are not the primary determinants of gene activation potentials and differentiation capacities of mesenchymal stem cells. Posttranslational histone modifications on promoters contribute to the establishment of a permissive states of differentiation, but based on current knowledge, they cannot predict transcriptional activation outcomes. Additional regulatory layers need to be examined to be able to explain cell fate commitment and ultimately predict cell fate.[50]

Cells of the early mammalian embryo, including pluripotent ESCs and PGCs, are epigenetically dynamic and heterogeneous. During early development, this heterogeneity of epigenetic states is associated with stochastic expression of lineage-determining transcription factors that establish an intimate cross-talk with epigenetic modifiers. Lineage-specific epigenetic modification of crucial transcription factor loci (e.g., methylation of the Elf5 promoter) leads to the restriction of transcriptional circuits and the fixation of lineage fate. The intersection of major epigenetic reprogramming and programming events in the early embryo creates plasticity, followed by commitment to the principal cell lineages of the early conceptus.[51]

The data suggest that our multipotent adult germline stem (maGS) cells are epigenetically stable for the H19 gene during *in vitro* modifications. Therefore, further studies on the epigenetic regulation and chromatin structure of maGS cells are necessary before their full potential is utilized in regenerative medicine.[52]

2.14 Epigenetics of the Gonads

Epigenetic regulation of gene expression is critical for oogenesis in mammals. In this study, a simple and efficient method was used to obtain the oocytes from cultured fetal mouse ovaries of 12.5 days postcoitum (DPC). The methylation pattern of these oocytes was examined. These results showed that the establishment of imprinting Igf2r and Peg3 in oocytes derived from cultured fetal mouse germ cells *in vitro*, follows a slower time course than that of oocytes *in vivo*. However, oocytes *in vitro* and *in vivo* share similar methylation patterns. Igf2r was gradually *de novo* methylated, and the methylation covers 80% CpG sites in oocytes cultured for 28 days. However, only 45% of the CpG sites get methylated in Peg3 at the same stage. Furthermore, it demonstrated that the degree of DNA methylation is positively correlated with the size of oocytes *in vitro* and *in vivo*, indicating a progressive methylation process during oocyte growth.[53]

The study of P transposable element repression in *Drosophila melanogaster* led to the discovery of the *trans*-silencing effect (TSE), a homology-dependent repression mechanism by which a P-transgene inserted in subtelomeric heterochromatin (telomere-associated sequences (TAS)) has the capacity to repress in trans, in the female germline, a homologous P-lacZ transgene located in euchromatin. Phenotypic and genetic analysis have shown that TSE exhibits variegation in ovaries, displays a maternal effect, as well as shows epigenetic transmission through meiosis, and involves heterochromatin (including HP1) and RNA silencing.[54]

Oocyte quality is one of the most important aspects of *in vitro* embryo development. Extensive epigenetic programming must occur during oocyte growth and maturation. A specific DNA methylation pattern of the imprinted genes must be established on differentially methylated regions (DMR). The insulin-like growth factor 2 (IGF2) gene is an important growth factor, and it is imprinted in several mammalian species. The aim of this study was to evaluate the methylation pattern on the DMR of the last exon of

IGF2 in immature and mature bovine oocytes with different developmental competencies. Mature oocytes from large follicles were less methylated (28.93%) than immature oocytes from large follicles (77.38%, $P = 0.002$), and there was also a tendency toward lower methylation in mature oocytes from large follicles (28.93%) compared with mature oocytes from small follicles (52.58%, $P = 0.07$). Immature oocytes from small and large follicles showed 53.85% (7/13) and 91.66% (11/12) hypermethylated sequences, respectively, whereas mature oocytes from small and large follicles showed 61.11% (11/18) and 40% (4/10), respectively. The hypomethylation pattern in mature oocytes from large follicles may be related to higher competence of these oocytes. Our results suggest that the methylation pattern in this DMR may be a useful parameter to investigate as a molecular marker for oocyte competence in cattle and as a model for studies in other species.[55]

The disturbances in oocyte genomic imprinting and modification of posttranslational histone and centromere architecture provide the first link between low bisphenol A (BPA) exposures and induction of *epigenetic changes, which may contribute to chromosome congressional failures* and meiotic errors, and to altered.[56]

The expression of mouse IC-transcript was studied, in brain, oocytes, in cultured neurons and glial cells by RT-PCR. *In situ* hybridization (ISH) was used in oocytes. The IC transcript was strongly expressed in brain (especially in neurons) and ovary (especially in oocytes and granulosa cells), while no expression was found in other tissues. This was confirmed by quantitative analysis and ISH. Expression levels in the brain were sevenfold higher compared to those in ovaries. ISH signals were observed in oocytes and granulosa cells of the secondary and developing follicles. These findings, together with previous data, suggest that the IC transcript may be associated with the establishment of PWS-IC methylation on the maternal chromosome as an AS-IC *cis*-acting element.[57]

Mammalian oocytes possess unique properties with respect to their ability to regulate and reprogram chromatin structure and epigenetic information. Proteins containing the conserved chromodomain motif, which is common to the Pc-G proteins, and the heterochromatin-associated protein HP1 plays essential roles in these processes, and more specifically in X-chromosome inactivation in female zygotes, extraembryonic tissues, and in the regulation of genomic imprinting. To characterize the potential role of these proteins in the regulation of epigenetic events during early human development, we utilized a degenerate PCR priming assay to assess the expression of mRNAs of chromodomain proteins in cDNA samples derived from the human female germline and preimplantation embryos. Expression of mRNAs of HP1 genes was observed in ovarian follicles (HP1 (HSα), HP1 (HSβ), and HP1 (HSγ)), mature oocytes (HP1 (HSα) and HP1 (HSβ)), cleavage stage preimplantation embryos (HP1 (HSα), HP1 (HSβ), and HP1 (HSγ)), and blastocysts (HP1 (HSα) and HP1 (HSγ)). Transcripts for three Pc-G genes, which are essential for early mammalian development (Yin-Yang 1 (YY1), enhancer of zeste-2 (EZH2), and embryonic ectoderm development (EED)) and that are essential for the regulation of X-inactivation and certain imprinted genes (EED) were revealed by gene-specific-PCR expression analysis of human ovarian follicles, oocytes, and preimplantation embryos. YY1 and EZH2 transcripts were additionally detected in metaphase II oocytes.[58]

Epigenetic reprogramming occurs during oocyte growth in mice, a stage where a number of important events are occurring, including transcription of maternal mRNAs for storage in the mature egg, global transcriptional silencing, and the acquisition of meiotic competence. Oocyte growth occurs in conjunction with follicular development over a period of many days. The signals involved in initiating different stages in oocyte and follicular development, and the concurrent epigenetic changes are poorly understood. Here, the role of stem cell factor (SCF or Kit ligand) has been examined on the early to mid stages of oocyte growth, on DNA methyltransferase expression, and function using a one-step *in vitro* culture system. Our results show that SCF promotes early oocyte growth and development to the multilaminar follicle stage. Oocyte growth is sufficient to trigger transcription of Dnmt1 and Dnmt3L from dedicated oocyte promoters, and we show that eggs undergoing growth in the absence of follicle development in Foxo3 mutants show elevated levels of Dnmt1. The methyltransferase proteins undergo sequential relocalization in the oocyte, with DNMT1 being exported from the nucleus at the bilaminar follicle stage, while DNMT3A is transported into the nucleus at the multilaminar stage, indicating an important role for trafficking in controlling imprinting. SCF is thought to signal partly through the phosphatidylinositol 3 (PI3) kinase pathway: inhibiting this path was previously shown to prevent FOXO3 nuclear export and we could show here that it also prevented DNMT1 export. Some oocytes reached full size (70 μm) in this *in vitro* system, but no secondary follicles were formed, most likely due to failure of the thecal layer to form properly. *De novo* methylation of imprinted genes was seen in some oocyte cultures, with methylation levels being highest for Snrpn and Igf2r, which are methylated early *in vivo*, while Peg1, which is methylated late, showed little or no methylation. SCF treatment did not increase the number of cultures showing methylation. We saw no evidence for *de novo* methylation of IAP repeats in our cultures. These results suggest that while methyltransferase loading is triggered by oocyte growth, in which SCF plays an important role, complete methylation probably requires progression to the secondary follicle stage and is unlikely to be affected by SCF.[59]

These results indicated that the differential expressions of paternally expressed genes occurred in testes *of intracytoplasmic sperm injection* (*ICSI*) mice, which may be mediated by methylation modification. ICSI procedure and mechanical stimulation can induce intergenerational transmission of the *epigenetic changes. In vitro* culture and mechanical stimulation were the main factors inducing the down regulation of paternally expressed imprinted genes in testes.[60]

Testis-derived male germline stem (GS) cells, the *in vitro* counterpart of spermatogonial stem cells (SSC), can acquire multipotency under appropriate culture conditions to become maGS cells, which upon testicular transplantation, produce teratoma instead of initiating spermatogenesis. Consequently, a molecular marker that can distinguish GS cells from maGS cells would be of potential value in both clinical and experimental research settings. Data suggest that GS cells have androgenetic DNA methylation and expression of imprinted miRNAs, which changes to ES cell-like pattern upon their conversion to maGS cells. Differential genomic imprinting of imprinted miRNAs may thus, serve as epigenetic miRNA signature or molecular marker to distinguish GS cells from maGS cells.[61]

Techniques such as ICSI and round spermatid injection (ROSI) may increase the incidence of imprinting disorders, and adversely affect embryonic development by using immature spermatozoa, which may not have established proper imprints or global methylation. Epigenetic changes, in contrast to genetic aberrations, may be less deleterious because they are potentially reversible. Further research could identify certain drugs capable of reversing epigenetic changes. Environmental toxins and drugs may cause sperm malformation, such as 5-aza-2′-deoxycytidine, vinclozolin, and methoxychlor.[62]

These results also suggest that the genomic DNA of *testicular tissues* from control rats is resistant to R.Msp1, while DNA from the experimental group testicular cells demonstrating high proliferation rate could not resist to R.Msp1 digestion due to DNA hypomethylation in CCGG sequence. In conclusion, it could be suggested that the reversal of gene imprinting in germinal cells may cause an increased cellular proliferation and R.Msp1 fragmentation, when induced by 5-aza-2′-deoxycytidine.[63]

Expression of imprinted genes is restricted to a single parental allele as a result of epigenetic regulation – DNA methylation and histone modifications. Igf2/H19 is a reciprocally imprinted locus exhibiting paternal Igf2 and maternal H19 expression. Their expressions are regulated by paternally methylated imprinting control regions (ICR) located between the two genes. Although the *de novo* DNA methyltransferases have been shown to be necessary for the establishment of ICR methylation, the mechanism by which they are targeted to the region remains unknown. It has been demonstrated that CTCFL/BORIS, a paralog of CTCF, is an ICR-binding protein expressed during embryonic male germ cell development, coinciding with the timing of ICR methylation. PRMT7, a protein arginine methyltransferase, with which CTCFL interacts, is also expressed during embryonic testis development. Symmetrical dimethyl arginine 3 of histone H4, a modification catalyzed by PRMT7, accumulates in germ cells during this developmental period. This modified histone is also found enriched in both H19 ICR and Gtl2 DMR chromatin of testis by chromatin immunoprecipitation (ChIP) analysis. *In vitro* studies demonstrate that CTCFL stimulates the histone-methyltransferase activity of PRMT7 via interactions with both histone and PRMT7. Finally, H19 ICR methylation is demonstrated by nuclear coinjection of expression vectors encoding CTCFL, PRMT7, and the *de novo* DNA methyltransferases, Dnmt3a, -b, and -L, in Xenopus oocytes. These results suggest that CTCFL and PRMT7 may play a role in male germline imprinted gene methylation.[64]

2.15 Role of the Innate Immune System

Now that stem cells are engaged in epigenetic programming, it can be proposed that stem cells will mutate in response to epigenetic influence. The biological function of the gonads is to transit the genome fatefully with all the old and new genes that might be there. However, the virtue of gonads is accuracy, so they would not participate in mutation.

One should note that stem cells reside in the bone marrow. This is a protected environment, the bone marrow is shielding for radiation, and possibly for much background noise in environmental impacts. So stem cells would respond only to significant impulses, which deserve attention. Stem cells respond to the epigenetic changes by cell proliferation. While proliferating stem cells make a lot of mistakes (1:25), so mutant stem cells will be generated. The mutant cells that find growth factors to stimulate them will survive. Mutant cells without stimulatory factors will die by apoptosis. For a proportion of surviving cells, there will not be appropriate function. Cells and organs without biological function are eliminated in Biology. Apparently functional activity is a *vital stimulus*. But what does it mean to survive in this situation? It means that the surviving cells have to outperform the old guard of cells, and be the best, this is the license for survival, and this principle is the driving force for evolution. Survival of the best, as Darwin pointed out. Now the mutant cells are new and need immunological protection. Since they come from stem cells in the bone marrow, the INIM system classify these cells as "self" and provide full protection for life. The immune system initiates the stress responses, which generates diversity, and the immune system is there to protect the progeny generated at the end of the evolutionary process. So, *the INIM system is the driving force of evolution.*

REFERENCES

1. Darwin C. *The origin of the species*. London: John Murray; 1859.
2. Szekeres Bartho J, Chaouhat G. Maternal fetal immune interaction. *Adv Neuroimmune Biol* 2011;**2**(1):2.
3. Babenko O, Kovalchuk I, Metz GA. Stress-induced perinatal and transgenerational epigenetic programming of brain development and mental health. *Neurosci Biobehav Rev* 2015;**48C**:70–91.
4. Ma Q, Zhang L. Epigenetic programming of hypoxic-ischemic encephalopathy in response to fetal hypoxia. *Prog Neurobiol* 2015;**124**:28–48.
5. Anacker C, O'Donnell KJ, Meaney MJ. Early life adversity and the epigenetic programming of hypothalamic-pituitary-adrenal function. *Dialogues Clin Neurosci* 2014;**16**: 321–33.
6. Schraut KG, Jakob SB, Weidner MT, Schmitt AG, Scholz CJ, Strekalova T, El Hajj Nijssen LM, Domschke K, Reif A, Haaf T, Ortega G, Steinbusch HW, Lesch KP, Van den Hove DL. Prenatal stress-induced programming of genome-wide promoter DNA methylation in 5-HTT-deficient mice. *Transl Psychiatry* 2014;**4**:e473.
7. Vaiserman AM. Epigenetic programming by early-life stress: evidence from human populations. *Dev Dyn* 2014;**10**(1002):24211.
8. Unterberger A, Andrews SD, Weaver IC, Szyf M. DNA methyltransferase 1 knockdown activates a replication stress checkpoint. *Mol Cell Biol* 2006;**26**:75–86.
9. Weaver IC. Epigenetic effects of glucocorticoids. *Semin Fetal Neonatal Med* 2009;**14**: 143–50.
10. Matthews SG, Phillips DI. Transgenerational inheritance of stress pathology. *Exp Neurol* 2012;**233**:95–101.
11. Bale TL. Sex differences in prenatal epigenetic programming of stress pathways. *Stress* 2011;**14**:348–56.
12. Hales CN, Barker DJP. Type 2 (non-insulin-dependent) diabetes mellitus: the thrifty phenotype hypothesis Diabetologia 1992; 35: 595–601.
13. Sebert S, Sharkey D, Budge H, Symonds ME. The early programming of metabolic health: is epigenetic setting the missing link? *Am J Clin Nutr* 2011;**94**(6 Suppl):1953S–8S.
14. Lucassen PJ, Naninck EF, van Goudoever JB, Fitzsimons C, Joels M, Korosi A. Perinatal programming of adult hippocampal structure and function; emerging roles of stress, nutrition and epigenetics. *Trends Neurosci* 2013;**36**:621–31.
15. Stroud LR, Papandonatos GD, Rodriguez D, McCallum M, Salisbury AL, Phipps MG, Lester B, Huestis MA, Niaura R, Padbury J, Marsit CJ. Maternal smoking during

pregnancy and infant stress response: test of a prenatal programming hypothesis. *Psychoneuroendocrinology* 2014;**48**:29–40.

16. Mueller BR, Bale TL. Sex-specific programming of offspring emotionality after stress early in pregnancy. *J Neurosci* 2008;**28**:9055–6.
17. Babenko O, Golubov A, Ilnytskyy Y, Kovalchuk I, Metz GA. Genomic and epigenomic responses to chronic stress involve miRNA-mediated programming. *PLoS ONE* 2012;**7**:e29441.
18. Thompson LP, Al-Hasan Y. Impact of oxidative stress in fetal programming. *J Pregnancy* 2012;582748.
19. Li Y, Gonzalez P, Zhang L. Fetal stress and programming of hypoxic/ischemic-sensitive phenotype in the neonatal brain: mechanisms and possible interventions. *Prog Neurobiol* 2012;**98**:145–65.
20. Wright RJ. Stress-related programming of autonomic imbalance: role in allergy and asthma. *Chem Immunol Allergy* 2012;**98**:32–47.
21. Harris, Seckl. Glucocorticoids, prenatal stress and the programming of disease. *Horm Behav* 2011;**59**:279–89.
22. Huang LT. Early-life stress impacts the developing hippocampus and primes seizure occurrence: cellular, molecular, and epigenetic mechanisms. *Front Mol Neurosci* 2014;**7**:8.
23. Murgatroyd C, Spengler D. Polycomb binding precedes early-life stress responsive DNA methylation at the Avp enhancer. *PLoS ONE* 2014;**9**:e90277.
24. Wu Y, Patchev AV, Daniel G, Almeida OF, Spengler D. Early-life stress reduces DNA methylation of the Pomc gene in male mice. *Endocrinology* 2014;**155**:1751–62.
25. Murgatroyd C, Spengler D. Epigenetic programming of the HPA axis: early life decides. *Stress* 2011;**14**:581–9.
26. McClelland S, Korosi A, Cope J, Ivy A, Baram TZ. Emerging roles of epigenetic mechanisms in the enduring effects of early-life stress and experience on learning and memory. *Neurobiol Learn Mem* 2011;**96**:79–88.
27. Vieau D. Perinatal nutritional programming of health and metabolic adult disease. *World J Diab* 2011;**2**:133–6.
28. Vaiserman AM. Epigenetic programming by early-life stress: evidence from human populations. *Dev Dyn* 2015;**244**:254–65.
29. Spaapen F, van den Akker GG, Caron MM, Prickaerts P, Rofel C, Dahlmans VE, Surtel DA, Paulis Y, Schweizer F, Welting TJ, Eijssen LM, Voncken JW. The immediate early gene product EGR1 and polycomb group proteins interact in epigenetic programming during chondrogenesis. *PLoS ONE* 2013;**8**:e58083.
30. Morgan CP, Bale TL. Early prenatal stress epigenetically programs dysmasculinization in second-generation offspring via the paternal lineage. *J Neurosci* 2011;**31**:11748–55.
31. Radtke KM, Ruf M, Gunter HM, Dohrmann K, Schauer M, Meyer A, Elbert T. Transgenerational impact of intimate partner violence on methylation in the promoter of the glucocorticoid receptor. *Transl Psychiatry* 2011;**1**:e21.
32. Schraut KG, Jakob SB, Weidner MT, Schmitt AG, Scholz CJ, Strekalova T, El Hajj Nijssen LM, Domschke K, Reif A, Haaf T, Ortega G, Steinbusch HW, Lesch KP, Van den Hove DL. Prenatal stress-induced programming of genome-wide promoter DNA methylation in 5-HTT-deficient mice. *Transl Psychiatry* 2014;**4**:e473.
33. Bertoldo MJ, Locatelli Y, O'Neill C, Mermillod P. Impacts of and interactions between environmental stress and epigenetic programming during early embryo development. *Reprod Fertil Dev*. 2014; doi: 10.1071/RD14049.
34. Gapp K, von Ziegler L, Tweedie-Cullen RY, Mansuy IM. Early life epigenetic programming and transmission of stress-induced traits in mammals: how and when can environmental factors influence traits and their transgenerational inheritance? *Bioessays* 2014;**36**:491–502.
35. Dinan TG, Cryan J, Shanahan F, Keeling PW, Quigley EM. IBS: an epigenetic perspective. *Nat Rev Gastroenterol Hepatol* 2010;**7**:465–71.

36. Babenko O, Golubov A, Ilnytskyy Y, Kovalchuk I, Metz GA Genomic and epigenomic responses to chronic stress involve miRNA-mediated programming. PLoS One 2012;7: e29441.
37. Darnaudéry M, Maccari S. Epigenetic programming of the stress response in male and female rats by prenatal restraint stress. *Brain Res Rev* 2008;**57**:571–85.
38. Bogdarina I, Welham S, King PJ, Burns SP, Clark AJ. Epigenetic modification of the renin-angiotensin system in the fetal programming of hypertension. *Circ Res* 2007;**100**:520–6.
39. Cottrell EC, Seckl JR. Prenatal stress, glucocorticoids and the programming of adult disease. *Front Behav Neurosci* 2009;**3**:19.
40. Gao Q, Tang J, Chen J, Jiang L, Zhu X, Xu Z. Epigenetic code and potential epigenetic-based therapies against chronic diseases in developmental origins. *Drug Discov Today* 2014;**19**:1744–50.
41. Butler JT, Hall LL, Smith KP, Lawrence JB. Changing nuclear landscape and unique PML structures during early epigenetic transitions of human embryonic stem cells. *J Cell Biochem* 2009;**107**:609–21.
42. Butler JT, Hall LL, Smith KP, Lawrence JB. Changing nuclear landscape and unique PML structures during early epigenetic transitions of human embryonic stem cells. *J Cell Biochem* 2009;**107**:609–21.
43. Eguizabal C, Shovlin TC, Durcova-Hills G, Surani A, McLaren A. Generation of primordial germ cells from pluripotent stem cells. *Differentiation* 2009;**78**:116–23.
44. Boquest AC, Noer A, Collas P. Epigenetic programming of mesenchymal stem cells from human adipose tissue. *Stem Cell Rev* 2006;**2**:319–29.
45. Delahaye F, Wijetunga NA, Heo HJ, Tozour JN, Zhao YM, Greally JM, Einstein FH. Sexual dimorphism in epigenomic responses of stem cells to extreme fetal growth. *Nat Commun* 2014;**5**:5187.
46. Lei J, Levin SA, Nie Q. Mathematical model of adult stem cell regeneration with cross-talk between genetic and epigenetic regulation. *Proc Natl Acad Sci USA* 2014;**111**: E880–7.
47. Magnúsdóttir E, Surani MA. How to make a primordial germ cell. *Development* 2014;**141**:245–52.
48. Gallagher KA, Joshi A, Carson WF, Schaller M, Allen R, Mukerjee S, Kittan N, Feldman EL, Henke PK, Hogaboam C, Burant CF, Kunkel SL. Epigenetic changes in bone marrow progenitor cells influence the inflammatory phenotype and alter wound healing in type 2 diabetes. *Diabetes* 2015;**64**:1420–30.
49. Vincent A, Van Seuningen I. On the epigenetic origin of cancer stem cells. *Biochim Biophys Acta* 2012;**1826**:83–8.
50. Collas P. Programming differentiation potential in mesenchymal stem cells. *Epigenetics* 2010;**5**:476–82.
51. Hemberger M, Dean W, Reik W. Epigenetic dynamics of stem cells and cell lineage commitment: digging Waddington's canal. *Nat Rev Mol Cell Biol* 2009;**10**:526–37.
52. Babenko O, Kovalchuk I, Metz GA. Stress-induced perinatal and transgenerational epigenetic programming of brain development and mental health. *Neurosci Biobehav Rev* 2015;**48**:70–91.
53. Oh SH, Jung YH, Gupta MK, Uhm SJ, Lee HT. H19 gene is epigenetically stable in mouse multipotent germline stem cells. *Mol Cells* 2009;**27**:635–40.
54. Song Z, Min L, Pan Q, Shi Q, Shen W. Maternal imprinting during mouse oocyte growth *in vivo* and *in vitro*. *Biochem Biophys Res Commun* 2009;**387**:800–5.
55. Todeschini AL, Teysset L, Delmarre V, Ronsseray S. The epigenetic trans-silencing effect in *Drosophila* involves maternally transmitted small RNAs whose production depends on the piRNA pathway and HP1. *PLoS ONE* 2010;**5**:e11032.
56. Fagundes NS, Michalczechen-Lacerda VA, Caixeta ES, Machado GM, Rodrigues FC, Melo EO, Dode MA, Franco MM. Methylation status in the intragenic differentially

methylated region of the IGF2 locus in *Bos taurus* indicus oocytes with different developmental competencies. *Mol Hum Reprod* 2011;**17**:85–91.
57. Trapphoff T, Heiligentag M, El Hajj N, Haaf T, Eichenlaub-Ritter U. Chronic exposure to a low concentration of bisphenol A during follicle culture affects the epigenetic status of germinal vesicles and metaphase II oocytes. *Fertil Steril* 2013;**100**(6):1758–67.
58. Mapendano CK, Kishino T, Miyazaki K, Kondo S, Yoshiura K, Hishikawa Y, Koji T, Niikawa N, Ohta T. Expression of the Snurf–Snrpn IC transcript in the oocyte and its putative role in the imprinting establishment of the mouse 7C imprinting domain. *J Hum Genet* 2006;**51**:236–43.
59. Hinkins M, Huntriss J, Miller D, Picton HM. Expression of Polycomb-group genes in human ovarian follicles, oocytes and preimplantation embryos. *Reproduction* 2005;**130**: 883–8.
60. Lees-Murdock DJ, Lau HT, Castrillon DH, De Felici M, Walsh CP. DNA methyltransferase loading, but not *de novo* methylation, is an oocyte-autonomous process stimulated by SCF signalling.. *Dev Biol* 2008;**321**:238–50.
61. Xu XR, Fu RG, Wang LY, Wang N, Zhang F, Le F, Li L, Li LJ, Liu XZ, Zheng YM, Lou HY, Jiang SW, Zhu XM, Zhu YM, Huang HF, Jin F. Epigenetic inheritance of paternally expressed imprinted genes in the testes of ICSI mice. *Curr Pharm Des* 2014;**20**:1764–71.
62. Shin JY, Gupta MK, Jung YH, Uhm SJ, Lee HT. Differential genomic imprinting and expression of imprinted microRNAs in testes-derived male germ-line stem cells in mouse. *PLoS ONE* 2011;**6**:e22481.
63. Rajender S, Avery K, Agarwal A. Epigenetics, spermatogenesis and male infertility. *Mutat Res* 2011;**727**:62–71.
64. Ozdemir O, Bulut HE, Korkmaz M, Onarlioglu B, Colak A. Increased cell proliferation and R.Msp1 fragmentation induced by 5-aza-2′-deoxycytidine in rat testes related to the gene imprinting mechanism. *Exp Toxicol Pathol* 2000;**52**:317–22.
65. Jelinic P, Stehle JC, Shaw P. The testis-specific factor CTCFL cooperates with the protein methyltransferase PRMT7 in H19 imprinting control region methylation. *PLoS Biol* 2006;**4**:e355.

CHAPTER 11

Prevention of Alcohol-Induced Inflammation of Murine Small Intestine by MicroRNA-155 Deficiency

Dora Lippai*, Gyongyi Szabo**
*Second Department of Internal Medicine, Semmelweis University, Budapest, Hungary
**Department of Medicine, University of Massachusetts Medical School, Worcester, MA, USA

1 ALCOHOL-INDUCED INFLAMMATION IN PROXIMAL SMALL INTESTINE – GENERAL CONSIDERATIONS (FIGURE 11.1)

World Health Organization indicated that alcoholism and related diseases pose major public health concerns across the world.[1] Alcohol use, both acute binge and chronic alcohol intake, has numerous negative health effects on different organs including the intestine.[2–4] Chronic alcohol consumption results in inflammation (pancreas, liver, brain, etc.) and enhances cancer risk (liver, colorectal, breast, etc.) in various organs, both in human and animal models.[5–7]

The pathogenesis of alcohol-induced changes in the small intestine involves bacterial overload, inflammatory changes, and impaired barrier function, that are partially mediated by innate immune responses.[4,5,8–12] Cellular changes, antibacterial protein production, proinflammatory cytokine production, apoptosis, and changes in microRNA (miRNA) expression are all part of the underlying mechanism.[4,5,8–12]

2 DIFFERENCES BETWEEN ACUTE AND CHRONIC DRINKING ON BARRIER FUNCTION AND SERUM–ENDOTOXIN LEVELS (TABLE 11.1)

Alcohol in the gastrointestinal tract impairs the barrier functions via epithelial cell apoptosis, necrosis, and inflammation.[4,9–12] As a result of alcohol consumption, the first part of the intestine, including the small intestine, is exposed to high concentrations of ethanol that may directly affect gut

Insights to Neuroimmune Biology
http://dx.doi.org/10.1016/B978-0-12-801770-8.00011-2

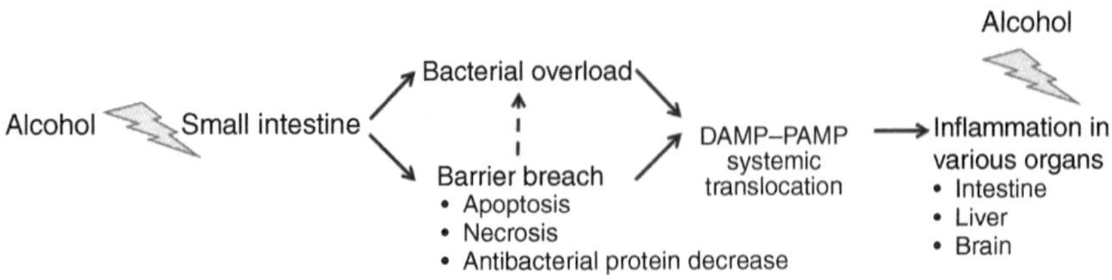

Figure 11.1 ***Alcohol Induced Inflammatory Changes.*** First, alcohol changes bacterial load, reduces barrier function and directly damages cells allowing PAMPs and DAMPs to increase in the circulation. Elevated DAMPs and PAMPs contribute to inflammatory changes in different organs including liver and brain.

Table 11.1 Acute versus chronic alcohol effects on murine small intestine[7]

	Acute binge	Chronic alcohol	
Factors (intestine)	**Wild type**	**Wild type**	**miRNA-155 deficiency**
Reg3b mRNA	Increase	Decrease	No change
Reg3b protein	Increase	Decrease	No change
Serum–endotoxin	Transient increase	Increase	No change
NF-κB activation	Increase	Increase	No change
TNF-α mRNA	Increase	Increase	No change
TNF-α protein	No change	Increase	No change
miRNA-155	No change	Increase	None detected
SHIP1 protein	No data	Decrease	No change

microbiome and barrier functions.[4] Chronic alcoholism is associated with bacterial overgrowth and impaired epithelial tight junctions in the proximal intestine.[8] The subsequently impaired barrier allows systemic translocation of pathogen- or danger-associated molecular patterns (PAMPs or DAMPs),[13] which activate proinflammatory pathways via pattern-recognition receptors (PRRs) in small intestine and other organs (liver, brain, etc.). Increased serum endotoxin level, observed in animal models of alcoholism, suggests intestinal barrier disruption and increased gut permeability.[14]

Acute binge gavage and chronic alcohol feeding results in increased serum–endotoxin levels in mice,[5] suggesting impairment in the barrier function. In acute alcohol gavage mouse model, the endotoxin increase is transient, lasting 12 h after intake.[5] These observations have been confirmed in humans, as both one-time binge drinkers and chronic alcoholic patients have increased serum–endotoxin levels.[15–17] Alcoholics, among other acute phase proteins, have higher levels of endotoxin-binding serum and lipopolysaccharide-binding protein (LBP) produced by hepatocytes.[18] Acute and chronic alcohol intake increase serum-LBP in mice.

LBP deficiency protects mice from alcohol or acetaminophen-induced liver injury.[19,20] These data suggest that endotoxin lipopolysaccharide (LPS) and LBP increase in both acute and chronic alcohol intake, would result in liver injury.

The proximal small intestine is in the first line of contact with alcohol. Important elements of the mucosal innate immune system are antibacterial proteins that are reportedly decreased in chronic alcoholics.[21] Antimicrobial peptides, such as regenerating islet-derived 3-beta and gamma (Reg3b and Reg3g), are part of the host defense mechanism against pathogens in the gut, enhancing the barrier function.[22] Acute alcohol binge increased Reg3b levels in murine small intestine.[5] Alcohol gavage prior to burn injuries in mice increases intestinal permeability and gut bacterial load, while decreasing intestinal interleukin-22 (IL-22), Reg3b, and Reg3g expression.[23] Considering this data, Reg3b increment in a three-day acute gavage model may represent a compensatory mechanism to maintain intact gut barrier after an acute alcohol binge,[5] which is supported by the fact that antibacterial protein increases in response to toll-like receptor (TLR) stimulation.[24] After repeated and sustained alcohol exposure, this mechanism is likely to be exhausted and results in reduced antibacterial protein levels.[5] Chronic alcohol consumption leads to decreased Reg3g and Reg3b expression both in animals and human proximal small intestine,[5,21] suggesting a decreased ability to protect against bacteria and subsequent bacterial overgrowth.

3 DIFFERENCES BETWEEN ACUTE AND CHRONIC DRINKING ON THE INTESTINAL INFLAMMATORY CHANGES (TABLE 11.1)

Apoptosis and necrosis, of small intestinal epithelial cells upon ethanol exposure, leads to impaired barrier integrity, consecutive trans-mural, and systemic translocation of PAMPs and DAMPs.[4,5,8–12] This process can facilitate inflammation in the intestinal wall, which can contribute to further loss of function. This might explain the clinical observation of common duodenal inflammation among alcoholic patients.[9]

Proinflammatory cytokine, tumor necrosis factor-alpha (TNF-α) secreting cells are increased in the inflamed mucosa during inflammatory bowel disease activation, which is associated with increased intestinal permeability.[25,26] Upregulation of TNF-α involves activation of transcription factors, such as nuclear factor-κB (NF-κB)[27] or activator protein-1 (AP-1).[28] NF-κB activation and TNF-α release play an important role in regulating intestinal epithelial function in inflammatory bowel disease.[29,30] Small intestinal NF-κB activation occurs after acute alcohol gavage and at a higher magnitude in chronic alcohol feeding in mice.[5] Acute and chronic alcohol consumption increase TNF-α mRNA level, however TNF-α protein increases only after chronic alcohol feeding in murine small intestine, despite the fact that alcohol binge only minimally increases NF-κB.[5] As NF-κB is not the sole mediator of TNF-α induction, there might be other transcription factors involved in the upregulation of TNF-α mRNA, such as AP-1.[5,28] Chronic alcohol feeding leads to a robust inflammatory response in murine small intestine which is characterized by increased TNF-α, NF-κB activation, and reduced Reg3b expression.[5] TNF-α antibody has been successfully used to treat inflammatory bowel diseases, however promising in alcoholic patients, studies with anti-TNF-α had to be suspended due to life-threatening opportunistic infections.[31,32]

4 miRNAs REGULATING INTESTINAL FUNCTIONS

miRNAs are evolutionally conserved small noncoding RNAs with regulatory functions involved in various biological processes such as development, differentiation, adaptive, and innate immune responses, including modulation of inflammation and cytokine production.[33–35] Mature miRNAs regulate posttranscriptional gene expression, mainly through repressing translation or inducing mRNA degradation.[36] Each miRNA can target various numbers of mRNAs to exert their regulatory effects and each mRNA can be regulated by different miRNAs.[35,37] Recently, other mechanisms, such as posttranslational stabilization of mRNA enabling increased translation, have also been proposed, however the exact mechanism is not fully understood.[38]

miRNAs are essential elements of proper intestinal homeostasis.[39] Several miRNAs have been implicated to play an important role in the fine-tuning of gene expressions in inflammatory processes.[33] Expressional patterns of miRNAs are altered in various intestinal disorders, including inflammatory bowel diseases, Celiac disease, colorectal cancer, etc.[40,41] Several studies show an association among changes of intestinal miRNA and infections, barrier function, inflammation, and tumor genesis.

Infections, including *Listeria monocytogenes* infection, can change the miRNA pattern in the small intestine of mice, and the host microbiota has an effect on the observed changes.[42] Deficiency of Dicer, which cleaves miRNAs into their mature forms, results in an impaired intestinal barrier function.[40] Alcohol-induced disruption of gut epithelial tight junctions is due to miRNA-212-mediated decrease in zonula occludens-1 expression.[43] miRNA-122a also plays a crucial role in the regulation of intestinal tight junctions. miRNA-122a overexpression leads to impaired barrier function as a result of decreased occludin expression, a target of miRNA-122a.[39,40]

Proinflammatory cytokine TNF-α induces increased intestinal permeability, which is associated with higher miRNA-122a levels.[39,40] miRNA-21 indirectly decreases occludin and barrier integrity gets upregulated in ulcerative colitis, moreover miRNA-21 deficiency protects from colitis.[40] Alcohol-induced gastric mucosal epithelial cell apoptosis is associated with c-Jun N-terminal kinase dependent decrease in miRNA-21.[44] Intestinal epithelial miRNA-146a can reduce inflammation during ischemia–reperfusion injury of intestine through decreasing TLR downstream signaling.[39] Furthermore, miRNA-146a single nucleotide polymorphism is associated with Crohn's disease and colorectal cancer susceptibility.[39]

Helicobacter pylori positive versus negative gastric cancer miRNA profiles differ suggesting a role of infection-associated miRNAs in the pathogenesis of tumor formation.[45] In neuroendocrine tumors, higher miRNA-21 expression is associated with high proliferation index and liver metastasis.[46] On the other hand, tumor suppressor gene miRNA-34 can induce apoptosis and its level is decreased in colon cancer.[47]

Some of the observed miRNA changes have not been connected to their functional relevance yet, but further work is warranted on this growing field.

5 miRNA-155 REGULATING INTESTINAL FUNCTIONS

miRNA-155, an important regulator of gene expression, has infection controlling, inflammatory, and oncological functions in the intestine. miRNA-155 deficient mice have immune deficiencies, including impaired T and B cell development, and antigen presentation by dendritic cells.[48,49] miRNA-155 deficient mice are susceptible to lung fibrosis and are less resistant to certain bacterial infections.[48,49]

Due to defective B cell function, miRNA-155 deficient mice have a higher pathogen burden and delayed pathogen clearing ability, which leads to the worsening of experimental colitis.[39] As an example, protection against

Citrobacter rodentium mucosal infection requires the production of miRNA-155 by lamina propria T cells.[39]

In an experimental pancreatitis model, upregulated intestinal miRNA-155 decreases rat sarcoma homolog gene family, member A (RhoA) levels, and is associated with disrupted intestinal barrier function.[50]

miRNA-155 is increased in inflammatory intestinal diseases, including ulcerative colitis,[51] acute graft versus host disease,[52] and also in the liver in alcoholic liver disease.[38] miRNA-155 is upregulated in active inflammatory bowel disease in the intestine as well as in the peripheral blood.[53] In the early phase of TLR4 signaling, miRNA-155 expression is upregulated and inhibits the expression of the negative regulators of sarcoma homology-2 domain containing inositol-5-phosphatase-1 (SHIP1) and suppressor of cytokine signaling-1 (SOCS1) allowing TLR signal transduction and proinflammatory cytokine production.[54,55] miRNA-155 is upregulated in macrophages and dendritic cells upon LPS, interferon or other TLR challanges.[53] miRNA-155 upregulation leads to decreased SOCS1 mRNA resulting in further inflammation via macrophages.[53] The diterpenoid epoxide triptolide prevents miRNA-155 induction in anastomositis after ileocecal resection in IL-10 deficient mice.[56] Triptolide can also inhibit miRNA-155 induction in LPS stimulated macrophages.[56]

miRNA-155 is induced in other inflammatory disorders, rheumatoid arthritis, and atopic dermatitis, and as an oncogene it is upregulated in different malignancies, colon cancer and B-cell lymphoma etc.[53,57] miRNA-155 can decrease the expression of mismatch repair genes, leading to increased mutation rates.[57]

6 miRNA-155 DEFICIENCY PROTECTS CHRONIC ALCOHOL-FED MICE FROM SERUM-ENDOTOXIN CHANGES (TABLE 11.1)

Based on one study, acute binge gavage does not change miRNA-155; however, chronic alcohol-feeding significantly increases its level in murine small intestine.[5] miRNA-155 deficiency protects against increase in serum endotoxin levels in chronic alcohol-fed mice.[5] Although serum–endotoxin is unchanged, serum-LBP gets increased in the same model.[5] LBP is mainly produced by hepatocytes and in a lesser degree by other cell types including intestinal epithelial cells.[58] Alcohol-induced monocytic transcription factors, signal transducer and activator of transcription-3 (STAT3) or AP-1,[59] can induce LBP.[60] These pathways might be independent from miRNA-155 in chronic alcohol feeding.

7 miRNA-155 DEFICIENCY PROTECTS CHRONIC ALCOHOL-FED MICE FROM INFLAMMATORY CHANGES IN THE SMALL INTESTINE (TABLE 11.1)

miRNA-155 as a positive regulator of TNF-α can increase its mRNA stability in macrophages.[38] TNF-α mRNA increases after acute alcohol gavage and chronic alcohol administration. However, TNF-α protein increases only after chronic alcohol feeding in murine small intestine.[5] This observation might be due to the lack of induction of intestinal miRNA-155 in the acute binge drinking in contrast to miRNA-155 induction in chronic alcohol feeding model.[5] Furthermore, miRNA-155 is upregulated in the colonic tissue of chronic alcohol-fed nonhuman primates, although it is associated with lower Th1 cytokine production, including TNF-α.[61]

Deficiency in miRNA-155 prevents Reg3b decrease in murine small intestine after chronic alcohol feeding.[5] Based on the regulatory effects of miRNA-155 on dendritic cells and B and T cell functions,[48] it is likely that some of these immune cells contribute to the protection of barrier function in miRNA-155 deficient mice. miRNA-155 deficient mice are prone to *Salmonella typhimurium* and *H. pylori* infection and are less likely to develop chronic gastritis or dextran sodium sulfate-induced colitis due to impaired B and T cell development.[62,63]

Amplification of TNF-α production in Kupffer cells, due to chronic alcohol exposure, involves alcohol-induced induction of NF-κB activation.[38] Deficiency in miRNA-155 prevents NF-κB activation and TNF-α increase in murine small intestine after chronic alcohol feeding.[5] miRNA-155 deficiency protects mice from the reduction of intestinal miRNA-155 target, SHIP1 upon chronic alcohol feeding, suggesting preservation of anti-inflammatory targets.[5]

Differences between the effects of acute binge and chronic alcohol administration in murine small intestine can be partially explained by miRNA-155, which is increased only after chronic alcohol feeding.[5] Alcoholics have intestinal inflammation, and a higher risk of colon tumors.[63] The role of miRNA-155 in intestinal inflammation and colon tumors have been previously linked,[64] and they deserve further investigation in alcoholics.

8 miRNA-155 DEFICIENCY PROTECTS CHRONIC ALCOHOL-FED MICE FROM INFLAMMATORY CHANGES IN THE CEREBELLUM (TABLE 11.2)

miRNA-155 and -132 are positive regulators of inflammation, while miRNA-125b and miRNA-146a have anti-inflammatory effects in most cell types.[65,66] Activation of the LPS-induced inflammatory pathway alters the expression of miRNA-125b, -132, -146a, and -155.[33,67] Chronic alcohol-fed mice have increased miRNA-155 and miRNA-132 whereas no changes in miRNA-125b or miRNA-146a levels have been observed in the cerebellum.[68]

Similar to small intestine, chronic alcohol induces inflammatory changes in cortical, hippocampal, and cerebellar brain regions.[2,69,70] A major proinflammatory regulator, NF-κB can upregulate multiple cytokines and chemokines, including TNF-α, monocyte chemotactic protein-1 (MCP1), and pro-IL-1β.[71,72] NF-κB activation mediates induction of miRNA-155 and in turn, it is suggested that miRNA-155 can decrease NF-κB activity.[38,73] Activation of NF-κB occurs by phosphorylation of p65, part of the p50/p65 NF-κB heterodimer and its translocation to the nucleus.[74] miRNA-155 deficiency, similar to TLR4 deficiency, protects chronic alcohol-fed mice from phosphorylation of p65, hence the NF-κB activation in the cerebellum.[68] Induction of miRNA-155 is TLR4-dependent in the cerebella of chronic alcohol-fed mice.[68]

Studies have confirmed the activation of microglia and astrocytes along with neuronal changes and cell death in alcoholic brains of humans and animals.[70,75] miRNA-155 is increased upon LPS stimulation in microglia cell line, reducing its target gene SOCS1.[35] Ethanol alone decreases miRNA-155 in murine microglia.[68] However, alcohol augments the TLR4

Table 11.2 Effects of chronic alcohol on murine cerebellum in different genotypes[10,12]

	Chronic alcohol		
Factors (cerebellum)	**Wild type**	**miRNA-155 deficiency**	**TLR4 deficiency**
miRNA-155	Increase	None detected	No change
NF-κB activation	Increase	No change	No change
TNF-α protein	Increase	No change	No change
MCP1 protein	Increase	No change	No change
Mature IL-lβ protein	Increase	Increase	Increase
Caspase-1 activation	Increase	Increase	Increase

ligand LPS-induced miRNA-155 production in murine microglia,[68] suggesting sensitization to PAMPs and possibly DAMPs.[68]

DAMPs and PAMPs are major inducers of inflammation via receptors, like the TLR family.[13] TLR4 can activate NF-κB[71] and can also induce miRNA-155 upregulation.[54,76] Alcohol-induced neuroinflammation can be triggered by TLR4.[70] There is less inflammation found in miRNA-155 deficient alcohol-fed mice in the cerebellum, based on the decreased NF-κB activation and a lack of proinflammatory cytokine production (TNF-α and MCP1).[68] Possibly, this attenuated effect of alcohol on the cerebellum is partially the result of less inflammation in the small intestine and lower systemic endotoxin translocation of alcohol-fed miRNA-155 deficient mice.[5] Additionally, miRNA-155 over-expressing mice showed increased TNF-α production upon LPS challenge.[66] However, some studies show that miRNA-155 decreases inhibitor of κB kinases (IKK) level which subsequently leads to the suppression of NF-κB pathways.[73,77]

Inflammasome activation and IL-1β increase is independent of miRNA-155 in alcohol-fed murine cerebellum,[68] as well as in LPS treated microglia cell line.[35] Pro-IL-1β mRNA levels are not increased in miRNA-155 deficient mice, most likely due to the lack of NF-κB activation. While IL-1β protein production is largely dependent on caspase-1 activation, pro-IL-1β mRNA induction is NF-κB mediated.[78] This mechanism has not been observed in the small intestine of alcoholics.

9 CONCLUSIONS

Alcohol, as one of the most commonly used substance of abuse in the world, is associated with various devastating health effects.[1] Recently, there has been a major ongoing interest in the mechanism of alcohol-induced changes in the intestine to guide future interventions. Short term, excessive, and prolonged alcohol use result in inflammatory changes in the small intestine and increased serum–endotoxin levels.[5] In contrast to chronic alcohol feeding, acute alcohol binge increases Reg3b and transiently increases serum endotoxin, which might be due to a rapid restoration of gut barrier function after disruption caused by acute alcohol binge.[5] miRNA-155 contributes to alcohol-induced intestinal damage and neuroinflammation.[5,68] miRNA-155 deficiency does not protect from cerebellar inflammasome activation and subsequent IL-1β release in alcohol-fed mice.[35,68] Conclusively, not all proinflammatory changes are dependent on miRNA-155 and can therefore be eliminated by solely blocking miRNA-155 but can greatly be modified. miRNA-155 silencing might have a therapeutic role in the improvement of alcohol-induced barrier dysfunction, intestinal wall inflammation, and neuroinflammation. Besides, further work on this field is warranted.[5]

REFERENCES

1. World Health Organization. Alcohol fact sheet 2015. Available from: http://www.who.int/mediacentre/factsheets/fs349/en/index.html
2. Lippai D, Bala S, Petrasek J, Csak T, Levin I, Kurt-Jones EA, et al. Alcohol-induced IL-1beta in the brain is mediated by NLRP3/ASC inflammasome activation that amplifies neuroinflammation. *J Leukoc Biol* 2013;**94**:171–82.
3. Szabo G, Bala S, Petrasek J, Gattu A. Gut-liver axis and sensing microbes. *Dig Dis* 2010;**28**:737–44.
4. Bode C, Bode JC. Effect of alcohol consumption on the gut. *Best Pract Res Clin Gastroenterol* 2003;**17**:575–92.
5. Lippai D, Bala S, Catalano D, Kodys K, Szabo G. Micro-RNA-155 deficiency prevents alcohol-induced serum endotoxin increase and small bowel inflammation in mice. *Alcohol Clin Exp Res* 2014;**38**:2217–24.
6. Roswall N, Weiderpass E. Alcohol as a risk factor for cancer: existing evidence in a global perspective. *J Prev Med Public Health* 2015;**48**:1–9.
7. Szabo G, Mandrekar P, Petrasek J, Catalano D. The unfolding web of innate immune dysregulation in alcoholic liver injury. *Alcohol Clin Exp Res* 2011;**35**:782–6.
8. Hauge T, Persson J, Danielsson D. Mucosal bacterial growth in the upper gastrointestinal tract in alcoholics (heavy drinkers). *Digestion* 1997;**58**:591–5.
9. Bhonchal S, Nain CK, Prasad KK, Nada R, Sharma AK, Sinha SK, et al. Functional and morphological alterations in small intestine mucosa of chronic alcoholics. *J Gastroenterol Hepatol* 2008;**23**:e43–8.
10. Tamai H, Kato S, Horie Y, Ohki E, Yokoyama H, Ishii H. Effect of acute ethanol administration on the intestinal absorption of endotoxin in rats. *Alcohol Clin Exp Res* 2000;**24**:390–4.
11. Beck IT, Dinda PK. Acute exposure of small intestine to ethanol: effects on morphology and function. *Dig Dis Sci* 1981;**26**:817–38.
12. Kirpich IA, Feng W, Wang Y, Liu Y, Beier JI, Arteel GE, et al. Ethanol and dietary unsaturated fat (corn oil/linoleic acid enriched) cause intestinal inflammation and impaired intestinal barrier defense in mice chronically fed alcohol. *Alcohol* 2013;**47**:257–64.
13. Menu P, Vince JE. The NLRP3 inflammasome in health and disease: the good, the bad and the ugly. *Clin Exp Immunol* 2011;**166**:1–15.
14. Mathurin P, Deng QG, Keshavarzian A, Choudhary S, Holmes EW, Tsukamoto H. Exacerbation of alcoholic liver injury by enteral endotoxin in rats. *Hepatology* 2000;**32**:1008–17.
15. Fujimoto M, Uemura M, Nakatani Y, Tsujita S, Hoppo K, Tamagawa T, et al. Plasma endotoxin and serum cytokine levels in patients with alcoholic hepatitis: relation to severity of liver disturbance. *Alcohol Clin Exp Res* 2000;**24**:48S–54S.
16. Bode C, Kugler V, Bode JC. Endotoxemia in patients with alcoholic and non-alcoholic cirrhosis and in subjects with no evidence of chronic liver disease following acute alcohol excess. *J Hepatol* 1987;**4**:8–14.
17. Bala S, Marcos M, Gattu A, Catalano D, Szabo G. Acute binge drinking increases serum endotoxin and bacterial DNA levels in healthy individuals. *PLoS ONE* 2014;**9**:e96864.
18. Summa KC, Voigt RM, Forsyth CB, Shaikh M, Cavanaugh K, Tang Y, et al. Disruption of the circadian clock in mice increases intestinal permeability and promotes alcohol-induced hepatic pathology and inflammation. *PLoS ONE* 2013;**8**:e67102.

19. Su GL, Gong KQ, Fan MH, Kelley WM, Hsieh J, Sun JM, et al. Lipopolysaccharide-binding protein modulates acetaminophen-induced liver injury in mice. *Hepatology* 2005;**41**:187–95.
20. Uesugi T, Froh M, Arteel GE, Bradford BU, Wheeler MD, Gabele E, et al. Role of lipopolysaccharide-binding protein in early alcohol-induced liver injury in mice. *J Immunol* 2002;**168**:2963–9.
21. Yan AW, Fouts DE, Brandl J, Starkel P, Torralba M, Schott E, et al. Enteric dysbiosis associated with a mouse model of alcoholic liver disease. *Hepatology* 2011;**53**:96–105.
22. Everard A, Belzer C, Geurts L, Ouwerkerk JP, Druart C, Bindels LB, et al. Cross-talk between *Akkermansia muciniphila* and intestinal epithelium controls diet-induced obesity. *Proc Natl Acad Sci USA* 2013;**110**:9066–71.
23. Rendon JL, Li X, Akhtar S, Choudhry MA. Interleukin-22 modulates gut epithelial and immune barrier functions following acute alcohol exposure and burn injury. *Shock* 2013;**39**:11–8.
24. Mowat AM, Agace WW. Regional specialization within the intestinal immune system. *Nat Rev Immunol* 2014;**14**:667–85.
25. Breese EJ, Michie CA, Nicholls SW, Murch SH, Williams CB, Domizio P, et al. Tumor necrosis factor alpha-producing cells in the intestinal mucosa of children with inflammatory bowel disease. *Gastroenterology* 1994;**106**:1455–66.
26. Gerova VA, Stoynov SG, Katsarov DS, Svinarov DA. Increased intestinal permeability in inflammatory bowel diseases assessed by iohexol test. *World J Gastroenterol* 2011;**17**:2211–5.
27. Liu H, Sidiropoulos P, Song G, Pagliari LJ, Birrer MJ, Stein B, et al. TNF-alpha gene expression in macrophages: regulation by NF-kappa B is independent of c-Jun or C/EBP beta. *J Immunol* 2000;**164**:4277–85.
28. Redhu NS, Saleh A, Halayko AJ, Ali AS, Gounni AS. Essential role of NF-kappaB and AP-1 transcription factors in TNF-alpha-induced TSLP expression in human airway smooth muscle cells. *Am J Physiol Lung Cell Mol Physiol* 2011;**300**:L479–85.
29. Suenaert P, Bulteel V, Lemmens L, Noman M, Geypens B, Van Assche G, et al. Anti-tumor necrosis factor treatment restores the gut barrier in Crohn's disease. *Am J Gastroenterol* 2002;**97**:2000–4.
30. Atreya I, Atreya R, Neurath MF. NF-kappaB in inflammatory bowel disease. *J Intern Med* 2008;**263**:591–6.
31. Tilg H, Jalan R, Kaser A, Davies NA, Offner FA, Hodges SJ, et al. Anti-tumor necrosis factor-alpha monoclonal antibody therapy in severe alcoholic hepatitis. *J Hepatol* 2003;**38**:419–25.
32. Naveau S, Chollet-Martin S, Dharancy S, Mathurin P, Jouet P, Piquet MA, et al. A double-blind randomized controlled trial of infliximab associated with prednisolone in acute alcoholic hepatitis. *Hepatology* 2004;**39**:1390–7.
33. Liu G, Abraham E. MicroRNAs in immune response and macrophage polarization. *Arterioscler Thromb Vasc Biol* 2013;**33**:170–7.
34. Ambros V. MicroRNA pathways in flies and worms: growth, death, fat, stress, and timing. *Cell* 2003;**113**:673–6.
35. Cardoso AL, Guedes JR, Pereira de Almeida L, Pedroso de Lima MC. miR-155 modulates microglia-mediated immune response by down-regulating SOCS-1 and promoting cytokine and nitric oxide production. *Immunology* 2012;**135**:73–88.
36. Valencia-Sanchez MA, Liu J, Hannon GJ, Parker R. Control of translation and mRNA degradation by miRNAs and siRNAs. *Genes Dev* 2006;**20**:515–24.
37. Szabo G, Sarnow P, Bala S. MicroRNA silencing and the development of novel therapies for liver disease. *J Hepatol* 2012;**57**:462–6.
38. Bala S, Marcos M, Kodys K, Csak T, Catalano D, Mandrekar P, et al. Up-regulation of microRNA-155 in macrophages contributes to increased tumor necrosis factor α (TNFα) production via increased mRNA half-life in alcoholic liver disease. *J Biol Chem* 2011;**286**:1436–44.

39. Runtsch MC, Round JL, O'Connell RM. MicroRNAs and the regulation of intestinal homeostasis. *Front Genet* 2014;**5**:347.
40. Ikemura K, Iwamoto T, Okuda M. MicroRNAs as regulators of drug transporters, drug-metabolizing enzymes, and tight junctions: implication for intestinal barrier function. *Pharmacol Ther* 2014;**143**:217–24.
41. Magni S, Comani GB, Elli L, Vanessi S, Ballarini E, Nicolini G, et al. miRNAs affect the expression of innate and adaptive immunity proteins in celiac disease. *Am J Gastroenterol* 2014;**109**:1662–74.
42. Archambaud C, Sismeiro O, Toedling J, Soubigou G, Becavin C, Lechat P, et al. The intestinal microbiota interferes with the microRNA response upon oral *Listeria* infection. *MBio* 2013;**4**:e00707–13.
43. Tang Y, Banan A, Forsyth CB, Fields JZ, Lau CK, Zhang LJ, et al. Effect of alcohol on miR-212 expression in intestinal epithelial cells and its potential role in alcoholic liver disease. *Alcohol Clin Exp Res* 2008;**32**:355–64.
44. Luo XJ, Liu B, Dai Z, Li TB, Li NS, Zhang XJ, et al. Expression of apoptosis-associated microRNAs in ethanol-induced acute gastric mucosal injury via JNK pathway. *Alcohol* 2013;**47**:481–93.
45. Chang H, Kim N, Park JH, Nam RH, Choi YJ, Lee HS, et al. Different microRNA expression levels in gastric cancer depending on *Helicobacter pylori* infection. *Gut Liver* 2015;**9**:188–96.
46. Vicentini C, Fassan M, D'Angelo E, Corbo V, Silvestris N, Nuovo GJ, et al. Clinical application of microRNA testing in neuroendocrine tumors of the gastrointestinal tract. *Molecules* 2014;**19**:2458–68.
47. Garzon R, Calin GA, Croce CM. MicroRNAs in Cancer. *Annu Rev Med* 2009;**60**:167–79.
48. Rodriguez A, Vigorito E, Clare S, Warren MV, Couttet P, Soond DR, et al. Requirement of bic/microRNA-155 for normal immune function. *Science* 2007;**316**:608–11.
49. Clare S, John V, Walker AW, Hill JL, Abreu-Goodger C, Hale C, et al. Enhanced susceptibility to *Citrobacter rodentium* infection in microRNA-155-deficient mice. *Infect Immun* 2013;**81**:723–32.
50. Tian R, Wang RL, Xie H, Jin W, Yu KL. Overexpressed miRNA-155 dysregulates intestinal epithelial apical junctional complex in severe acute pancreatitis. *World J Gastroenterol* 2013;**19**:8282–91.
51. Takagi T, Naito Y, Mizushima K, Hirata I, Yagi N, Tomatsuri N, et al. Increased expression of microRNA in the inflamed colonic mucosa of patients with active ulcerative colitis. *J Gastroenterol Hepatol* 2010;**25**(Suppl 1):S129–33.
52. Ranganathan P, Heaphy CE, Costinean S, Stauffer N, Na C, Hamadani M, et al. Regulation of acute graft-versus-host disease by microRNA-155. *Blood* 2012;**119**:4786–97.
53. Chapman CG, Pekow J. The emerging role of miRNAs in inflammatory bowel disease: a review. *Therap Adv Gastroenterol* 2015;**8**:4–22.
54. Tili E, Croce CM, Michaille JJ. miR-155: on the crosstalk between inflammation and cancer. *Int Rev Immunol* 2009;**28**:264–84.
55. He X, Jing Z, Cheng G. MicroRNAs: new regulators of toll-like receptor signalling pathways. *Biomed Res Int* 2014;**2014**:945169.
56. Wu R, Li Y, Guo Z, Gong J, Zhu W, Li N, et al. Triptolide ameliorates ileocolonic anastomosis inflammation in IL-10 deficient mice by mechanism involving suppression of miR-155/SHIP-1 signaling pathway. *Mol Immunol* 2013;**56**:340–6.
57. Hutchison J, Cohen Z, Onyeagucha BC, Funk J, Nelson MA. How microRNAs influence both hereditary and inflammatory-mediated colon cancers. *Cancer Genet* 2013;**206**:309–16.
58. Vreugdenhil AC, Dentener MA, Snoek AM, Greve JW, Buurman WA. Lipopolysaccharide binding protein and serum amyloid A secretion by human intestinal epithelial cells during the acute phase response. *J Immunol* 1999;**163**:2792–8.

59. Norkina O, Dolganiuc A, Shapiro T, Kodys K, Mandrekar P, Szabo G. Acute alcohol activates STAT3, AP-1, and Sp-1 transcription factors via the family of src kinases to promote IL-10 production in human monocytes. *J Leukoc Biol* 2007;**82**:752–62.
60. Schumann RR, Kirschning CJ, Unbehaun A, Aberle HP, Knope HP, Lamping N, et al. The lipopolysaccharide-binding protein is a secretory class 1 acute-phase protein whose gene is transcriptionally activated by APRF/STAT/3 and other cytokine-inducible nuclear proteins. *Mol Cell Biol* 1996;**16**:3490–503.
61. Asquith M, Pasala S, Engelmann F, Haberthur K, Meyer C, Park B, et al. Chronic ethanol consumption modulates growth factor release, mucosal cytokine production, and microRNA expression in nonhuman primates. *Alcohol Clin Exp Res* 2014;**38**: 980–93.
62. Oertli M, Engler DB, Kohler E, Koch M, Meyer TF, Muller A. MicroRNA-155 is essential for the T cell-mediated control of *Helicobacter pylori* infection and for the induction of chronic gastritis and colitis. *J Immunol* 2011;**187**:3578–86.
63. Bujanda L. The effects of alcohol consumption upon the gastrointestinal tract. *Am J Gastroenterol* 2000;**95**:3374–82.
64. Bakirtzi K, Hatziapostolou M, Karagiannides I, Polytarchou C, Jaeger S, Iliopoulos D, et al. Neurotensin signaling activates microRNAs-21 and -155 and Akt, promotes tumor growth in mice, and is increased in human colon tumors. *Gastroenterology* 2011;**141** 1749-61.e1.
65. Taganov KD, Boldin MP, Chang KJ, Baltimore D. NF-kappaB-dependent induction of microRNA miR-146, an inhibitor targeted to signaling proteins of innate immune responses. *Proc Natl Acad Sci USA* 2006;**103**:12481–6.
66. Tili E, Michaille JJ, Cimino A, Costinean S, Dumitru CD, Adair B, et al. Modulation of miR-155 and miR-125b levels following lipopolysaccharide/TNF-alpha stimulation and their possible roles in regulating the response to endotoxin shock. *J Immunol* 2007;**179**:5082–9.
67. Pedersen I, David M. MicroRNAs in the immune response. *Cytokine* 2008;**43**:391–4.
68. Lippai D, Bala S, Csak T, Kurt-Jones EA, Szabo G. Chronic alcohol-induced microRNA-155 contributes to neuroinflammation in a TLR4-dependent manner in mice. *PLoS ONE* 2013;**8**:e70945.
69. Qin L, Crews FT. NADPH oxidase and reactive oxygen species contribute to alcohol-induced microglial activation and neurodegeneration. *J Neuroinflamm* 2012;**9**: 5-2094-9-5.
70. Alfonso-Loeches S, Pascual-Lucas M, Blanco AM, Sanchez-Vera I, Guerri C. Pivotal role of TLR4 receptors in alcohol-induced neuroinflammation and brain damage. *J Neurosci* 2010;**30**:8285–95.
71. Zhang G, Ghosh S. Toll-like receptor-mediated NF-kappaB activation: a phylogenetically conserved paradigm in innate immunity. *J Clin Invest* 2001;**107**:13–9.
72. Thompson WL, Van Eldik LJ. Inflammatory cytokines stimulate the chemokines CCL2/MCP-1 and CCL7/MCP-3 through NFkB and MAPK dependent pathways in rat astrocytes. *Brain Res* 2009;**1287**:47–57.
73. O'Neill LA, Sheedy FJ, McCoy CE. MicroRNAs: the fine-tuners of Toll-like receptor signalling. *Nat Rev Immunol* 2011;**11**:163–75.
74. Yang F, Tang E, Guan K, Wang CY. IKK beta plays an essential role in the phosphorylation of RelA/p65 on serine 536 induced by lipopolysaccharide. *J Immunol* 2003;**170**:5630–5.
75. Crews FT, Nixon K. Mechanisms of neurodegeneration and regeneration in alcoholism. *Alcohol Alcohol* 2009;**44**:115–27.
76. Ceppi M, Pereira PM, Dunand-Sauthier I, Barras E, Reith W, Santos MA, et al. MicroRNA-155 modulates the interleukin-1 signaling pathway in activated human monocyte-derived dendritic cells. *Proc Natl Acad Sci USA* 2009;**106**:2735–40.

77. Verstrepen L, Bekaert T, Chau TL, Tavernier J, Chariot A, Beyaert R. TLR-4, IL-1R and TNF-R signaling to NF-kappaB: variations on a common theme. *Cell Mol Life Sci* 2008;**65**:2964–78.
78. He Y, Franchi L, Nunez G. TLR agonists stimulate Nlrp3-dependent IL-1beta production independently of the purinergic P2X7 receptor in dendritic cells and *in vivo*. *J Immunol* 2013;**190**:334–9.

PART 6

Neuroprotection

CHAPTER 12

Neuroprotection

Istvan Berczi
Department of Immunology, Faculty of Medicine, The University of Manitoba, Winnipeg, Manitoba, Canada; Departamento de Fisiología y Farmacología, Centro de Ciencias Básicas, Universidad Autónoma de Aguascalientes, Aguascalientes, Estado Libre y Soberano de Aguascalientes, Mexico City, Mexico

1 INTRODUCTION

Today, it is fashionable to think that the brain needs protection. However, no argument determines the best way to protect the central nervous system (CNS). While, some authors recommend stimulation, others would use drugs. Here, stimulation of the CNS by gut microbiota has been discussed. The evidence indicates that gut microbiota provides vital stimulus for the immune system and also for the CNS. This stimulus maintains the development of immune system and brain in higher animals and humans.

2 GUT MICROBIOTA AND THE BRAIN

In zoology textbooks, we may distinguish carnivorous as meat eating, herbivorous as plant eating, and omnivorous as those animals that can eat everything. In anatomy, we may find that the structure of digestive tract differs according to what the animal is eating. Herbivorous animals have an additional stomach, which is called the rumen; in ruminants and other herbivorous animals the well-developed cecum does the same job. Animal physiology teaches that the function of the cecum and rumen is to ferment cellulose, and during this process, a significant amount of microorganisms grow that will thrive on the material that the animal has consumed. This may be cellulose, fibrous, or even woody material, which has very poor nutritional value. However, the microbes which pass from the rumen or cecum to the regular stomach, and into the digestive tract, supplements the consumed food with valuable nutrients, such as protein, fat, vitamins, and possibly other nutrients. This is how a moose survives in Northern Canada during winter eating only tree bark for nutrition. As a student of veterinary medicine, I often thought that my mother's cow, which got only dry corn leaves and a little barely to eat, would have died quickly if she knew what the books of animal nutrition are saying about good nutrition.

Insights to Neuroimmune Biology
http://dx.doi.org/10.1016/B978-0-12-801770-8.00012-4

So, Mother Nature provides the first example as to what microbiota can do in animal welfare. All herbivorous animals use microbes to supplement their food with essential nutrients. This is nature's life saving experiment for the supplementation of substandard food. That it will be adequate for survival.

The ancestors of humans used to have a cecum as they were herbivorous. As the food habit moved in the omnivorous direction, the cecum became redundant. So it regressed and gradually became an appendix. Our feeding habits elicited a change in our gastrointestinal (GI) system. Since our nutrition normally has everything, we do not need a cecum for supplementation. Hence, we lost this part of our digestive system because of the alterations in our food intake.

Is this an abnormality? Are there health costs of this change? No, this was unlikely, even though our food was going in the direction of a more concentrated, omnivorous type style, for a long time. Northern people consume almost exclusively meat, fat, and little or no carbohydrates. The northern native people in Canada were healthy with their diet for millennia, until they started to eat the Southern Canadian diet, which is rich in starch, other carbohydrates, and sugar. With this diet 60–80% of the native people become diabetic.[1] Clearly, the sugar in the starch, carbohydrates, and purified sugars are major pathogenic agents, and they should be avoided in the diet. There could be little doubt that sugar-rich diet is pathogenic even for white people, even though some individuals exert tolerance. Diabetes is a major problem in the "western" nations. Our omnivorous system poorly tolerates large amounts of starch, carbohydrates, and sugar, all of which have been introduced artificially into our diet. So it makes good sense to stick to vegetables, dairy products, including butter, some lean meat, poultry, fish, and a little whole wheat bread, as they are permissible. This is the recommendation from a scientist with type II diabetes. Being lean also helps to stay healthy. However, fat has physiological functions and the mediators produced are necessary for a healthy organism. So fat is not an inert tissue, which is for energy storage only, it is an integral part of the neuroimmune system and its functions are vital.[2–4] What the immune system does with sugar metabolism and glucose resistance, will be discussed later. Inheritance also plays an important role in diabetes.

2.1 Germ-Free Animals

The current debate addresses the role of microbes in development and health. The aforementioned story proves that microbes contribute a lot to the well being of some animal species. Our body would hang on to the favorable effects as a rule. So the present situation will be discussed here.

Germ-free rats were first described by Gustaffson.[5] He also discovered that germ-free rats did not develop their immune systems normally. Normal intestinal microflora were required for the development of a normal immune function.

2.2 Naive Lymphocytes

Later, it was learnt that naive or antigen-sensitive lymphocytes have to meet antigen in order to achieve mature effector cell status. The gut microflora were readily available to provide a final antigenic stimulus for most naive lymphocytes. However, naive lymphocytes also play a role in the secondary immune response. Memory cells need to recruit naive lymphocytes in order to generate a secondary immune response. So naive lymphocytes are needed continuously for normal immune function. After exposure to antigen, the naive lymphocytes become mature effector cells, which are ready to react with specific antigens.[6] Young HYPOX rats (100 g) could not develop a secondary antibody response because they lacked naive T lymphocytes.[7] This situation suggests that intestinal microbiota serve as enormous antigenic stimulus and most of these naive lymphocytes participate in the defense effort against the microbiota in the host organism. By this process, the immune system must grow tremendously to take care of the rapidly growing mass of microbes. The immune system develops into a strong and very powerful system, while responding to this massive microbial stimulus. Since the intestinal microbiota elicits a massive immune response of the host, it seems obvious that this immune response would play a role in the control of intestinal microbiota. The situation is analogous to normal antigenic stimulation, in which clearly protective immune responses are elicited.

2.3 The Gut–Microbiome–Brain Axis

In recent papers, some authors argue that there is a *gut–microbiome–brain axis*, by the application of proper commensal bacteria (pre- or probiotic agents) will correct the abnormal brain function, which is observed with unsuitable microbes (autism, depression, irritable bowel syndrome (IBS), etc.). Others talk about a metabolic effect of microbes and an effect on physiology.[8–16]

One paper mentions hormones, which control gut function and another paper mentions immune activation, but then they turn back to the beneficial effects of proper microbiota even for the treatment of *allergy*, which is clearly an immunological disease as only proper microbiota can help.[17,18]

Others talk about the possible involvement of gut hormones, especially of the *gut–microbiome–vagus–brain axis*. Cholecystokinin and glucagon-like peptide 1 excite vagal-afferent neurons to activate an ascending pathway, leading to the inhibition of food intake. *Ghrelin* inhibits vagal-afferent neurons, leading to an increased food intake. Gut–brain signaling is also influenced by *leptin*.[19]

2.4 Gut Microbiota – GI Nerves and Hormones

Commensal intestinal microbiota were necessary for normal excitability of gut *sensory neurons* and they provide a potential mechanism for the transfer of information between the microbiota and nervous system.[20] CNS neurotransmission can be profoundly disturbed by the absence of a normal gut microbiota in early life. This aberrant neurochemical, but not behavioral, profile is resistant to restoration of a normal gut flora in later life.[21]

2.5 Gut Microbiota – Immune

Between 500 and 1,000 microbial species colonize the mammalian colon to a density of approximately 1,000,000,000,000 bacteria per gram of content, comprising of 100 times more cells in total, than those that make up the host itself. Studies about germ-free and gnotobiotic animals revealed that intestinal microbiota influence many aspects of immunity including development and function of mucosal immunity, development of oral tolerance, and generation of the antibody repertoire. A recent study demonstrated that capsular polysaccharide produced by the commensal bacteria, *Bacteroides fragilis*, stimulates maturation of the developing immune system by contributing to the generation of CD4+ T cells and the proper T helper 1/T helper 2 cytokine balance.[22]

Thus pharmaceutical treatment has failed to inhibit the tsunami of endemic diseases spreading around the world, and no new tools are in sight. Dramatic alterations, in direction of a Paleolithic-like lifestyle and food habits, seem to be the only alternatives with the potential to control the present escalating crisis.[23]

The GI tract of adult mammals is colonized by a complex and dynamic community of microorganisms. Maximum protection against potential pathogens occurs via a mucosal immune system involving mechanisms of innate immunity, as well as a secondary lymphoid organ, the gut-associated lymphoid tissue (GALT). However, the bacterial community also supports its host against invasion by potential pathogens, by a mechanism called *colonization resistance*. Young animals need time to develop a complex bacterial community and their immature GI tract immune systems, and until such developments have taken place, they are vulnerable in the presence of potential pathogens in their GI tract. Initial protection against invading pathogens is provided by *milk* and *colostrum*, which contain antibodies and other bioactive components. At weaning, with the introduction of solid food and deprivation of the mother's milk, the young must also cope with a rapidly changing microbiota. The colonizing microbiota not only provide *colonization resistance* to potentially pathogenic bacteria, but also has a major role in the development of the intestinal immune system, in terms of GALT development, mucosal immunity, and the induction of oral tolerance. Studies using gnotobiotic animal models have revealed that the presence of even limited numbers of the indigenous microbiota may influence the GI tract immune system. Regulation of the composition of the GI tract microbiota, for example, by the use of pre- and probiotics, offers the possibility to influence the development of mucosal and systemic immunity.[24]

Perturbations in the GI microbiota composition that occur as a result of antibiotics and diet in "westernized" countries are strongly associated with allergies and asthma (hygiene hypothesis). The microbiota (microflora) plays a crucial role in the development of mucosal tolerance, including the airways. Significant attention has been focused on the role of the microbiota in GI development, immune adaptation, and initiation of GI inflammatory diseases.

This chapter covers the postdevelopmental functions that the microbiota plays in the regulation of immunological tolerance to allergen exposure outside the GI tract and proposes the question: *is the microbiota a major regulator of the immune system?*[25]

Organisms live in continuous interaction with their environment. This interaction is of vital importance but at the same time, it can be life threatening. The largest and most important between the organism and its environment is represented by surfaces covered with epithelial cells. In these surfaces, mucosae comprise of approximately 300 m^2, and the skin covers approximately 1.8 m^2 surface of the human body. Mucosal tissues contain two effector arms of the immune system, *innate* and *adaptive*, which operate in synergy. Interaction with commensal bacteria, which outnumber the nucleated cells of our body, occurs physiologically on epithelial surfaces. This interaction could pose the risk of inflammation. The mucosal immune system has developed a complex network of regulatory signaling cascades that is a prerequisite not only for proper activation, but also for a timely inactivation of the pathway. As demonstrated in gnotobiotic animal models of human diseases, impaired regulation of mucosal responses to commensal bacteria plays an important role in the development of several inflammatory and autoimmune diseases.[26]

An association exists between bacterial load and IgA subclass production. An increase in IgA2 may enhance mucosal protection and probably reflects immunomodulation caused by lipopolysaccharides.[27]

2.6 Gut–Microbiota Immune and Nerve

Although the gut microbiota has been previously studied in the context of inflammatory diseases, it has recently become clear that this microbial environment has a beneficial role during normal homeostasis, by modulating the immune system, or bowel motor function.[28]

Key physiological functions of the intestine are governed by nerves and neurotransmitters. This complex control relies on two neuronal systems: an extrinsic innervation supplied by the two branches of the autonomic nervous system and an intrinsic innervation provided by the enteric nervous system. As a result of constant exposure to commensal and pathogenic microflora, the intestine developed a *tightly regulated immune system.*[29]

2.7 Early Life Stress Alters Behavior, Immunity, and Microbiota

Early life stress results in an altered brain–gut axis and is therefore an important model for the investigation of potential mechanistic insights into stress-related disorders, including depression and IBS.[30] Maternal probiotic intervention-induced activation of neonatal stress pathways and an imbalance in gut microflora. However, more importantly, it improved the immune environment of stressed animals and in part, protected against stress-induced disturbances in adult gut microflora.[31]

2.8 Neuroprotection

Effector T cells specific for CNS antigens can assist recovery from an array of environmental insults ranging from nerve injury to psychological stress, while in contrast, regulatory T cells (Treg) oppose recovery. Indeed, some reported effects of this so-called *protective autoimmunity* seem to have a direct relevance in depressive disorders. Even in SPF models, the effectors of beneficial effects are not the proinflammatory autoimmune cells themselves, but rather unidentified regulatory cells.[32]

Protective and autoreactive T cells are involved in neuroprotection (T_{NEP})*. These are nursing cells as they secrete NGF and other growth factors. Such nursing T cells* *may be distributed and recirculated throughout the body, and may be distributed as required. They may play an important role in brain development and contribute to the nursing and healing power of the immune system.*

In rats having experimental autoimmune encephalitis (EAE), it was observed that clinically the legs got paralyzed, which correlated with the inflammatory cellular infiltration in the spinal cord. However, the brain had just as much infiltrating cells, but no clinical disease! It was suggested that the brain was protected by neuroprotective T cells, which were seen in other systems.[33]

2.9 Diseases

Gut mucosal immunology is thought to be relevant with immunological changes reflected as peripheral blood cytokine level changes. Molecular technology advances suggest a role for microbiota by activating the gut immunological system.[34]

Emerging evidence suggests that alterations in the GI microbiota and altered immune functions may play a role in the pathogenesis of IBS. These recent data suggest a plausible model in which *changes in the intestinal microbiota and activation of the enteric immune system* may impinge upon the brain–gut axis, causing the alterations in GI function and clinical symptoms observed in patients with IBS.[35]

In the past few years, intestinal microbiota has emerged as a novel target for the treatment of gut–brain axis alterations. These include functional GI disorders, such as IBS, which can be comorbid with stress-related psychiatric conditions. Thus, modulation of the microbiota (e.g., with the use of probiotics) could be proposed as a novel strategy not only for the treatment of IBS, but also as an adjuvant for psychiatric treatment of anxiety and depression.[36]

A number of food groups appear to have a protective effect with regard to IGT and diabetes, including vegetables (odds ratio (OR) = 0.41; confidence interval (CI) = 0.18–0.91), breakfast foods (OR = 0.41; CI = 0.18–0.93), and hot meal foods (OR = 0.29; CI = 0.11–0.78). Most of these foods are relatively high in fiber and low in fat.[37]

3 CLINICAL RELEVANCE OF NEUROIMMUNE REGULATION

3.1 Sleep

Although sleep deprivation did not potentiate EAT growth, it decreased the survival of female tumor-bearing mice.[38]

Our study provided original evidence concerning the immune outcomes of PSD, indicating that cellular decrease caused by PSD is not restricted to circulation, but also to immune sites. Taken together, our results could help shed light on the physiological mechanisms of leukocyte trafficking.[39]

IL-6 elevations in the context of inflammatory disease affects the organism, with a focus on *sleep*-related symptoms *and fatigue*; and conversely, how alterations in sleep duration and quality stimulate the increase in concentrations of IL-6 in the circulation. Research showing that acute as well as chronic psychological stress also increases concentrations of IL-6, supports the notion of a close link between an organism's response to *physiological and psychological perturbations*.[40]

3.2 Glucose Homeostasis

Evidence showing the capacity of IL-1β to mediate adjustments of glucose homeostasis, can be added to the already well-known pleiotropic effects of this cytokine, has been discussed here. Such adjustments, which are necessary for satisfying the high energy demands of immune/inflammatory responses, can be mediated by effects of endogenous IL-1 exerted at peripheral and brain levels, in a concerted action with other cytokines and neuroendocrine mechanisms.[41]

3.3 Physical Exercise

Like any other stressor, the body's response to exercise triggers a systematic series of neuroendocrine and immune events directed at bringing the system back to a state of homeostasis. Physical exercise presents a unique physiological stress, where the neuroendocrine and immune systems contribute to accommodate the increase in physiological demands. These systems of the body also adapt to chronic overloads or exercise training. Such adaptations alleviate the magnitude of subsequent stress or minimize the exercise challenge within homeostatic limits. This adaptive capacity of collaborating systems resembles the acquired, or adaptive, branch of the immune system, characterized by the memory capacity of the cells involved. Specific to the adaptive immune response, once a specific antigen is encountered, memory cells, or lymphocytes, mount a response that reduces the magnitude of immune response to subsequent encounters of the same stress. In each case, the endocrine response to physical exercise and the adaptive branch of the immune system share the ability to adapt to a stressful encounter. Moreover, each of these systemic responses to stress is influenced by gender.[42]

Physical exercise diminishes inflammation and elevates agents and factors involved in immunomodulatory functions. The alleviatory effects of life-long physical activity upon multiple cancer forms and the palliative effects of physical activity for individuals afflicted by cancer, offer advantages in health intervention. Chronic conditions of stress and affective dysregulation are associated with neuroimmunological insufficiency and inflammation, contributing to health risk and mortality. Physical exercise regimes have induced the manifestation of anti-inflammatory benefits, mediated possibly by brain-derived neurotrophic factors. The epidemic proportions of metabolic disorders, obesity, and diabetes demand attention. Several variants of exercise regimes have been found to repeatedly induce prevention and improvement under laboratory and clinical conditions. Physical exercise offers a unique nonpharmacological intervention, incorporating multiple activity regimes, for example, endurance versus resistance exercises that may be adapted to conform to the particular demands of diagnosis, intervention, and prognosis inherent to the staging of autoimmune disorders and related conditions.[43]

Physical inactivity plays a negative role in several aspects of health status. A sedentary lifestyle represents a predisposing base for the development of obesity, type 2 diabetes, cardiovascular disease, osteoporosis, depression, and cancer. On the other hand, an active life is distinguished by physical exercise and the beneficial effects it brings to health, thereby reducing the risk of contracting diseases. The general state of exercise-related psychophysical health is regulated by a variety of mechanisms that affect the highly integrated and synergistic responses of the CNS and efficiency of the overall immune system. There is positive evidence concerning the influence of neuroimmune systems on the status of health, during moderate and regular physical exercise.[44]

Daily exercise potentially improves cognition in aging rats by modulating hippocampal neurogenesis, immune, and neuroimmune cytokine signaling.[45]

3.4 Pain

Adults who have experienced neonatal injury display increased pain and injury-induced hyperalgesia in the affected region, but mild injury can also induce widespread baseline hyposensitivity across the rest of the body surface, suggesting the involvement of several underlying mechanisms, depending upon the type of early life experience. The stress/HPA axis has been suggested to mediate this widespread hypoalgesia.[46]

Cold pressor test (CPT) acute pain may affect proinflammatory pathways, possibly through mechanisms related to adrenergic activation.[47]

Pain, fatigue, altered sleep pattern, anorexia, and fever are common symptoms in sick animals and humans. Many of these physiological changes represent adaptive responses that are considered to promote animal survival, and this constellation of events results in sickness behavior.[48]

Mediators released by immune cells, such as cytokines, sensitize nociceptive signaling in the peripheral and CNS. Preclinical data suggests an immune pathogenesis of *neuropathic pain*, but clinical evidence of a central role of the immune system is less clear.[49]

Following peripheral nerve injury, linking focal changes in Schwann cell activation and release of the proinflammatory cytokine tumor necrosis factor-alpha (TNF-α) with subsequent activation and sensitization of ascending sensory neurons and glia, which culminate in the neuropathic pain state. The pathophysiology, neuropathology, and molecular biology of Wallerian degeneration are also reviewed from a perspective that links it to upregulation of proinflammatory cytokines and the development of neuropathic pain states.[50]

Chronic neuropathic and inflammatory pain is a major public health problem. Nociceptors undergo sensitization, first in peripheral tissues and then in the CNS, via neuroimmune interactions linking neurons, glial cells (microglia and astrocytes), and immune cells. These interactions may either exacerbate or attenuate the pain and inflammation, which normally reach a state of equilibrium. With more powerful or longer lasting stimuli, specific profiles of microglial, and subsequently, astrocytic activation in the dorsal horn, play a key role in neuronal plasticity and transition to chronic pain.[51]

Adult pain responses in the rat are "primed" by tissue injury in the neonatal period. Rats that experience hindpaw incision injury at the age of 3 days, display an increased magnitude and duration of hyperalgesia following incision in adulthood when compared with those having no early life pain experience. Early life injury predisposes individuals to enhanced sensitivity in painful events.[52]

Neural immune signaling modulates the susceptibility to migraine, modeled *in vitro* using *spreading depression* (*SD*), as a means to develop novel therapeutic targets for episodic and chronic migraine. SD is the probable cause of migraine aura and migraine pain. It is a paroxysmal loss of neuronal function triggered by initially increased neuronal activity, which slowly propagates within susceptible brain regions. *SD activates trigeminal nociceptive pathways* and *immune signaling in mature hippocampal slice cultures.* Since the cultures closely grow parallel to their *in vivo* counterpart with *mature trisynaptic function,* they *show quiescent astrocytes, microglia, and cytokine levels.* Hence, *SD is easily induced in an unanaesthetized preparation.* Furthermore, the slices are long lasting, and *SD can be induced on consecutive days without injury*, making this preparation the sole means till date capable of modeling the neuroimmune consequences of chronic SD, and thus perhaps chronic migraine. Normal brain function is exquisitely sensitive to, and relies on, coincident low-level immune signaling.[53]

In patients with functional abdominal pain syndromes, conscious perception of interoceptive information from the GI tract, or recall of interoceptive memories of such input, can occur in the form of constant or recurrent discomfort or pain. This is often associated with alterations in autonomic nervous system output and with emotional changes. A model is proposed that incorporates reported peripheral and central abnormalities in patients with IBS, extrapolates similar alterations in brain-gut interactions to patients with other chronic abdominal pain syndromes, and provides novel treatment targets.[54]

3.5 Stress

IgA could be a potential biomarker for extreme environmental conditions, such as the winter in Antarctica.[55]

It was found that circulating levels of interferon-γ (IFN-γ), which can affect brain functions and induce depressive symptoms, were lower in the high-happiness group than in the low happiness group. A negative correlation between the levels of perceived happiness and IFN-γ concentrations was also observed. Furthermore, we also found that experimentally induced happiness could reduce peripheral IFN-γ levels.[56]

IL-1β and IL-6, enzyme cyclooxygenase 2, and the chemokines, CXCL1 and MCP-1, in plasma, hypothalamus, and prefrontal cortex were studied. There was no effect of chronic intermittent cold stress (CIC stress; 4°C; 6 h/day; 14 days) stress on basal expression of these markers, 24 h after the termination of stress. However, CIC stress enhanced the acute induction of the proinflammatory cytokines, IL-1β and particularly IL-6, and the chemokines, CXCL1 and MCP-1, in plasma, hypothalamus and prefrontal cortex in response to LPS, and also sensitized the hypothalamic IL-1β response to acute footshock. Thus, sensitization of acute proinflammatory responses in the brain could potentially mediate some of the CIC-dependent changes in HPA and cognitive function.[57]

The stress response, although theoretically stereotyped in nature, may lead to slightly different outcomes according to several modifiers. Our group has compared the effects of several stressors over different parameters of brain activity, behavior, immunity, and glucocorticoid levels. Although, these data show that during an increased turnover of noradrenaline in the hypothalamus, along with anxiety-like behaviors, and an increase in serum corticosterone are present very often, the magnitude of changes in immunity may vary considerably. Thus, data generated over the past decade was reviewed, and it was found to support the effects of stressors on immunity and behavior highly depend on their specifics, animal model, frequency, duration, intensity, perception, and coping by the stressed animal.[58]

3.6 Aging

Although aging is an inexorable component of life, its progress depends on how cumulative disruptions of homeostasis are compensated. Cumulative oxidative and inflammatory processes must be controlled to maintain successful aging. Heat shock proteins, such as those of the Hsp70 family, can be considered a danger signal, and their effects can either support longevity by neutralizing danger or can become detrimental, when their production is not balanced. Here, we discuss evidence indicating that these highly conserved proteins can favor longevity when such balance is achieved. Emphasis was given to the mechanisms affected by Hsp72 that can interfere with the effects of excessive oxidative stress, subtle inflammation, and acts either directly or by affecting neuroimmune–endocrine interactions, can mediate metabolic, neuroprotective, and behavioral adjustments during the aging process.[59]

During cerebral development, vitamin D may act like a neurosteroid hormone in the areas of neurotransmission, neuroprotection, and neuroimmunomodulation. Moreover, vitamin D deficiency has been associated with neurological and psychiatric disorders. In older adults, hypovitaminosis D has been associated with neuromuscular disorders, dementia, and Parkinson's disease. Thus, vitamin D supplementation might have a protective effect against these neurological disorders.[60]

Cholinergic anti-inflammatory modulation provides therapeutic advantages in sepsis. The vagus nerve controls inflammation in healthy, but not in septic mice. Likewise, vagus nerve and cholinergic agonists fail to control inflammation in splenectomized and nude animals. Unlike typical suppressor CD25+ cells, CD4+, CD25− lymphocytes re-establish the anti-inflammatory potential of the vagus nerve and cholinergic agonists in immunocompromised and septic animals.[61]

4 CONCLUSIONS

Although the ADIM system shows autonomy, this does not mean independence from the host organism. Stem cells reside in the host's bone marrow for both the immune and neuroendocrine systems. The CNS is a part of the INIM system. INIM provides the first antigen presenting cells for ADIM in the form of macrophages. Further, the suppressor/regulatory T cells (Tsr) are innate immune cells, as they grow in response to glucocorticoids and catecholamines, which are stimulators of innate immunity. A type of T cell that recognizes the CNS and exert a protective effect against dangerous insults.[62] We name these unique T cells as *neuroprotective* (T_{NEP}) cells.

The fact that the existence of the immune–neuroendocrine circuitry has been acknowledged, it has been stated that the immune system belongs to this circuitry in the body, and fulfills important systemic regulatory function. This is not an optional function, rather this is the central regulatory circuitry of the organism and participation is obligatory in the neuroimmune supersystem (NISS).[62] Malfunction of any participant in NISS has a series of consequences, which let disease to develop. So the immune system is a part of the central regulatory establishment, which coordinates and regulates the entire host organism. Participation in this system rules out autonomy.

Neuroimmune mechanisms play a role in reproduction, menstruation, neonatal development, brain development and function, decision making; early life stress; and neuroprotection (neuroprotective T cells).

An interesting recent development is the relationship between gut microbiota and brain functions. The proposal is that, abnormal microbes lead to abnormal brain function. When this happens, it may be corrected by probiotic organisms. Some experiments support these claims, but the underlying mechanisms remain unclear.

It is known for a long time that germ-free rats do not develop a normal immune system. If the gut microflora get restored, normal immune system develops. So the gut microflora stimulate immune function, which in turn *could affect the brain.*

Naive lymphocytes are fully mature cells, but have not been exposed to antigenic stimulation. They need antigenic stimulus to mature into fully immunocompetent effector cells. The gut microflora is readily available for stimulation. The result is that a robust immune system develops, which is ready to restrain the intestinal army of microbes and any other intruder that might come around. The rule is that, when the immune system responds, it will protect. This rule is valid for intestinal microbes as well.

Nerve cells in the CNS and the nerve fibers express TLR (toll-like antigen receptors are expressed by both the brain and immune systems, so both are immune organs). For this reason, it was suggested that the CNS belongs to the innate immune system.[64] So, it is clear that the CNS would be able to sense the intestinal microbes as already have been confirmed. The brain is the most powerful immunoregulatory organ, which commands the entire neuroimmune orchestra. With this kind of mobilization, the gut microbes must behave, or else.

However, the first cells that encounter microbes are the mucosal epithelial cells. Could they respond to the challenge? Yes indeed, mucosal epithelial cells possess TLR and are found to respond with activation, cytokine, and chemokine production to bacterial invasion.[64–67] Given the vast number of epithelial cells, their role must not be discounted.

A question arises here, that is, how do gut microbiota affect the brain? Directly via TLR, which are present on the nerves. These nerves sense the bacteria directly. Could this sensing contribute to beneficial or malicious effects? This can be found out from future studies.

The immune system is directly stimulated by the microbiome and communicates this to the brain. An activated immune system has lots to do with brain physiology. Again future studies will tell the significance of this phenomenon.

Mucosal epithelial cells have tremendous number and were found already to possess healing power in pneumonia. This is perhaps the most powerful defense protecting us from microbial domination.

Moreover, mucosal microbiota should be considered as fully integrated into the physiology of the host organism and it plays a fundamental role in the maintenance of health. These microbes belong to us, we need them commensally, to nourish, stimulate, and maintain essential bodily functions for us. Clearly our body is not complete without these microbes.

The rapidly growing microbes pose a formidable challenge to the host. However, the entire neuroimmune supersystem is mobilized for host defense, which does a very good defense indeed!

REFERENCES

1. Gittelsohn J, Wolever TM, Harris SB, Harris-Giraldo R, Hanley AJ, Zinman B. Specific patterns of food consumption and preparation are associated with diabetes and obesity in a Native Canadian community. *J Nutr* 1998;**128**:541–7.
2. Romacho T, Elsen M, Röhrborn D, Eckel J. Adipose tissue and its role in organ crosstalk. *Acta Physiol* 2014;**210**(4):733–53.

3. Andrade JM, Paraíso AF, Garcia ZM, Ferreira AV, Sinisterra RD, Sousa FB, Guimarães AL, Paula AM, Campagnole-Santos MJ, Santos RA, Santos SH.
4. Proença AR, Sertié RA, Oliveira AC, Campaãa AB, Caminhotto RO, Chimin P, Lima FB. New concepts in white adipose tissue physiology. *Braz J Med Biol Res.* 2014;**47**: 192–205.
5. Gustaffson B. Germ-free rearing of rats. *Acta Pathol Microbiol Scand* 1948;LXXIII.
6. Berczi I, Nagy E, Kovacs K, Horvath E. Regulation of humoral immunity in rats by pituitary hormones. *Acta Endocrinol* 1981;**98**:506–13.
7. Hsiao EY, McBride SW, Hsien S, Sharon G, Hyde ER, McCue T, Codelli JA, Chow J, Reisman SE, Petrosino JF, Patterson PH, Mazmanian SK. Microbiota modulate behavioral and physiological abnormalities associated with neurodevelopmental disorders. *Cell* 2013;**155**(7):1451–63.
8. Ray K. Gut microbiota: microbial metabolites feed into the gut-brain-gut circuit during host metabolism. *Nat Rev Gastroenterol Hepatol* 2014;**11**:76.
9. De Vadder F, Kovatcheva-Datchary P, Goncalves D, Vinera J, Zitoun C, Duchampt A, Bäckhed F, Mithieux G. Microbiota-generated metabolites promote metabolic benefits via gut-brain neural circuits. *Cell* 2014;**156**(1-2):84–96.
10. Chen X, D'Souza R, Hong ST. The role of gut microbiota in the gut-brain axis: current challenges and perspectives. *Protein Cell* 2013;**4**(6):403–14.
11. Ait-Belgnaoui A, Colom A, Braniste V, Ramalho L, Marrot A, Cartier C, Houdeau E, Theodorou V, Tompkins T. Probiotic gut effect prevents the chronic psychological stress-induced brain activity abnormality in mice. *Neurogastroenterol Motil* 2013;**26**(3).
12. Montiel-Castro AJ, González-Cervantes RM, Bravo-Ruiseco G, Pacheco-López G. The microbiota-gut-brain axis: neurobehavioral correlates, health and sociality. *Front Integr Neurosci* 2013;**7**:70.
13. Umu OC, Oostindjer M, Pope PB, Svihus B, Egelandsdal B, Nes IF, Diep DB. Potential applications of gut microbiota to control human physiology. *Antonie Van Leeuwenhoek.* 2013;**104**(5):609–18.
14. Berczi I, Nagy E, Kovacs K, Horvath E. Regulation of humoral immunity in rats by pituitary hormones. *Acta Endocrinol* 1981;**98**:506–13.
15. Dinan TG, Cryan JF. Melancholic microbes: a link between gut microbiota and depression? *Neurogastroenterol Motil* 2013;**25**(9):713–9.
16. Dockray GJ. Gastrointestinal hormones and the dialogue between gut and brain. *J Physiol* 2014.
17. Lanning DK, Knight KL. Intestinal bacteria and development of the antibody repertoire. *Discov Med* 2005;**5**(28):393–8.
18. McVey Neufeld KA, Mao YK, Bienenstock J, Foster JA, Kunze WA. The microbiome is essential for normal gut intrinsic primary afferent neuron excitability in the mouse. *Neurogastroenterol Motil* 2013;**25**(2):183-e88.
19. Clarke G, Grenham S, Scully P, Fitzgerald P, Moloney RD, Shanahan F, Dinan TG, Cryan JF. The microbiome-gut-brain axis during early life regulates the hippocampal serotonergic system in a sex-dependent manner. *Mol Psychiatry* 2013;**18**:666–73.
20. Bengmark S. Gut microbiota, immune development and function. *Pharmacol Res* 2013;**69**:87–113.
21. Bauer E, Williams BA, Smidt H, Verstegen MW, Mosenthin R. Influence of the gastrointestinal microbiota on development of the immune system in young animals. *Curr Issues Intest Microbiol* 2006;**7**:35–51.
22. Noverr MC, Huffnagle GB. Does the microbiota regulate immune responses outside the gut? *Trends Microbiol* 2004;**12**(12):562–8.
23. Tlaskalova-Hogenova H, Tuckova L, Mestecky J, Kolinska J, Rossmann P, Stepankova R, Kozakova H, Hudcovic T, Hrncir T, Frolova L, Kverka M. Interaction of mucosal microbiota with the innate immune system. *Scand J Immunol* 2005;**62**(Suppl 1):106–13.

24. Kett K, Baklien K, Bakken A, Kral JG, Fausa O, Brandtzaeg P. Intestinal B-cell isotype response in relation to local bacterial load: evidence for immunoglobulin A subclass adaptation. *Gastroenterology* 1995;**109**:819–25.
25. Kanauchi O, Andoh A, Mitsuyama K. Effects of the modulation of micro- biota on the gastrointestinal immune system and bowel function. *J Agric Food Chem* 2013;**61**(42): 9977–83.
26. Costes LM, Boeckxstaens GE, de Jonge WJ, Cailotto C. Neural networks in intestinal immunoregulation. *Organogenesis* 2013;**9**:216–23.
27. O'Mahony SM, Marchesi JR, Scully P, Codling C, Ceolho AM, Quigley EM, Cryan JF, Dinan TG. Early life stress alters behavior, immunity, and microbiota in rats: implications for irritable bowel syndrome and psychiatric illnesses. *Biol Psychiatry* 2009;**65**:263–7.
28. Barouei J, Moussavi M, Hodgson DM. Effect of maternal probiotic intervention on HPA axis, immunity and gut microbiota in a rat model of irritable bowel syndrome. *PLoS ONE* 2012;**7**(10):e46051.
29. Rook GA, Lowry CA, Raison CL. Lymphocytes in neuroprotection, cognition and emotion: is intolerance really the answer? *Brain Behav Immun* 2011;**25**(4):591–601.
30. Arebi N, Gurmany S, Bullas D, Hobson A, Stagg A, Kamm M. Review article: the psychoneuroimmunology of irritable bowel syndrome an exploration of interactions between psychological, neurological and immunological observations. *Aliment Pharmacol Ther* 2008;**28**(7):830–40.
31. Ringel Y, Maharshak N. Intestinal microbiota and immune function in the pathogenesis of irritable bowel syndrome. *Am J Physiol Gastrointest Liver Physiol* 2013;**305**(8): G529–41.
32. Berczi I, Villalobos Hernández Egina C, Quintanar Stephano A, Campos R, Kovacs K. The healing power of vasopressin. *Adv Neuroimmune Biol* 2012;**2**:217–24.
33. Bravo JA, Julio-Pieper M, Forsythe P, Kunze W, Dinan TG, Bienenstock J, Cryan JF. Communication between gastrointestinal bacteria and the nervous system. *Curr Opin Pharmacol* 2012;**12**:667–72.
34. Maragno-Correa JM, Patti CL, Zanin KA, Wuo-Silva R, Ruiz FS, Zager A, Sá-Nunes A, Tufik S, Andersen ML, Frussa-Filho R. Sleep deprivation increases mortality in female mice bearing Ehrlich ascetic tumor. *Neuroimmunomodulation* 2013;**20**(3):134–40.
35. Zager A, Ruiz FS, Tufik S, Andersen ML. Immune outcomes of paradoxical sleep deprivation on cellular distribution in naive and lipopolysaccharide-stimulated mice. *Neuroimmunomodulation* 2012;**19**(2):79–87.
36. Rohleder N, Aringer M, Boentert M. Role of interleukin-6 in *stress, sleep, and fatigue*. *Ann NY Acad Sci* 2012;**1261**:88–96.
37. Besedovsky HO, Del Rey A. Interleukin-1 resets glucose homeostasis at central and peripheral levels: relevance for immunoregulation. *Neuroimmunomodulation* 2010;**17**(3): 139–41.
38. Fragala MS, Kraemer WJ, Denegar CR, Maresh CM, Mastro AM, Volek JS. Neuroendocrine-immune interactions and responses to exercise. *Sports Med* 2011;**41**:621–39.
39 Archer T, Fredriksson A, Schütz E, Kostrzewa RM. Influence of physical exercise on neuroimmunological functioning and health: aging and stress. *Neurotox Res* 2011;**20**:69–83.
40. Verratti V. Neuroimmune biology of physical exercise. *J Biol Regul Homeost Agents* 2009;**23**:203–6.
41. Speisman RB, Kumar A, Rani A, Foster TC, Ormerod BK. Daily exercise improves memory, stimulates hippocampal neurogenesis and modulates immune and neuroimmune cytokines in aging rats. *Brain Behav Immun* 2013;**28**:25–43.
42. Schwaller F, Fitzgerald M. The consequences of pain in early life: injury-induced plasticity in developing pain pathways. *Eur J Neurosci* 2014;**39**:344–52.
43. Griffis CA, Crabb Breen E, Compton P, Goldberg A, Witarama T, Kotlerman J, Irwin MR. Acute painful stress and inflammatory mediator production. *Neuroimmunomodulation* 2013;**20**:127–33.

44. McCusker RH, Kelley KW. Immune-neural connections: how the immune system's response to infectious agents influences behavior. *J Exp Biol* 2013;**216**(Pt 1):84–98.
45. Calvo M, Dawes JM, Bennett DL. The role of the immune system in the generation of *neuropathic pain. Lancet Neurol* 2012;**11**:629–42.
46. Myers RR, Shubayev VI. The ology of neuropathy: an integrative review of the role of neuroinflammation and TNF-α axonal transport in neuropathic pain. *J Peripher Nerv Syst* 2011;**16**:277–86.
47. Guillot X, Semerano L, Decker P, Falgarone G, Boissier MC. Pain and immunity. *Joint Bone Spine* 2012;**79**:228–36.
48. Beggs S, Currie G, Salter MW, Fitzgerald M, Walker SM. Priming of adult pain responses by neonatal pain experience: maintenance by central neuroimmune activity. *Brain* 2012; **135**(Pt 2):404–17.
49. Pusic AD, Grinberg YY, Mitchell HM, Kraig RP. Modeling neural immune signaling of episodic and chronic migraine using spreading depression *in vitro*. *J Vis Exp* 2011; **13**(52):2910.
50. Mayer EA, Tillisch K. The brain-gut axis in abdominal pain syndromes. *Annu Rev Med* 2011;**62**:381–96.
51. Yadav AP, Mishra KP, Ganju L, Singh SB. Wintering in Antarctica: impact on immune response of Indian expeditioners. *Neuroimmunomodulation* 2012;**19**(6):327–33.
52. Matsunaga M, Isowa T, Yamakawa K, Tsuboi H, Kawanishi Y, Kaneko H, Kasugai K, Yoneda M, Ohira H. Association between perceived happiness levels and peripheral circulating pro-inflammatory cytokine levels in middle-aged adults in Japan. *Neuro Endocrinol Lett* 2011;**32**(4):458–63.
53. Girotti M, Donegan JJ, Morilak DA. Chronic intermittent cold stress sensitizes neuro-immune reactivity in the rat brain. *Psychoneuroendocrinology* 2011;**36**:1164–74.
54. Costa-Pinto FA, Palermo-Neto J. Neuroimmune interactions in stress. *Neuroimmunomodulation* 2010;**17**(3):196–9.
55. Ortega E, Bote ME, Besedovsky HO, del Rey A. Hsp72, inflammation, and aging: causes, consequences, and perspectives. *Ann NY Acad Sci.* 2012;**1261**:64–71.
56. Annweiler C, Schott AM, Berrut G, Chauviré V, Le Gall D, Inzitari M, Beauchet O. Vitamin D and ageing: neurological issues. *Neuropsychobiology* 2010;**62**(3):139–50.
57. Peña G, Cai B, Ramos L, Vida G, Deitch EA, Ulloa L. Cholinergic regulatory lymphocytes re-establish neuromodulation of innate immune responses in sepsis. *J Immunol* 2011;**187**:718–25.
58. Latorre E, Mendoza C, Layunta E, Alcalde AI, Mesonero JE. TLR2, TLR3, and TLR4 activation specifically alters the oxidative status of intestinal epithelial cells. *Cell Stress Chaperones* 2014;**19**:289–93.
59. Fűri I, Sipos F, Germann TM, Kalmár A, Tulassay Z, Molnár B, Műzes G. Epithelial toll-like receptor 9 signaling in colorectal inflammation and cancer: clinico-pathogenic aspects. *World J Gastroenterol* 2013;**19**:4119–26.
60. Cleaver JO, You D, Michaud DR, Pruneda FA, Juarez MM, Zhang J, Weill PM, Adachi R, Gong L, Moghaddam SJ, Poynter ME, Tuvim MJ, Evans SE. Lung epithelial cells are essential effectors of inducible resistance to pneumonia. *Mucosal Immunol* 2014;**7**:78–88.
61. Mayer EA, Tillisch K. The brain-gut axis in abdominal pain syndromes. *Annu Rev Med* 2011;**62**:381–96.
62. Petitto JM, Meola D, Huang Z. Interleukin-2 and the brain: dissecting central versus peripheral contributions using unique mouse models. *Methods Mol Biol* 2012;**934**: 301–11.
63. Kokaia Z, Martino G, Schwartz M, Lindvall O. Cross-talk between neural stem cells and immune cells: the key to better brain repair? *Nat Neurosci* 2012;**15**:1078–87.
64. Gruol DL. Neuroimmune regulation of neurophysiology in the cerebellum. *Cerebellum* 2013;**12**:307–9.

65. Kipnis J, Gadani S, Derecki NC. Pro-cognitive properties of T cells. *Nat Rev Immunol* 2012;**12**(9):663–9.
66. Chiu CT, Wen LL, Pao HP, Wang JY. Cortistatin is induced in brain tissue and exerts neuroprotection in a rat model of bacterial meningoencephalitis. *J Infect Dis* 2011; **204**(10):1563–72.
67. Rosas-Ballina M, Olofsson PS, Ochani M, Valdés-Ferrer SI, Levine YA, Reardon C, Tusche MW, Pavlov VA, Andersson U, Chavan S, Mak TW, Tracey KJ. Acetylcholine-synthesizing T cells relay neural signals in a vagus nerve circuit. *Science* 2011;**334**(6052): 98–101.

INDEX

I

N

FSC
www.fsc.org
MIX
Papier | Fördert
gute Waldnutzung
FSC® C083411

Kolibri 360 GmbH
Ferdinand-Jühlke-Straße 7
99095 Erfurt, Deutschland
produktsicherheit@kolibri360.de